Commemorative Issue in Honor of Professor Karlheinz Schwarz on the Occasion of his 80th Birthday

Commemorative Issue in Honor of Professor Karlheinz Schwarz on the Occasion of his 80th Birthday

Editor

Henry Chermette

MDPI • Basel • Beijing • Wuhan • Barcelona • Belgrade • Manchester • Tokyo • Cluj • Tianjin

Editor
Henry Chermette
Institut des Sciences
Analytiques, UMR 5280,
CNRS
Université Claude Bernard
Lyon 1
Villeurbanne
France

Editorial Office
MDPI
St. Alban-Anlage 66
4052 Basel, Switzerland

This is a reprint of articles from the Special Issue published online in the open access journal *Computation* (ISSN 2079-3197) (available at: www.mdpi.com/journal/computation/special_issues/Commem_KS).

For citation purposes, cite each article independently as indicated on the article page online and as indicated below:

LastName, A.A.; LastName, B.B.; LastName, C.C. Article Title. *Journal Name* **Year**, *Volume Number*, Page Range.

ISBN 978-3-0365-4974-3 (Hbk)
ISBN 978-3-0365-4973-6 (PDF)

Cover image courtesy of Karlheinz Schwarz

Contents

computation MDPI

Editorial

Dedication: Commemorative Issue in Honor of Professor Karlheinz Schwarz on the Occasion of His 80th Birthday

Peter Blaha [1,*] and Henry Chermette [2,*]**

[1] Institute of Materials Chemistry, Vienna University of Technology, Getreidemarkt 9/165-TC, 1060 Vienna, Austria
[2] Institut des Sciences Analytiques, UMR 5280, CNRS, Université Claude Bernard Lyon1, Université de Lyon, 5 rue de la Doua, F-69100 Villeurbanne, France
* Correspondence: peter.blaha@tuwien.ac.at (P.B.); henry.chermette@univ-lyon1.fr (H.C.)

Citation: Blaha, P.; Chermette, H. Dedication: Commemorative Issue in Honor of Professor Karlheinz Schwarz on the Occasion of His 80th Birthday. *Computation* **2022**, *10*, 78. https://doi.org/10.3390/computation10050078

Received: 17 May 2022
Accepted: 19 May 2022
Published: 23 May 2022

Karlheinz Schwarz was born in January 1941 in Vienna (Austria), and he married Christine Schwarz in 1969. They have one child (Caroline), and he is a happy grandfather of four grandchildren. Besides his passion for science, he likes classical music. In fact, when he was young, he was not sure whether he should study chemistry or music, and maybe he would have become a virtuoso pianist instead of a famous scientist. Another passion was skiing; he even became a skiing instructor and impressed (and taught) numerous colleagues at various winter conferences in the mountains with his elegant style.

He studied Chemistry at the University Vienna and obtained his PhD under the supervision of A. Neckel in 1968, where he performed the first theoretical solid-state calculations in Austria using the APW method. I still remember his stories about all the problems he had to solve due to the limited computer resources at that time, including sleeping at the computing center because he had to remount tapes several times during the night, when such a calculation was running.

After his PhD, he worked as a postdoc in John C. Slater's group in Gainesville, FL, and later with Frank Herman at the IBM Research Lab in San Jose, CA. With his programming skills, developed during his PhD in Austria, he could impress the leading solid-state theory groups and speed up their codes by an order of magnitude. He came into contact with the X-alpha method—a predecessor of modern DFT—and optimized the alpha parameter for all atoms in a clever and unique way. The corresponding paper [1] was one of the most cited theory papers at that time.

Coming back from the US, he completed his Habilitation in "Quantum Chemistry" at the Technical University of Vienna, where he became a professor in 1976. His scientific reputation and active scientific life led to many invitations, and he was a guest professor at the Universities of Bochum, Uppsala, Montreal, Duisburg, Gainesville and Paris. He was a research fellow in Tsukuba, Japan, and six times a summer faculty member at the IBM T.J. Watson Research Center, where he worked intensively with Art Williams and Vic Moruzzi. During these times, his interest turned towards magnetism [2] and concepts such as "covalent magnetism" [3] and the "fixed-spin method" (FSM) [4], which laid the foundations to explain the Invar alloy problem [5].

Together with S.B. Trickey, P. Sorantin and myself, he formed the basis for the WIEN code, a full-potential all-electron augmented plane-wave-based computer program. This code, first published in 1990 [6], developed over the years into the WIEN2k code [7,8], a powerful and widely used package able to calculate many different properties of solids. In particular, it is famous for its unprecedented accuracy in solving the Kohn–Sham equations, often referred to as the "DFT limit". This user-friendly program is licensed by more than 3500 groups worldwide in academia and industry and still serves as a benchmark for all newly developed solid-state codes.

Although science was his passion, as a professor, he also served in many management positions. He was department head for 4 years and vice-head/head of the "FG Chemie" for more than 8 years. He served as president of the "Chemisch-Physikalische gesellschaft" and of the Austrian Fulbright Alumni Association, paying back the support he received from the Fulbright Association as a young postdoc. Realizing how important a good computer infrastructure for a modern university will be, he became head of the computing center of the Technical University of Vienna (TU Wien) in 1990 and later on head of their advisory board.

Karlheinz Schwarz realized early how important dissemination and discussions with colleagues are, and besides giving excellent invited talks at many conferences, he organized several conferences or served in the corresponding advisory boards. He was chairman of two Gordon conferences ("Phase transitions in non-metallic solids", Volterra, Italy, 1994, and "Electron densities and chemical bonding", Oxford, UK, 1998), and he founded and chaired the famous conference series "Workshop on Novel Materials and Superconductors" at Planneralm and Obertraun for an amazing 32 years (1989–2017). Similarly, to promote our WIEN2k code, he organized 26 WIEN workshops all over the world (Italy, France, Poland, the USA, Canada, Iran, Singapore, Japan and Austria). He was a very good friend of Walter Kohn [9], and his invitation to two conferences organized by Schwarz in Vienna (International Conference on DFT 1997 and Applied DFT2001) helped to settle the relation of Walter Kohn with his country of birth.

Personally, I met Karlheinz Schwarz in 1979 when I was a young chemistry student at the Technical University of Vienna and attending a course on "Solid state theory" given by him. Although most students had problems at the beginning to digest "strange concepts" such as the "reciprocal lattice" or "periodic boundary conditions", our passionate teacher explained this so well and with unique enthusiasm that this topic really started to catch my interest. When I finally realized that K. Schwarz was a scientist with the highest international experience and reputation, I was very happy to accept his offer to start a PhD under his supervision. He introduced me to the LAPW method, and this was the beginning of a deep friendship and, for me, due to his never-ending support, the start of my career. (P.B.)

I met Karlheinz Schwarz for the first time in Montreal, in 1981, where we crossed as postdocs in the Dennis Salahub group. For me, he was the famous author of the optimization of the α statistical exchange parameter [1] I cited and used in all my MS-Xα calculations, e.g., [10,11]. Later, at the beginning of the 1990s, he participated for the first time in the series of DFT congresses we initiated with the passed Alessandro Bencini and Annick Goursot. He organized the meeting in 1997 in Vienna and became more or less the leading member of the scientific committee of the series of these meetings. I am glad that a respectable number of friends (18) accepted to write a paper in this dedicated issue. We can notice the large panel of scientific topics covered by Karlheinz's knowledge. We deeply acknowledge the following contributions related to spectroscopy by Manuel Yañez et al. [12], Juan-Carlos Sancho-García and Emilio San-Fabián [13]; excited states by Ágnes Nagy [14], Kalidas Sen et al. [15] and Fabrizia Negri et al. [16]; DFT developments by Fabio Della Sala et al. [17], Mathias Rapacioli and Nathalie Tarrat [18], Emmanuel Fromager et al. [19], José Manuel García de la Vega et al. [20] and Harry Ramanantoanina [21]; results analysis by Andreas Savin et al. [22] and Manuel Richter et al. [23]; and, of course, the solid state and surfaces by Leila Kalantari and Fabien Tran et al. [24], Denis Salahub et al. [25], Peter Blaha et al. [26], Samuel B. Trickey [27], William Lafargue-Dit-Hauret and Xavier Rocquefelte [28], Tzonka Mineva and Hazar Guesmi et al. [29]. (H.C.)

Conflicts of Interest: The authors declare no conflict of interest.

References

1. Schwarz, K. Optimization of the Statistical Exchange Parameter α for the Free Atoms H through Nb. *Phys. Rev. B* **1972**, *5*, 2466. [CrossRef]
2. Williams, A.R.; Moruzzi, V.L.; Gelatt, C.D., Jr.; Kübler, J.; Schwarz, K. Aspects of transition-metal magnetism. *J. Appl. Phys.* **1982**, *53*, 2019. [CrossRef]

3. Schwarz, K.; Mohn, P.; Blaha, P.; Kubler, J. Electronic and magnetic structure of BCC Fe-Co alloys from band theory. *J. Phys. F Met. Phys.* **1984**, *14*, 2659. [CrossRef]
4. Schwarz, K.; Mohn, P. Itinerant metamagnetism in YCO2. *J. Phys. F Met. Phys.* **1984**, *14*, L129. [CrossRef]
5. Entel, P.; Hoffmann, E.; Mohn, P.; Schwarz, K.; Moruzzi, V.L. First-principles calculations of the instability leading to the Invar effect. *Phys. Rev. B* **1993**, *47*, 8706. [CrossRef] [PubMed]
6. Blaha, P.; Schwarz, K.; Sorantin, P.; Trickey, S.B. Full-potential, linearized augmented plane wave programs for crystalline systems. *Comput. Phys. Commun.* **1990**, *59*, 399. [CrossRef]
7. WIEN2k. Available online: http://www.wien2k.at (accessed on 17 May 2022).
8. Blaha, P.; Schwarz, K.; Tran, F.; Laskowski, R.; Madsen, G.K.H.; Marks, L.D. WIEN2k: An APW + lo program for calculating the properties of solids. *J. Chem. Phys.* **2020**, *152*, 074101. [CrossRef]
9. Schwarz, K.; Sham, L.J.; Mattsson, A.E.; Scheffler, M. Obituary for Walter Kohn (1923–2016). *Computation* **2016**, *4*, 40. [CrossRef]
10. Chermette, H.; Pedrini, C. MSX-Alpha Calculation of Adiabatic Potential-Energy Surfaces of Cu+ in Sodium-Chloride Lattice in the A1g Subspace—Incidence of the Copper Chlorine Distance on the Electronic-Structure. *J. Chem. Phys.* **1982**, *77*, 2460–2465. [CrossRef]
11. Chermette, H. 20 Years of the Ms-X-Alpha Method. *New J. Chem.* **1992**, *16*, 1081–1088.
12. Lamsabhi, A.M.; Montero-Campillo, M.M.; Mó, O.; Yañez, M. A Theoretical Survey of the UV–Visible Spectra of Axially and Peripherally Substituted Boron Subphthalocyanines. *Computation* **2022**, *10*, 14. [CrossRef]
13. Sancho-García, J.-C.; San-Fabián, E. Organic Emitters Showing Excited-States Energy Inversion: An Assessment of MC-PDFT and Correlation Energy Functionals Beyond TD-DFT. *Computation* **2022**, *10*, 13. [CrossRef]
14. Nagy, Á. Density Functional Theory of Coulombic Excited States Based on Nodal Variational Principle. *Computation* **2021**, *9*, 93. [CrossRef]
15. Doma, S.B.; Salem, M.A.; Sen, K.D. Plasma Confined Ground and Excited State Helium Atom: A Comparison Theorem Study Using Variational Monte Carlo and Lagrange Mesh Method. *Computation* **2022**, *9*, 138. [CrossRef]
16. Dai, Y.; Zubiria-Ulacia, M.; Casanova, D.; Negri, F. Impact of Charge-Resonance Excitations on CT-Mediated J-Type Aggregation in Singlet and Triplet Exciton States of Perylene Di-Imide Aggregates: A TDDFT Investigation. *Computation* **2022**, *10*, 18. [CrossRef]
17. Fabiano, E.; Sarcinella, F.; Constantin, L.A.; Della Sala, F. Energy Density Functionals Based on a Generalized Screened Coulomb Potential: Linear Response and Future Perspectives. *Computation* **2022**, *10*, 30. [CrossRef]
18. Rapacioli, M.; Tarrat, N. Periodic DFTB for Supported Clusters: Implementation and Application on Benzene Dimers Deposited on Graphene. *Computation* **2022**, *10*, 39. [CrossRef]
19. Sekaran, S.; Saubanère, M.; Fromager, E. Local Potential Functional Embedding Theory: A Self-Consistent Flavor of Density Functional Theory for Lattices without Density Functionals. *Computation* **2022**, *10*, 45. [CrossRef]
20. Ema, I.; Ramírez, G.; López, R.; García de la Vega, J.M. Generation of Basis Sets for Accurate Molecular Calculations: Application to Helium Atom and Dimer. *Computation* **2022**, *10*, 65. [CrossRef]
21. Ramanantoanina, H. LFDFT—A Practical Tool for Coordination Chemistry. *Computation* **2022**, *10*, 70. [CrossRef]
22. Pernot, P.; Savin, A. Should We Gain Confidence from the Similarity of Results between Methods? *Computation* **2022**, *10*, 27. [CrossRef]
23. Richter, M.; Kim, S.-J.; Koepernik, K. Helge Rosner and Arnulf M?bius, Accuracy and Precision in Electronic Structure Computation: Wien2k and FPLO. *Computation* **2022**, *10*, 28. [CrossRef]
24. Kalantari, L.; Tran, F.; Blaha, P. Density Functional Theory Study of Metal and Metal-Oxide Nucleation and Growth on the Anatase TiO_2(101) Surface. *Computation* **2022**, *9*, 125. [CrossRef]
25. Naseri, M.; Jalilian, J.; Salahub, D.R.; Lourenço, M.P.; Rezaei, G. Hexatetra-Carbon: A Novel Two-Dimensional Semiconductor Allotrope of Carbon. *Computation* **2022**, *10*, 19. [CrossRef]
26. Rubel, O.; Blaha, P. Length-Gauge Optical Matrix Elements in WIEN2k. *Computation* **2022**, *10*, 22. [CrossRef]
27. Zhang, L.; Kozhevnikov, A.; Schulthess, T.; Cheng, H.P.; Trickey, S.B. Performance Enhancement of APW + lo Calculations by Simplest Separation of Concerns. *Computation* **2022**, *10*, 43. [CrossRef]
28. Lafargue-Dit-Hauret, W.; Rocquefelte, X. Influence of the chemical pressure on the magnetic properties of the mixed anion cuprates Cu_2OX_2 (X = Cl, Br, I). *Computation* **2022**, *10*, 73. [CrossRef]
29. Gaumard, R.; Dragún, D.; Pedroza-Montero, J.N.; Alonso, B.; Guesmi, H.; Malkin Ondík, I.; Mineva, T. Regression Machine Learning Models Used to Predict DFT-Computed NMR Parameters of Zeolites. *Computation* **2022**, *10*, 74. [CrossRef]

computation

MDPI

Article

A Theoretical Survey of the UV–Visible Spectra of Axially and Peripherally Substituted Boron Subphthalocyanines †

Al Mokhtar Lamsabhi, M. Merced Montero-Campillo, Otilia Mó and Manuel Yáñez *

Departamento de Química, Módulo 13, Facultad de Ciencias and Institute for Advanced Research in Chemical Sciences (IAdChem), Universidad Autónoma de Madrid, 28049 Madrid, Spain; mokhtar.lamsabhi@uam.es (A.M.L.); mm.montero@uam.es (M.M.M.-C.); otilia.mo@uam.es (O.M.)
* Correspondence: manuel.yanez@uam.es

† This paper is dedicated to Prof. Karlheinz Schwarz on the occasion of his 80th birthday.

Abstract: The UV–visible spectra of a series of subphthalocyanines (SubPcs) characterized by three different axial substituents (A_n) in combination with H, F, NO_2, SO_2H and SO_2CH_3 peripheral substituents (R_i) were predicted and analyzed by means of time-dependent DFT calculations, including chloroform as a solvent. In this analysis, we paid particular attention to the Q band, which remained almost unchanged regardless of the nature of the axial substituent. For the same axial substituent, changes in the Q band were also rather small when hydrogens at the periphery were replaced by $R_1 = SO_2H$ and $R_1 = R_2 = SO_2H$. However, the shifting of the Q band was almost 10 times larger when $R_1 = NO_2$ and $R_1 = R_2 = NO_2$ due to the participation of this substituent in the π SubPc cloud. In most cases, the characteristics of the spectra can be explained considering only the transitions involving the HOMO-1, HOMO, LUMO and LUMO + 1 orbitals, where the Q band can be decomposed into two main contributions, leading to charge separation. Only for SubPc(A_3,F,F,H) would one of the two contributions from the deepest orbital involved not lead to charge transfer. For this latter case, the HOMO-2 orbital must also be taken into account. In summary, the results obtained with the analysis of the MO indicate that the studied SubPcs are appropriate for photochemical devices.

Keywords: subphthalocyanines; UV–visible spectra; axial substituents; peripheral substituents; time-dependent DFT

Citation: Lamsabhi, A.M.; Montero-Campillo, M.M.; Mó, O.; Yáñez, M. A Theoretical Survey of the UV–Visible Spectra of Axially and Peripherally Substituted Boron Subphthalocyanines. *Computation* **2022**, *10*, 14. https://doi.org/10.3390/computation10020014

Academic Editors: Henry Chermette and Giuseppe Zollo

Received: 20 December 2021
Accepted: 14 January 2022
Published: 18 January 2022

1. Introduction

Theoretical and computational chemistry has played a very important role in the development of chemistry since the late 20th century to the present, providing tools to understand experiments and ways to predict the behavior of many different systems. In this sense, the work developed by K. Schwarz is a good example. Suffice it to mention his contributions to the development of functionals to achieve a good description of solid-state materials [1], the WIEN code being a paradigmatic example of this potentiality [2]. In particular, computational chemistry has been equally useful for shedding light on the mechanisms behind the photoactivity of various organic compounds, such as subphthalocyanines (SubPcs) [3].

SubPcs are a typical porphyrin species (see Scheme 1) discovered in the last half of the 20th century [4]. They present a non-planar, 14 π-electron macrocyclic ring that accommodates a B(III) ion at its binding core.

The most frequently used method to incorporate macrocycles for the axial functionalization of SubPcs is substitution at its central boron atom, with the added advantage that the π-conjugated macrocycle maintains its electronic characteristics [5]. This synthetic versatility is behind a very rich and interesting SubPc metallosupramolecular chemistry. It also allows for the preparation of energy donors or energy acceptors based on SubPcs, which lends them ideal properties as photoactive entities through their excited states. SubPcs typically display intense Q-band transitions in the 550–650 nm region [5]. More importantly, their excellent light-harvesting ability, with high extinction coefficients, can be modulated through the nature of the peripheral substituents, which are important factors regarding SubPc photostability [5]. Interestingly, due to the fact that SubPcs have a cone-shaped aromatic structure, the effect of peripheral substituents can be intricate. It has been found [6] that the profile of the absorption spectra depends on whether both substituents are on the same or different sides of the molecular cone.

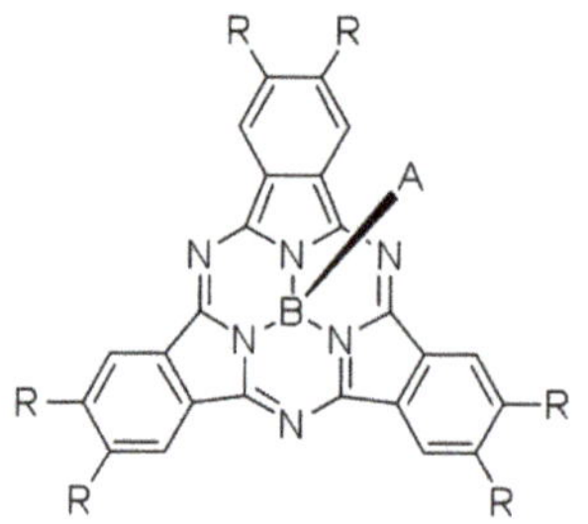

Scheme 1. Subphthalocyanines (SubPcs).

An adequate combination with axial substituents allows for even more fine tuning of SubPcs with excellent properties for photochemical devices, such as a lower tendency to aggregate [5]. A paradigmatic example of photochemical tuning is dodecafluorosubphthalocyanine, an excellent electron-acceptor unit. When this unit is covalently bound to a triphenylamine moiety, which is an excellent electron donor, the resulting derivative becomes an efficient electron donor-acceptor system [3]. Further relevant examples of what can be achieved through axial functionalization have been reported. For instance, SubPcs with axial alkoxo substituents form optically active inclusion complexes with β-cyclodextrin [7], SubPcs dyads with an axial ferrocenyl substituent can exhibit efficient fluorescence quenching [8], and axial aryl-substituted SubPcs exhibit very similar photoluminescence behavior regardless of the nature of the substituent at the para position of the aryl group [9]. SubPcs with a ferrocenylethynyl unit in the axial position exhibit interesting optical and redox properties [10], and rylene-annulated SubPcs also present an excellent photovoltaic performance [11]. SubPcs with ferrocenecarboxylic acid as an axial ligand also combine very interesting redox and photophysical properties, as the first reversible oxidations take place at the axial ligand, while the second oxidation is located in the SubPc ring [12]. Very recently, SubPc hydride derivatives with a high reactivity have been prepared for the first time as hydroboration reagents of aldehydes [13]. Another interesting case is that of the axial-phenoxylated SubPcs, which show spectroscopical properties for the singly reduced and singly oxidized species that can be tuned with changes in axial and peripheral substituents [14], whereas the reversibility of the redox reactions seems to be correlated with some characteristics of the boron-to-axial ligand bond [14]. In regard to larger axial substituents, it has been shown that a novel series of SubPcs with macrocycle axial substitution show interesting behavior as fluorescent probes and photodynamic therapy agents, their biological activity being determined by the nature of the macrocycle [15]. Finally, it was found that SubPc polymers have similar photoluminescence characteristics to those of SubPc units, although the emission of the polymer depends on the excitation wavelength [16].

In our group, we have previously explored, on theoretical grounds, the UV–visible spectra of SubPcs, in which the central boron atom was replaced by aluminum or gallium by using time-dependent DFT calculations, including chloroform as a solvent [17]. This replacement led to a redshift of the Q band, which is not very dependent on either the nature of the peripheral substituents or the nature of the central atom. More recently, the behavior of boron chloride SubPcs with a core-expanded six-membered ring was theoretically investigated, and the compound with three six-membered rings was found to be a promising organic solar cell donor material due to its small exciton binding energy [18].

As indicated above, there are many studies on SubPcs that include a variety of peripheral substituents, but the amount of information is very scarce when dealing with axial substituents. The aim of the present study is to determine the effects on the photochemical behavior of SubPcs of three different axial substituents (An) combined with some of the most common peripheral R_1, R_2 and R_3 substituents, as shown in Scheme 2.

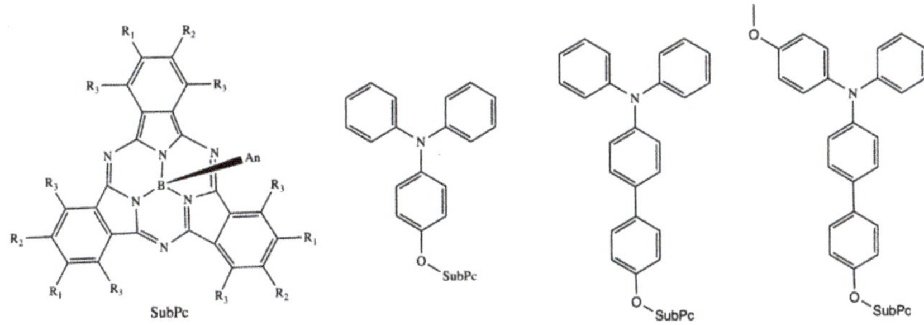

A1 **A2** **A3**

R_1: H, SO_2H, SO_2CH_3, NO_2, F R_2: H, SO_2CH_3, NO_2, F R_3: H, F

Scheme 2. SubPcs(An,R_1,R_2,R_3) envisaged in this study, where An = {A1, A2, A3} are the axial substituents and R_1, R_2 and R_3 are the peripheral substituents.

2. Computational Details

The geometries of all SubPcs were fully optimized in the ground state at the B3LYP/6-31G(d) computational level. The stationary points found were assessed as local minima of the corresponding potential energy surface through the calculation of their harmonic frequencies at the same level of accuracy. To obtain the corresponding UV–Vis absorption spectra, we employed a linear response time-dependent density functional theory (TD-DFT) formalism, using the optimized geometries mentioned above for the ground state. Accordingly, the excitation energies (both in vacuum and using chloroform as a solvent), the oscillator strengths f and the dominant electronic transitions correspond to vertical excitations. These values for all these magnitudes were obtained by single-point calculations at the B3LYP/6-31 + G(d,p) level.

Since all the experimental data are usually obtained in chloroform and not in the gas phase, solvent effects were evaluated by means of the polarizable continuum model (PCM) as implemented in Gaussian 09 programs. More precisely, solvent shifts of the excitation bands were obtained by the nonequilibrium implementation of the PCM through single-point calculations on equilibrium geometries obtained in vacuum. This methodology has been shown to be suitable to reproduce the experimental values in systems similar to those considered in the present publication [3,6,17].

The absorption spectra were plotted using the SWizard program package, represented as a sum of Gaussian functions that follows Equation (1):

$$\varepsilon(\omega) = 2.174 * 10^8 \sum_i \frac{f_i}{\Delta_{1/2}} \exp(-2.773 \frac{(\omega - \omega_i)^2}{\Delta^2_{1/2}}) \tag{1}$$

where ε is the molar absorbance (mol^{-1} cm^{-1} L^{-1}), which depends on the computed excitation energies, ω_i (in cm^{-1}), and their corresponding oscillator strengths, f_i. The half-weight bandwidth is represented by $\Delta^2{}_{1/2}$, and it is assumed to be constant (1200 cm^{-1}).

3. Results and Discussion

For the remainder of this paper, we use the notation SubPc(An,R_1,R_2,R_3) to identify the different SubPcs included in our study (see Scheme 2). For the largest A2 and A3 axial substituents, calculations with SO_2H, SO_2CH_3 and NO_2 as peripheral substituents may become prohibitively expensive. Hence, for A3 only, the parent compound and the all-fluor-substituted derivative are considered, whereas for A2, we add to these two cases the derivative in which R_1 = NO_2. The optimized geometries are reported in Table S1 of Supplementary Materials. Although we do not discuss them in detail, it is important to emphasize that all of them are nonplanar, a basic structural characteristic of these compounds. A good measure of the pyramidalization of the structure is the sum of the bond angles around the boron atom, with a value close to 312 degrees. This value is practically independent of the nature of the peripheral and the axial substituents.

3.1. Calculated UV–Visible Spectra

Since comparisons between the simulated and experimental spectra for the particular case of SubPcs are almost inexistent, we thought it was important, as the first step of our survey, to check the performance of our theoretical scheme. For this purpose, we compare in Figure S1 of Supplementary Materials the available experimental spectrum for SubPc(A2,F,F,F) [3] with our calculated one. The agreement is rather satisfactory, with the maximum displacement being 40 nm for the B band and, most importantly, 10 nm smaller for the Q band.

Although the calculated spectra for the SubPcs to be discussed always correspond to those in $CHCl_3$ solution, we considered it of interest to know whether the solvent effect is important. To check this aspect, we calculated the UV–visible spectra for SubPc(A2,H,H,H) in both the gas phase and in $CHCl_3$ solution. As shown in Figure S2 of Supplementary Materials, the solvent leads to a rather small shifting (~10 nm) of the Q band to lower frequencies, whereas the effects on the B band are almost negligible.

As the agreement with the experimental data and the effect of the solvent are revised, we now focus our attention on the analysis of the impact of the axial and peripheral substituents on the optical properties. Figure 1 shows the effect of replacing the axial substituents in two different structures, namely, SubPc(An,H,H,H) and SubPc(An,F,F,F). The first important and conspicuous fact is that, in all cases, the Q band is not sensitive to the nature of the axial substituent. Instead, the peripheral substituents have a clear effect: in the spectra shown in the (SubPc(An,H,H,H)) case, the maximum for the Q band is at 516 nm, which is shifted to 528 nm for (SubPc(An,F,F,F)). The situation is different as far as the B band is concerned. When there are no peripheral substituents, the maxima appear at 310 nm independently of the nature of the axial substituent, though for A2 and A3, a second secondary maximum is predicted at longer wavelengths (350 and 364 nm, respectively). The situation changes slightly when all the peripheral substituents are F atoms. Now, the maximum of the B band appears at a shorter wavelength for the A1 axial substituent (320 nm) than for A2 and A3 (359 and 369 nm, respectively).

Figure 2 shows a comparison between SubPc(A1,H,H,H), SubPc(A1,SO_2H,H,H) and SubPc(A1,SO_2H,SO_2H,H), where the axial substituent remains unchanged. As evidenced by the picture, the position of the B and Q bands is affected very little by the presence of the sulfonate groups. The maximum of the B band shifts from 316 to 323 nm in the first case and to 328 nm in the second, whereas the shiftings for the Q band from 520 to 524 and 528 nm, respectively, though some changes in their intensity are also observed.

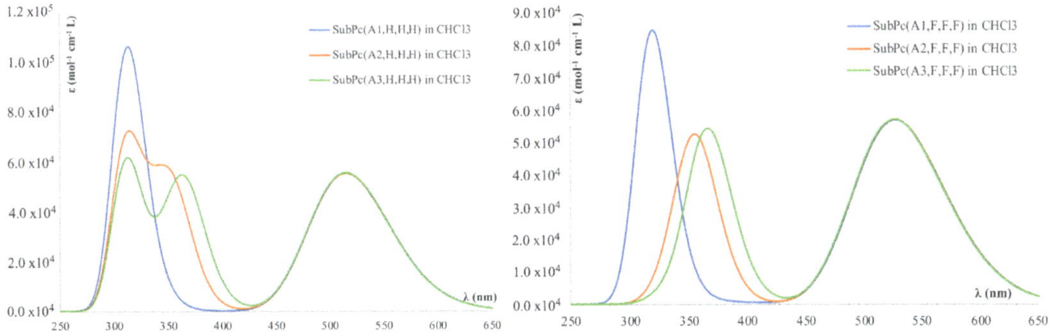

Figure 1. Calculated UV–visible spectra for SubPc(An,R_1,R_2,R_3), with n = 1, 2, 3; $R_1 = R_2 = R_3$ = H, F, dissolved in $CHCl_3$.

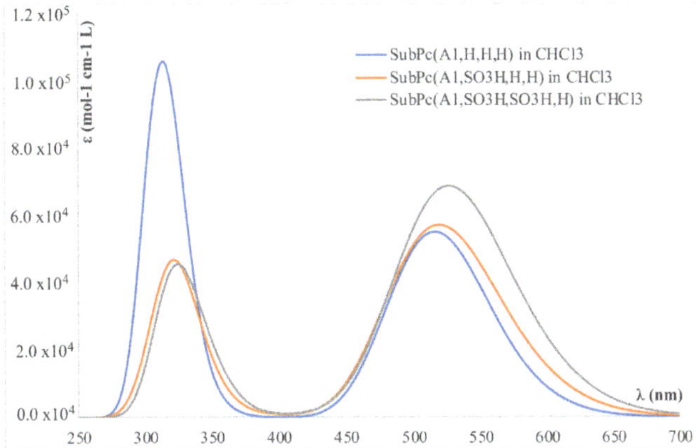

Figure 2. Calculated UV–visible spectra for SubPc(A1,R_1,R_2,R_3) with $R_1 = R_2 = R_3$ = H; $R_1 = SO_2H$, $R_2 = R_3$ = H; $R_1 = R_2 = SO_2H$, R_3 = H, dissolved in $CHCl_3$.

Figure 3 shows the situation when the peripheral substituents are nitro groups rather than SO_2H. In this case, it can be observed that the shifting of the Q band is now significantly larger (from 519 to 576 nm), but, again, it is rather similar for $R_1 = NO_2$ and $R_1 = R_2 = NO_2$. The effects on the B band are also more significant than those observed for SO_2H substituents, in both the intensity of the band and its position. When only R_1 substitutions take place, the B band splits into two peaks of much lower intensity and shifts from 314 to 345 and 427 nm. When R_2 substitutions also take place, only a shoulder at about 433 nm remains. This is very likely due to the ability of the NO_2 groups to participate through their electron lone pair into the aromatic cloud of the SubPc six-membered rings, which does not occur for substituents such as F and SO_2H. We come back to this point in the next section.

We saw above that, when the axial substituent is A1, the replacement of the substitution of R_1 by NO_2 splits the B band into two sub-bands. Figure 4 shows that a similar splitting is observed for axial substituent A2, though the intensity of the first sub-band is much higher than for A1.

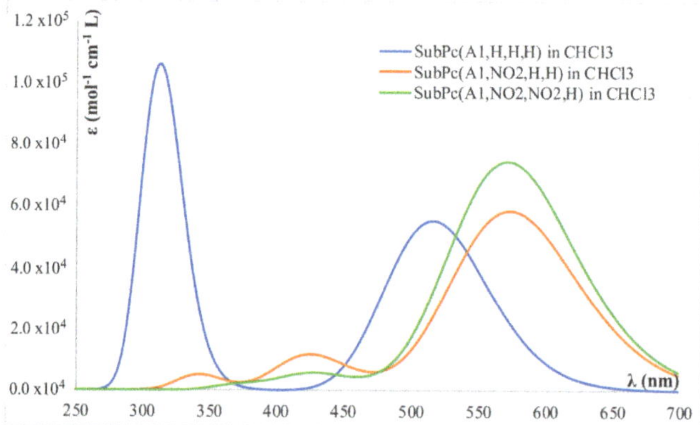

Figure 3. Calculated UV–visible spectra for SubPc(A1,R_1,R_2,R_3), with $R_1 = R_2 = R_3$ = H; $R_1 = NO_2$, $R_2 = R_3$ = H; $R_1 = R_2 = NO_2$, R_3 = H, dissolved in $CHCl_3$.

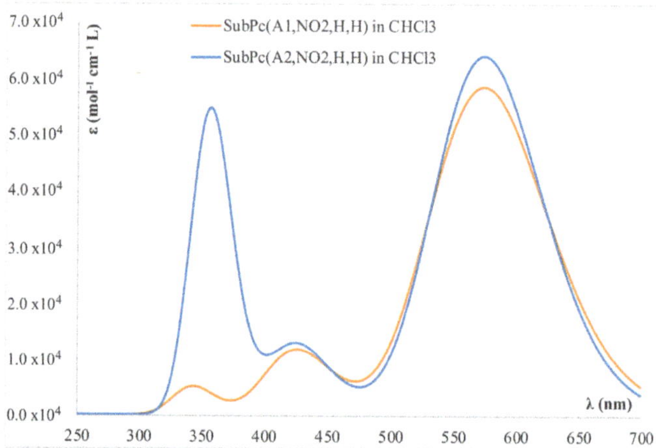

Figure 4. Comparison between the calculated UV-visible spectra of SubPc(A1,NO_2,H,H) and SubPc(A1,NO_2,H,H).

3.2. MO Analysis

The calculated spectra obtained for the SubPcs under analysis can be rationalized through the use of the well-known four-orbital model of Martin Gouterman, who, back in the 1960s, described the spectra of porphyrins [19] and proposed that the characteristics of the Q band observed at about 550 nm for these compounds could be related to the properties of the two top filled and two lowest empty π orbitals involved in the weak excitation to the first excited state (S0 $\rightarrow$ S1). Two years later, this assumption was consistent with rather simple molecular orbital calculations [20], showing, indeed, that the Q band could be associated with transitions involving the HOMO-1, HOMO, LUMO and LUMO + 1 orbitals. The rest of the absorptions of the spectra that form the Soret or B band at lower wavelengths (around 400 nm) are associated with S0 —> S2 electronic excitations. The same model was also successfully used to rationalize the Q bands of subporphyrins [21], subporphyrazines and SubPcs [6,17,22–25], as well as Be-SubPcs [26] complexes. As explained below, the results show that this model is also valid for the transitions observed in our set, but it needs to be expanded in some specific cases to include the HOMO-2 orbital.

In Table 1, we summarize the most relevant information for the Q band for the species investigated in solution. Similar information in vacuum is provided in Table S2 of Supplementary Materials.

Table 1. Q-band (λ_{max}) excitation energies in chloroform, oscillator strengths *f* and dominant electronic transitions for the SubPcs included in this study [a].

An	R_1	R_2	R_3	lmax (eV)	nm	*f*	Main Configuration
A1	H	H	H	2.40	516.5	0.4107	H-1→L + 1(+87%)
				2.41	515.5	0.4403	H-1→L + 2(+87%)
A1	F	F	F	2.35	527.9	0.4230	H-1→L + 1(+99%)
				2.36	526.3	0.4534	H-1→L + 2(+99%)
A1	SO_2H	H	H	2.38	520.8	0.4749	H-1→L + 0(+97%)
				2.39	519.2	0.5087	H-1→L + 1(+97%)
A1	SO_2CH_3	SO_2CH_3	H	2.35	527.2	0.5735	H-1→L + 0(+97%)
				2.36	525.6	0.6106	H-1→L + 1(+97%)
A1	SO_2CH_3	H	H	2.39	519.8	0.4762	H-1→L + 0(+97%)
				2.39	518.1	0.5091	H-1→L + 1(+97%)
A1	NO_2	H	H	2.16	573.7	0.4387	H-1→L + 0(+98%)
				2.17	572.2	0.4614	H-1→L + 1(+97%)
A1	NO_2	NO_2	H	2.16	572.7	0.5569	H-1→L + 0(+98%)
				2.18	569.9	0.588	H-1→L + 1(+97%)
A2	H	H	H	2.4	515.9	0.4108	H-1→L + 0(+97%)
				2.41	514.8	0.4432	H-1→L + 1(+97%)
A2	NO_2	H	H	2.16	573.6	0.3952	H-1→L + 0(+86%)
				2.17	572.6	0.4249	H-1→L + 1(+85%)
A2	F	F	F	2.35	527.5	0.4215	H-1→L + 0(+97%)
				2.36	526.5	0.449	H-1→L + 1(+96%)
A3	H	H	H	2.4	515.8	0.4102	H-1→L + 0(+97%)
				2.41	514.8	0.4452	H-1→L + 1(+97%)
A3	F	F	F	2.34	529.7	0.1861	H-2→L + 0(+55%); H-1→L + 1(39%)
				2.35	527.9	0.3587	H-1→L + 0(+82%); H-2→L + 1(+13%)
				2.37	523.2	0.2743	H-1→L + 1(+56%) H-2→L + 0(+41%)

[a] H-2, H-1, L + 0, L + 1 stands for HOMO-2, HOMO-1, LUMO and LUMO + 1, respectively.

The results in Table 1 show that, for all the systems investigated, there are two main transitions responsible for the Q band at very similar wavelengths. The first one, Q_1, is associated with the HOMO-1 → LUMO transition, and the second one, Q_2, is associated with the HOMO-1 → LUMO + 1 transition, which indicates that, in practically all the systems investigated, the LUMO and LUMO + 1 orbitals are almost degenerate. It can also be observed that SubPc(A3,F,F,F) is the only exception to this general behavior, as we discuss later.

The transitions mentioned above as being responsible for the Q band are illustrated in Figure 5 for the particular case of SubPc(A2,H,H,H), which is taken as a suitable example, but this same scheme is also observed for SubPc(A1,H,H,H), SubPc(A1,NO_2,H,H), SubPc(A1,NO_2,NO_2,H), SubPc(A1,SO_2H,H,H), SubPc(A1,SO_2CH_3,H,H), SubPc(A1,SO_2CH_3, SO_2CH_3,H), SubPc(A2,NO_2,H,H) and SubPc(A2,F,F,F). Very importantly, this figure clearly illustrates that, for all these systems, the HOMO-1 orbital is located at the axial substituent, whereas both the LUMO and LUMO + 1 orbitals are localized at the SubPc moiety, with the obvious consequence that the Q band is associated with a charge transfer phenomenon from the axial substituent toward the SubPc moiety.

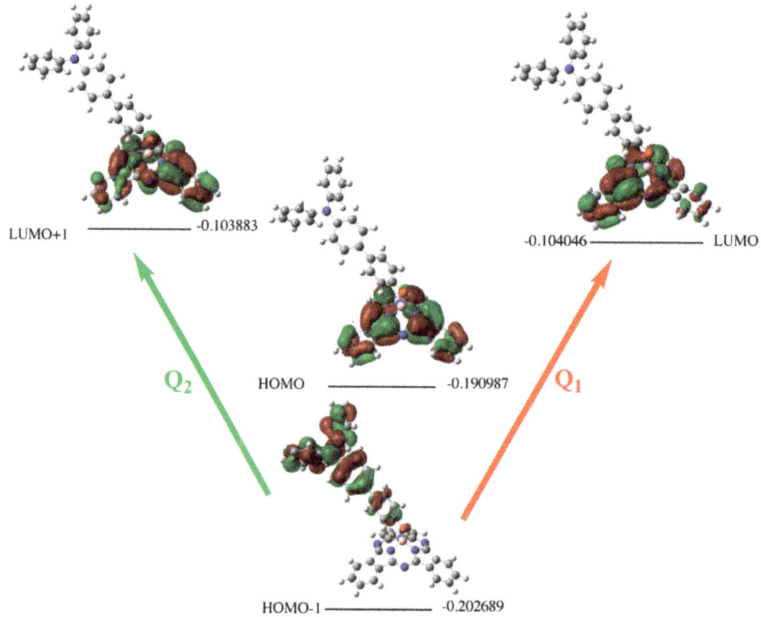

Figure 5. MO diagram showing, for SubPc(A2,H,H,H), the main transitions Q_1 and Q_2 contributing to the Q band, which are accompanied by a charge transfer phenomenon. Molecular orbital energies are in hartrees.

The fact that the LUMO and the LUMO + 1 orbitals are localized at the SubPc moiety also permits the explanation as to why, as already mentioned in the previous section, the shifting of the Q band is much larger for $R_1 = NO_2$ and $R_1 = R_2 = NO_2$ with respect to the parent compound ($R_1 = R_2 = H$) than it is for $R_1 = SO_2H$ and $R_1 = R_2 = SO_2H$. Indeed, as illustrated in Figure 5, the two main components of the Q band correspond to transitions that reach these two empty orbitals, whose energies change significantly when the H atoms are replaced by NO_2 groups. These substituents, as shown in Figure S3 of Supplementary Materials, conjugate with the aromatic π-system through their N lone pair. This interaction cannot occur when R_1 = H or $R_1 = SO_2H$, as evidenced in Figure S3, as these groups participate neither in the LUMO nor in the LUMO + 1 orbitals.

Rather interestingly, when A2 is replaced by A3 as an axial substituent, the dominant transitions are still HOMO-1 $\rightarrow$ LUMO for the Q_1 component and HOMO-1 $\rightarrow$ LUMO + 1 for the Q_2 one. However, the presence of the methoxy group at the axial substituent stabilizes the orbital located at the SubPc subunit, which becomes the HOMO-1 orbital, whereas the orbital localized at the axial substituent becomes the HOMO orbital (see Figure 6). As a consequence, the Q absorption band of SubPc(A3,H,H,H) is not accompanied by a charge transfer phenomenon, because the two Q_1 and Q_2 components involve electron densities that are always localized at the SubPc subunit.

The situation becomes somewhat more complicated when moving from the parent compound to the derivative, in which all peripheral substituents are F atoms. In this case, some transitions have their origin not only in the HOMO-1 orbital but also in the HOMO-2 orbital, so the Q band has three rather than two components (see Figure 7).

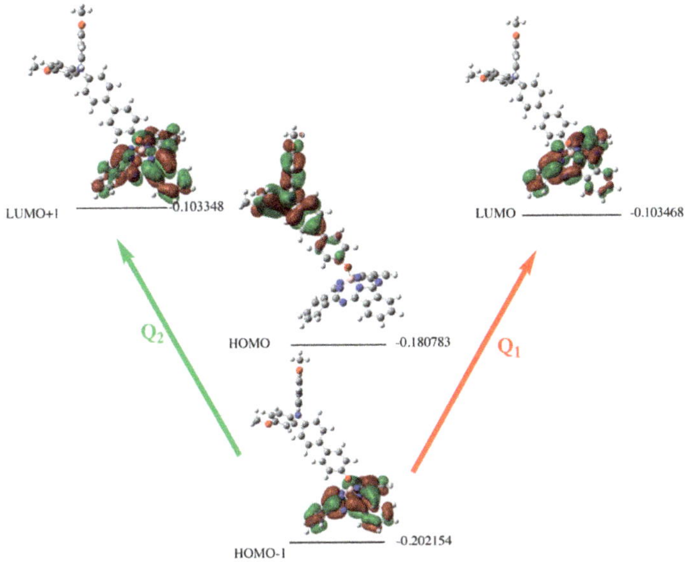

Figure 6. MO diagram of SubPc(A3,H,H,H), showing the main transitions Q_1 and Q_2 contributing to the Q band, which is not accompanied by a charge transfer phenomenon. Molecular orbital energies are in hartrees.

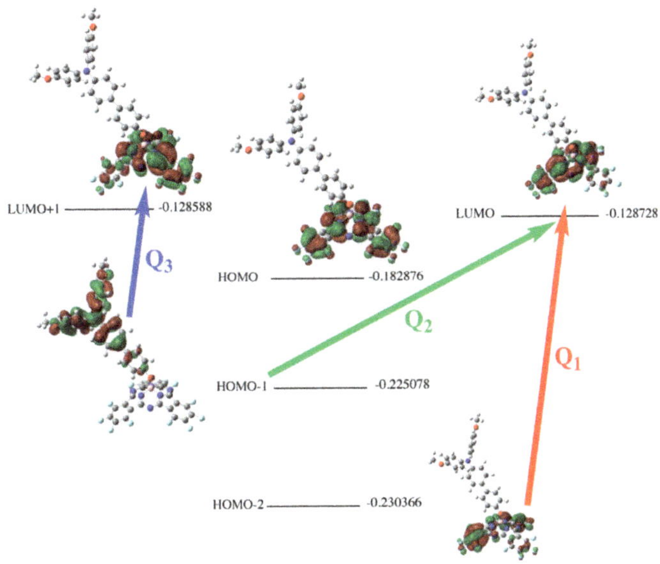

Figure 7. MO diagram showing, for SubPc(A3,F,F,H), the main transitions Q_1, Q_2 and Q_3 contributing to the Q band. The Q_2 and Q_3 components are accompanied by a charge transfer phenomenon. Molecular orbital energies are in hartrees.

More specifically, the following transitions are detected:

Q_1 component: HOMO-2 → LUMO(55%); HOMO-1 → LUMO + 1(39%)

Q_2 component: HOMO-1 → LUMO(82%); HOMO-2 → LUMO + 1(13%)

Q_3 component: HOMO-1 → LUMO + 1(56%); HOMO-2 → LUMO(41%)

Hence, globally, the Q band is accompanied by a charge transfer phenomenon but only through the Q_2 and Q_3 components.

4. Conclusions

The electronic spectra of SubPc are typically characterized by two main absorption bands: one typically in the visible region, called the Q band, and another at lower wavelengths, named the B band or the Soret band. In our theoretical survey, we found that the Q band is not sensitive to axial replacement for SubPcs with H or F peripheral substituents. The situation is not very different when it is the axial substituent that remains unchanged (A1) and peripheral hydrogens that are replaced by $R_1 = SO_2H$ and $R_1 = R_2 = SO_2H$. Indeed, in this case, the Q band is only slightly shifted by 4 and 8 nm. The shiftings when $R_1 = NO_2$ and $R_1 = R_2 = NO_2$ are rather similar but almost 10 times larger than those observed for SO_2H.

In general, the effects on the characteristics of the B band are more apparent. For instance, the Q band is not altered when all peripheral substituents are fluorine atoms and when the axial substituent A1 is replaced by A2 or A3, but the B band is shifted 40 nm to longer wavelengths. Similarly, the B band splits into two peaks of much lower intensity and shifts from 31 and 113 nm to longer wavelengths when $R_1 = H$ is replaced by $R_1 = NO_2$, keeping the A1 axial substituent unchanged. These effects are slightly stronger when both R_1 and R_2 are replaced.

In most cases, the characteristics of the spectra can be explained by considering only the transitions involving the HOMO-1, HOMO, LUMO and LUMO + 1 orbitals, as in the well-known Gouterman model [19]. Nevertheless, when the axial substituent is A3, we found that the spectra can only be accounted for if transitions from the HOMO-2 orbital are also taken into account.

From a global perspective, the results collected in this work and those previously reported for other substitution patterns indicate that the optical properties of boron SubPcs regarding the Q band are very robust toward axial and peripheral substitutions.

Supplementary Materials: The following supporting information can be downloaded at: https://www.mdpi.com/article/10.3390/computation10020014/s1, Figure S1: Comparison between the calculated and the experimental UV-Vis spectra for SubPc (A2,F,F,F) dissolved in $CHCl_3$; Figure S2: Comparison between the UV-Vis spectrum of SubPc (A2,H,H,H) in the gas-phase and in CHCl3 solution; Figure S3: LUMO of SubPc (A1,H,H,H), SubPc (A1,SO_2CH_3, SO_2CH_3,H) and SubPc (A1,NO_2,H,H); Table S1: Cartesian coordinates; Table S2: Q-Band excitation energies in vacuum, oscillator strengths and dominant electronic transitions.

Author Contributions: Conceptualization, M.Y. and A.M.L.; original draft preparation, M.Y.; writing, review and editing, M.M.M.-C., O.M., M.Y. and A.M.L.; formal analysis, M.M.M.-C., O.M., M.Y. and A.M.L.; methodology, M.M.M.-C. and A.M.L.; investigation, M.M.M.-C., O.M., M.Y. and A.M.L.; project administration, O.M.; funding acquisition, O.M. and M.Y. All authors have read and agreed to the published version of the manuscript.

Funding: This work was carried out with financial support from the projects PGC2018-094644-B-C21 and PID2019-110091GB-I00 funded by the Ministerio de Ciencia, Innovación y Universidades of Spain (MICIN/AEI) and the PRIES-CM project Ref: Y2020/EMT-6290 from the Comunidad Autónoma de Madrid.

Institutional Review Board Statement: Not applicable.

Acknowledgments: Thanks are also given to the Centro de Computación Científica of the UAM (CCC-UAM) for the generous allocation of computer time and for their continued technical support.

Conflicts of Interest: The authors declare no conflict of interest.

References

1. Blaha, P.; Schwarz, K.; Sorantin, P.; Trickey, S.B. Full-potential, linearized augmented plane-wave programs for crystalline systems. *Comput. Phys. Commun.* **1990**, *59*, 399–415. [CrossRef]
2. *IAEA Wien Automatic System Planning (WASP) Package A Computer Code for Power Generating System Expansion Planning Version WASP-IV*; Computer Manual Series no. 16; International Atomic Energy Agency: Vienna, Austria, 2001.
3. Medina, A.S.; Claessens, C.G.; Rahman, G.M.A.; Lamsabhi, A.M.; Mó, O.; Yáñez, M.; Guldi, D.M.; Torres, T. Accelerating charge transfer in a triphenylamine-subphthalocyanine donor-acceptor system. *Chem. Commun.* **2008**, *15*, 1759–1761. [CrossRef] [PubMed]
4. Meller, A.; Ossko, A. Triisindolo 1,2,3-cd-1′,2′,3′-gh-1″,2″,3″-kl 2,3a,5,6a,8,9a,9b-Hena-azaboraphenalene. *Monatsh. Chem.* **1972**, *103*, 150–153. [CrossRef]
5. Claessens, C.G.; González-Rodríguez, D.; Rodríguez-Morgade, M.S.; Medina, A.; Torres, T. Subphthalocyanines, Subporphyrazines, and Subporphyrins: Singular Nonplanar Aromatic Systems. *Chem. Rev.* **2014**, *114*, 2192–2277. [CrossRef] [PubMed]
6. Lamsabhi, A.; Yáñez, M.; Mó, O.; Trujillo, C.; Blanco, F.; Alkorta, I.; Elguero, J.; Caballero, E.; Rodríguez-Morgade, M.S.; Claessens, C.G.; et al. TDDFT study of the UV-vis spectra of subporphyrazines and subphthalocyanines. *J. Porphyr. Phthalocyanines* **2011**, *15*, 1220–1230. [CrossRef]
7. Adachi, K.; Watarai, H. Site-selective formation of optically active inclusion complexes of alkoxo-subphthalocyanines with beta-cyclodextrin at the toluene/water interface. *Chem. Eur. J.* **2006**, *12*, 4249–4260. [CrossRef] [PubMed]
8. Solntsev, P.V.; Spurgin, K.L.; Sabin, J.R.; Heikal, A.A.; Nemykin, V.N. Photoinduced Charge Transfer in Short-Distance Ferrocenyl-subphthalocyanine Dyads. *Inorg. Chem.* **2012**, *51*, 6537–6547. [CrossRef]
9. Bonnier, C.; Josey, D.S.; Bender, T.P. Aryl-Substituted Boron Subphthalocyanines and their Application in Organic Photovoltaics. *Austral. J. Chem.* **2015**, *68*, 1750–1758. [CrossRef]
10. Muñoz, A.V.; Gotfredsen, H.; Jevric, M.; Kadziola, A.; Hammerich, O.; Nielsen, M.B. Synthesis and Properties of Subphthalocyanine-Tetracyanobutadiene-Ferrocene Triads. *J. Org. Chem.* **2018**, *83*, 2227–2234. [CrossRef]
11. Huang, T.D.; Chen, H.; Feng, J.J.; Zhang, A.D.; Jiang, W.; He, F.; Wang, Z.H. Rylene Annulated Subphthalocyanine: A Promising Cone-Shaped Non-Fullerene Acceptor for Organic Solar Cells. *Acs Mater. Lett.* **2019**, *1*, 404–409. [CrossRef]
12. Swarts, P.J.; Conradie, J. Redox and Photophysical Properties of Four Subphthalocyanines Containing Ferrocenylcarboxylic Acid as Axial Ligands. *Inorg. Chem.* **2020**, *59*, 7444–7452. [CrossRef]
13. Tejerina, L.; Labella, J.; Martínez-Fernández, L.; Corral, I.; Martínez-Diaz, M.V.; Torres, T. Subphthalocyaninato Boron (III) Hydride: Synthesis, Structure and Reactivity. *Chem. Eur. J.* **2021**, *27*, 12058–12062. [CrossRef]
14. Sampson, K.L.; Jiang, X.Q.; Bukuroshi, E.; Dovijarski, A.; Raboui, H.; Bender, T.P.; Kadish, K.M. A Comprehensive Scope of Peripheral and Axial Substituent Effect on the Spectroelectrochemistry of Boron Subphthalocyanines. *J. Phys. Chem. A* **2018**, *122*, 4414–4424. [CrossRef] [PubMed]
15. van de Winckel, E.; Mascaraque, M.; Zamarron, A.; de la Fuente, A.J.; Torres, T.; de la Escosura, A. Dual Role of Subphthalocyanine Dyes for Optical Imaging and Therapy of Cancer. *Adv. Funct. Mater.* **2018**, *28*, 1705938. [CrossRef]
16. Lessard, B.H.; Sampson, K.L.; Plint, T.; Bender, T.P. Boron subphthalocyanine polymers: Avoiding the small molecule side product and exploring their use in organic light-emitting diodes. *J. Polym. Sci. A* **2015**, *53*, 1996–2006. [CrossRef]
17. Montero-Campillo, M.M.; Lamsabhi, A.M.; Mó, O.; Yáñez, M. UV/Vis Spectra of Subporphyrazines and Subphthalocyanines with Aluminum and Gallium: A Time-Dependent DFT Study. *Chemphyschem* **2013**, *14*, 915–922. [CrossRef] [PubMed]
18. Peng, S.P.; Zheng, S.H. A computational investigation on core-expanded subphthalocyanines. *Int. J. Quant. Chem.* **2019**, *119*, e25942. [CrossRef]
19. Gouterman, M. Spectra of porphyrins. *J. Mol. Spect.* **1961**, *6*, 138–163. [CrossRef]
20. Gouterman, M.; Wagnière, G.H.; Snyder, L.C. Spectra of porphyrins: Part II. Four orbital model. *J. Mol. Spect.* **1963**, *11*, 108–127. [CrossRef]
21. Takeuchi, Y.; Matsuda, A.; Kobayashi, N. Synthesis and characterization of meso-triarylsubporphyrins. *J. Am. Chem. Soc.* **2007**, *129*, 8271–8281. [CrossRef]
22. Shimizu, S.; Yamazaki, Y.; Kobayashi, N. Tetrathiafulvalene-Annulated Subphthalocyanines. *Chem. Eur. J.* **2013**, *19*, 7324–7327. [CrossRef]
23. Liang, X.; Shimizu, S.; Kobayashi, N. Sizeable red-shift of absorption and fluorescence of subporphyrazine induced by peripheral push and pull substitution. *Chem. Commun.* **2014**, *50*, 13781–13784. [CrossRef]
24. Liu, Q.; Shimizu, S.; Kobayashi, N. Cyclophanes Containing Bowl-Shaped Aromatic Chromophores: Three Isomers of anti- 2.2 (1,4)Subphthalocyaninophane. *Angew. Chem. Int. Ed.* **2015**, *54*, 5187–5191. [CrossRef] [PubMed]
25. Esteso, V.; Calio, L.; Espinos, H.; Lavarda, G.; Torres, T.; Feist, J.; García-Vidal, F.J.; Bottari, G.; Míguez, H. Light-Harvesting Properties of a Subphthalocyanine Solar Absorber Coupled to an Optical Cavity. *Sol. Rrl* **2021**, *5*, 2100308. [CrossRef]
26. Montero-Campillo, M.M.; Lamsabhi, A.; Mó, O.; Yáñez, M. Photochemical Behavior of Beryllium Complexes with Subporphyrazines and Subphthalocyanines. *J. Phys. Chem. A* **2016**, *120*, 4845–4852. [CrossRef] [PubMed]

computation

MDPI

Article

Organic Emitters Showing Excited-States Energy Inversion: An Assessment of MC-PDFT and Correlation Energy Functionals Beyond TD-DFT

Juan-Carlos Sancho-García *,† and Emilio San-Fabián *,†

Department of Physical Chemistry, University of Alicante, E-03080 Alicante, Spain
* Correspondence: jc.sancho@ua.es (J.-C.S.-G.); sanfa@ua.es (E.S.-F.)
† These authors contributed equally to this work.

Abstract: The lowest-energy singlet (S_1) and triplet (T_1) excited states of organic conjugated chromophores are known to be accurately calculated by modern wavefunction and Time-Dependent Density Functional Theory (TD-DFT) methods, with the accuracy of the latter heavily relying on the exchange-correlation functional employed. However, there are challenging cases for which this cannot be the case, due to the fact that those excited states are not exclusively formed by single excitations and/or are affected by marked correlation effects, and thus TD-DFT might fall short. We will tackle here a set of molecules belonging to the azaphenalene family, for which research recently documented an inversion of the relative energy of S_1 and T_1 excited states giving rise to a negative energy difference (ΔE_{ST}) between them and, thereby, contrary to most of the systems thus far treated by TD-DFT. Since methods going beyond standard TD-DFT are not extensively applied to excited-state calculations and considering how challenging this case is for the molecules investigated, we will prospectively employ here a set of non-standard methods (Multi-Configurational Pair Density Functional Theory or MC-PDFT) and correlation functionals (i.e., Lie–Clementi and Colle–Salvetti) relying not only on the electronic density but also on some modifications considering the intricate electronic structure of these systems.

Keywords: TD-DFT; MC-PDFT; Lie–Clementi; Colle–Salvetti; OLEDs

Citation: Sancho-García, J.C.; San-Fabián,E. Violation of Hund's Rule in Organic Emitters: An Assessment of MC-PDFT and Correlation Energy Functionals Beyond TD-DFT. *Computation* **2022**, *10*, 13. https://doi.org/10.3390/computation10020013

Academic Editor: Henry Chermette

Received: 23 December 2021
Accepted: 12 January 2022
Published: 18 January 2022

Publisher's Note: MDPI stays neutral with regard to jurisdictional claims in published maps and institutional affiliations.

1. Introduction

The violation of Hund's rule in molecules [1], analogously to atoms, is commonly ascribed to an inversion of the excitation energies of the lowest states of spin-singlet (S_1) or spin-triplet (T_1) multiplicity. In common situations, the energy difference between S_1 and T_1 excited states, that is $\Delta E_{ST} = E(S_1) - E(T_1)$ is positive, contrary to what happens if Hund's rule is altered (in that case, ΔE_{ST} would be negative). Note that the negative sign contradicts the fact that the exchange energy (K) is normally thought to be positive, historically $\Delta E_{ST} \approx 2K$, thus, implying that the lowest singlet excited state lies energetically above the lowest triplet excited state and not the opposite.

That exchange energy is known to be of the order of hundreds of meV for common organic chromophores, and this becomes a key parameter for photophysics and related applications [2–4]. However, it has been demonstrated that strong correlation effects can decrease the ΔE_{ST} value [5,6] and even invert the energies of S_1 and T_1 excited states due to a more marked stabilization of the former vs. the latter state [7], although very few molecules are discovered up to now displaying such an excited-state energy inversion. Additionally, fast environmental effects (whenever they are reliably introduced) could also lead to negative ΔE_{ST} values.

Among those environmental effects, we mention thermal fluctuations of molecular conformations or microscopic electronic polarization effects in amorphous films of carbazole derivatives [8], thus, opening a whole world of future studies and applications around this unexpected issue. However, for solvation effects, current implementations of continuum solvation models should be employed with caution since it could lead to spurious excited-state energy inversion [9].

The physical origin of this inversion and its possible practical implications should not merely be considered as an academic questions. As an example of practical use, due to the spin statistics, triplet excitons (dark) are known to be formed upon a 3:1 ratio with respect to singlet excitons (bright) thus limiting the efficiency of electroluminiscent processes. Therefore, the mentioned energy inversion could be further exploited in the, e.g., recovery of triplet excitons created upon electroluminiscence to increase the internal quantum efficiency or quantum yields of Organic Light-Emitting Diodes (OLEDs).

Many other applications in related fields (photocatalysts, covalent organic frameworks, liquid crystals, etc.) have also been envisioned and recently reviewed [10]. However, from an experimental point of view, the range of disclosed molecules showing that excited-state energy inversion is very limited, and this goes back to the discovery in the 1980s of some azaphenalene molecules candidates [11,12] for such a violation of Hund's rule. Those initial molecules were also extended to other (not-yet-synthesized) candidates after a massive screening of compounds recently performed [13], showing that the topic is still open and under active investigation.

Therefore, theoretical methods based on one-electron excitations (i.e., Time-Dependent Density Functional Theory or TD-DFT) are questioned in its current implementation to recognize that excited-state energy inversion, due to the lack of inclusion of higher-than-single excitations into their formulation. In contrast, wavefunction methods have been shown in recent studies [14–21] to provide reasonably accurate values for that ΔE_{ST} energy difference, although at a cost certainly higher than TD-DFT.

That limitation of TD-DFT is known to occur independently of the underlying exchange-correlation functional and basis sets chosen. However, since excited-state wavefunction methods can capture double and higher-order excitations by definition, depending of the truncation done for the excitation operator, those methods are able to predict the excited-state energy inversion while concomitantly providing accurate individual (i.e., S_1 and T_1) energies for the excited-states involved.

Based on these findings, our goal here is to investigate if methods going beyond standard (TD-)DFT could predict that excited-state energy inversion and thus compete in accuracy with wavefunction results. To assess that, we will employ methods merging wavefunction and correlation functionals, in the hope of including both kind of correlation effects (dynamical and non-dynamical) for any of the electronic states involved. These results will also serve to confirm the key role played by marked correlation effects, as well as to invigorate more research of DFT methods out of the most commonly found implementations.

2. Systems, Methods, and Computational Details

2.1. *Choice of the Target Systems*

The set of systems selected is exclusively motivated by the previous experimental and theoretical findings mentioned above. The set of azaphenalene molecules shown in Figure 1 is known to display negative ΔE_{ST} values at various wavefunction levels, from pioneering studies [14–16] later extended to related and/or larger systems [17–21], which clearly constitutes a challenge for any theoretical method.

Interestingly, chemical substitution of the heptazine core $-C_6H_7H_3-$ with chlorine $-C_6H_7Cl_3-$, cyano $-C_6H_7(CN)_3-$, or p-methoxyphenylene $-C_6N_7(p\text{-}C_6H_4OCH_3)_3-$ groups preserved the negative ΔE_{ST} value, showing that the chemical structure of the core is indeed responsible as well as that more potential and synthetically viable molecules could soon be theoretically disclosed and/or experimentally achieved.

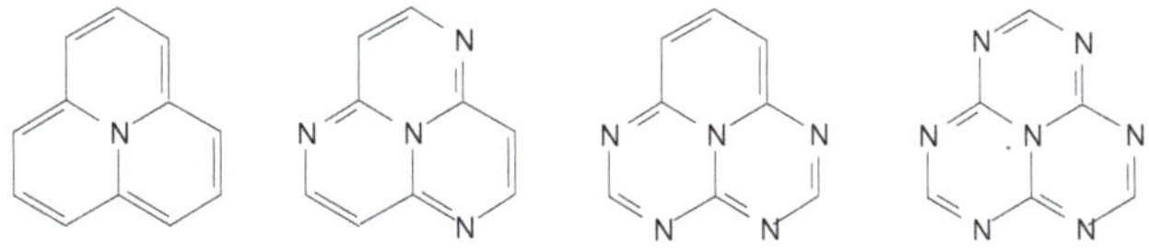

MAP 4AP 5AP 7AP

Figure 1. Chemical structures (from left to right) of the molecules MAP (monoazaphenalene or cyclazine), TAP (tetraazaphenalene), 5AP (pentaazaphenalene), and 7AP (heptaazaphenalene or heptazine). Hydrogen atoms are omitted for clarity.

From the experimental point of view, some studies that were performed in the 1980s also concluded with a real possibility of having an excited-state inversion for some of these azaphenalenes [11,12]. More recently, some heptazine derivatives have been successfully employed as emitters in real devices [22,23], with an exceptionally high quantum yield reported and explained by the (partial) conversion of triplet into singlet excitons possibly assisted by a negative ΔE_{ST} value.

Note also that other related cores could be also potential candidates, after the conclusions reached by a massive screening of thousands of potential azaphenalene candidates by Pollice et al. [13]. However, we are more interested in assessing the reliability of theories going beyond (TD-)DFT and we will, thus, restrict this work to the compounds shown in Figure 1 for which reference results are available as well.

2.2. Physical Meaning of Reduced Density Matrices

While the electronic density, or first-order reduced density matrix, is given by:

$$\rho(\mathbf{r}) = N \int \Psi^{\star}(\mathbf{r}, \mathbf{r}_2, \ldots, \mathbf{r}_N) \Psi(\mathbf{r}, \mathbf{r}_2, \ldots, \mathbf{r}_N) d\mathbf{r}_2 \ldots d\mathbf{r}_N, \tag{1}$$

and thus integrates over the number of electrons N, $\int \rho(\mathbf{r}) d\mathbf{r} = N$, the corresponding spinless second-order reduced density matrix integrates to the total number of interacting electron pairs:

$$\iint \rho_2(\mathbf{r}_1, \mathbf{r}_2) d\mathbf{r}_1 d\mathbf{r}_2 = \frac{N(N-1)}{2}, \tag{2}$$

and represents the probability density of finding a particle at point $\mathbf{r}_1$ and simultaneously another particle at point $\mathbf{r}_2$. The explicit form is given by:

$$\gamma_2(\mathbf{x}_1', \mathbf{x}_2'; \mathbf{x}_1, \mathbf{x}_2) = \frac{N(N-1)}{2} \int \cdots \int \gamma_N(\mathbf{x}_1', \mathbf{x}_2', \mathbf{x}_3 \ldots, \mathbf{x}_N; \mathbf{x}_1, \mathbf{x}_2, \mathbf{x}_3, \ldots, \mathbf{x}_N) d\mathbf{x}_3 \ldots d\mathbf{x}_N, \tag{3}$$

or better its reduced form $\xi_2(\mathbf{r}_1', \mathbf{r}_2'; \mathbf{r}_1, \mathbf{r}_2) = \sum_{s_1, s_2} = \gamma_2(\mathbf{x}_1', \mathbf{x}_2'; \mathbf{x}_1, \mathbf{x}_2)|_{s_i' = s_i}$. Finally, the diagonal element or $\rho_2(\mathbf{r}_1, \mathbf{r}_2)$ would account from any correlation effect arising from interparticle interaction, as it can be easily seen from the electron–electron mean value as a function of this new magnitude:

$$\langle \hat{V}_{ee} \rangle = \iint \frac{\rho_2(\mathbf{r}_1, \mathbf{r}_2)}{|\mathbf{r}_1 - \mathbf{r}_2|} d\mathbf{r}_1 d\mathbf{r}_2. \tag{4}$$

Note also that $\rho(\mathbf{r})$ and $\rho_2(\mathbf{r}_1, \mathbf{r}_2)$ are related through:

$$\rho(\mathbf{r}) = \frac{2}{N-1} \int \rho_2(\mathbf{r}_1, \mathbf{r}_2) d\mathbf{r}_2. \tag{5}$$

In the following, we will denote $\rho_2(\mathbf{r}) = \rho_2(\mathbf{r}_1, \mathbf{r}_2)|_{\mathbf{r}_1 = \mathbf{r}_2}$ as the function at the two-electron coalescence point, whose modelling has been extensively pursued in the past [24–26], as well as its integration into excited-state formalisms [27,28], as the next step for the description of electronic structure beyond the use of merely the electronic density $\rho(\mathbf{r})$.

2.3. Theories Going beyond (TD-)DFT

The methods included in this study can be categorized in popular language as methods going beyond (standard) DFT, in the sense that they are based on the on-top second-order reduced density matrix $\rho_2(\mathbf{r})$, and not only on the first-order density $\rho(\mathbf{r})$, or post-MCSCF, in the sense that a Multi-Configurational Self-Consistent Field (MCSCF) calculations needs to be done first from which the magnitude $\rho_2(\mathbf{r})$ is obtained.

Note that the on-top second-order reduced density matrix represents the probability that two opposite-spin electrons are found at point **r** and integrates to the total number of interacting pairs. Multiconfiguration Pair-Density Functional Theory (MC-PDFT [29–32]) can be thus viewed as a post-MCSCF method that evaluates the energy of any state with on-top pair-density function theory. Basically, for a MCSCF wavefunction, $|\Psi^{\text{MCSCF}}\rangle = \sum_\mu C_\mu |\Psi_\mu\rangle$, one can obtain the total electronic energy as:

$$E = \langle \Psi^{\text{MCSCF}} | \hat{T} | \Psi^{\text{MCSCF}} \rangle + \int v(\mathbf{r})\rho(\mathbf{r})d\mathbf{r} + \int\int \frac{\rho(\mathbf{r}_1)\rho(\mathbf{r}_2)}{|\mathbf{r}_1 - \mathbf{r}_2|} d\mathbf{r}_1 d\mathbf{r}_2 + E_{xc}[\rho(\mathbf{r}), \rho_2(\mathbf{r})], \quad (6)$$

with the different terms being the kinetic, potential, Coulomb, and exchange-correlation energies, respectively, the latter relying on a modification of a common DFT functional to be employed together with Equation (1). The most extensively tested on-top density functional is called tPBE and will, thus, be used here consequently.

On the other hand, one can also directly employ a functional explicitly depending upon the design of the on-top second-order reduced density matrix, such as the Colle–Salvetti (CS [33–35]) correlation functional. We will denote, in the following, the CS expression as a two-body correlation functional, cast as $E_c[\rho(\mathbf{r}), \rho_2(\mathbf{r})]$, in contrast with conventional or one-body functionals commonly used for standard DFT calculations, or simply $E_c[\rho(\mathbf{r})]$. Note that the famous Lee–Yang–Parr (LYP [36]) correlation functional is a reformulation of the Colle–Salvetti expression to avoid the explicit dependence on $\rho_2(\mathbf{r})$ at the price of neglecting its use with, e.g., MCSCF wavefunctions. For this case, the total electronic energy is calculated by a two-step procedure,

$$E = \langle \Psi^{\text{MCSCF}} | \hat{T} + \hat{V}_{Ne} + \hat{V}_{ee} | \Psi^{\text{MCSCF}} \rangle + E_c[\rho(\mathbf{r}), \rho_2(\mathbf{r})], \quad (7)$$

after adding the correlation (mainly dynamic) energy to the energy calculated by the underlying MCSCF procedure, which already includes the non-dynamic (or static) correlation energy, with $\hat{T}$, $\hat{V}_{Ne}$, $\hat{V}_{ee}$ the kinetic, nuclear-electron, and electron–electron operators.

Another not-so-common approximation is given by the Lie–Clementi (LC [37,38]) correlation functional, with an explicit dependence on the natural (fractional) occupation numbers, if a MCSCF calculation is done first. A modified density is built, such as

$$\rho_m(\mathbf{r}) = \sum_i n_i e^{-(2-n_i)^2/2} \tilde{\rho}(\mathbf{r}), \quad (8)$$

depending on the density built from the natural orbitals, $\tilde{\rho}(\mathbf{r})$, and the corresponding natural orbital occupation numbers (n_i). That density is, thus, inserted into the reparameterized correlation functional of Gombas et al., generally denoting this class of functionals as $E_c[\rho_m(\mathbf{r})]$. Interestingly, those orbitals with $n_i < 2$ do not contribute to the correlation energy as much as those doubly occupied ($n_i = 2$), thus, describing both ground- and excited-states independently.

Therefore, MC-PDFT, $E_{xc}[\rho(\mathbf{r}), \rho_2(\mathbf{r})]$, and $E_c[\rho_m(\mathbf{r})]$ exchange-correlation functionals will be all based here on a Complete Active Space Self-Consistent Field (CASSCF) wavefunction with an active space of N electrons housed in M orbitals, or simply (N, M), to incorporate non-dynamic (or static) correlation effects in a consistent way. The active space chosen, (6,6) or (12,12), is indeed based on the occupancy (and degeneracy) or molecular orbitals found at the uncorrelated level.

Note that: (i) MC-PDFT could be instead used with many MCSCF wavefunction types, such as GVB, CASSCF, RASSCF, CAS-CI, and RAS-CI, as is the case for the other two-body functionals too. We will, however, limit this work to the same wavefunction type for both schemes for the sake of coherence. (ii) Since these theories can be applied to any state of interest, independently of its spin, there is no need to invoke a linear-response regime as it happens for TD-DFT. (iii) Any of these methods will incorporate all correlation effects, be they static or dynamic, thus, possibly disentangling the importance of any of these contributions into the final results.

2.4. Computational Details

The ground-state (S_0) geometry of all the compounds was optimized by the B97-3c method [39], without any imaginary frequency obtained. The energy difference between the lowest-energy spin-singlet (S_1) and spin-triplet (T_1) excited-states is denoted as ΔE_{ST}, which is normally positive unless for an energy inversion of the S_1 and T_1 energies, thus, giving rise to $\Delta E_{ST} < 0$. The def2-SVP and def2-TZVP basis sets [40] are used for all the calculations, with the auxiliary def2/JK and def2-TZVP/C basis sets [41] to reduce the computational cost.

For some control TD-DFT calculations, we will employ the ωB97, ωB97X [42], and ωB97X-2 [43], which form a set of range-separated exchange-correlation functionals belonging, respectively, to non-hybrid (ωB97), hybrid (ωB97X), and double-hybrid (ωB97X-2) rungs to infer if the addition of exact-exchange (for a hybrid) or pertubation-like (for a double-hybrid) brings any difference to the results.

We used the following quantum-chemical packages for the calculations performed here: ORCA 5.0 [44] for the (standard) TD-DFT with hybrid and double-hybrid functionals, GAMESS [45] for the MC-PDFT method with the tPBE functional, and an in-house program [46,47] (interfaced with GAMESS) for the $E_c[\rho_m(\mathbf{r})]$ and $E_c[\rho(\mathbf{r}), \rho_2(\mathbf{r})]$ calculations employing the Lie–Clementi and the Colle–Salvetti correlation functionals.

3. Results and Discussion

3.1. Reference Results Available

We compare, in the following, the results obtained here not only with previously applied wavefunction methods but also with respect to the experimental information available: Leupin et al. obtained [11] for MAP a $S_1 \leftarrow S_0$ and $T_1 \leftarrow S_0$ excitation energies of 0.972 and between 0.972 and 0.984, respectively, and thus with a ΔE_{ST} energy difference possibly negative. For 4AP, Leupin et al. obtained [12] excitation energies below 2.39 for $S_1 \leftarrow S_0$ (although the final value could not be completely determined experimentally) and 2.29 eV for $T_1 \leftarrow S_0$, respectively, again not excluding a $\Delta E_{ST} < 0$ value depending on how low $S_1 \leftarrow S_0$ was in reality.

Finally, for 7AP, Leonard et al. obtained [48] a $S_1 \leftarrow S_0$ excitation energy of 2.60 eV. From the theoretical point of view, a large number of previous studies are available. However, some of them (DLPNO-STEOM-CCSD, NEVPT2, SCS-CC2, and SCS-ADC(2)) used the def2-TZVP basis set, as done here, but others (EOM-CCSD, DLPNO-NEVPT2, ADC(2), and ADC(3)) employed the smaller cc-pVDZ or def2-SVP basis sets instead.

In all cases, all these correlated wavefunction methods were able to predict a $\Delta E_{ST} < 0$ value for all the MAP, 4AP, 5AP, and 7AP systems. Choosing one of these methods as reference, necessarily motivated by the completeness of the values found in the literature [17] together with the high accuracy demanded, and the SCS-CC2/def2-TZVP results are listed next for cross-comparison along the study.

MAP: $S_1 \leftarrow S_0$ ($T_1 \leftarrow S_0$) value of 1.110 (1.334) eV, with $\Delta E_{ST} = -0.22$ eV
4AP: $S_1 \leftarrow S_0$ ($T_1 \leftarrow S_0$) value of 2.258 (2.342) eV, with $\Delta E_{ST} = -0.08$ eV
5AP: $S_1 \leftarrow S_0$ ($T_1 \leftarrow S_0$) value of 2.308 (2.541) eV, with $\Delta E_{ST} = -0.23$ eV
7AP: $S_1 \leftarrow S_0$ ($T_1 \leftarrow S_0$) value of 2.847 (3.226) eV, with $\Delta E_{ST} = -0.38$ eV

3.2. TD-DFT Calculations

First, we also illustrate here if a standard TD-DFT calculation is able or not to provide a $\Delta E_{ST} < 0$ value and (if so) with what accuracy. For this purpose, we will choose not a selection of different functionals but a set of models of increasing complexity (i.e., semi-local, hybrid, and double-hybrid functionals). Additionally, since range-separation was before invoked as an accurate tool for the modelling of organic emitters of this type [49], Table 1 gathers the results obtained by the ωB97, ωB97X, and ωB97X-2 range-separated exchange-correlation functionals.

Strikingly, ωB97 and ωB97X are unable to predict the excited-state energy inversion, providing values for individual excitation energies not differing greatly between both methods, which is not often the case since those values are known to depend on the amount of (short-range) exact-exchange introduced (0 % in ωB97 vs. $\approx$16 % in ωB97X) and/or the range-separation parameter ($\omega = 0.4$ in ωB97 vs. $\omega = 0.3$ in ωB97X).

Furthermore, we note how the $S_1 \leftarrow S_0$ excitation energies are overestimated (severely in the case of 7AP) by both methods with respect to experimental or SCS-CC2 reference results. Note that many previous publications [14–21] already demonstrated how this was the case for all exchange-correlation functionals assessed up to now, which is also confirmed here.

We additionally assessed if the use of the ωB97X-D3 or ωB97X-D4 models, which are reparameterized to include the dispersion correction, would lead to any difference: the ΔE_{ST} value only changed only 0.01 eV with respect to the value calculated by the original ωB97X.

We next analysed the performance of the ωB97X-2 double-hybrid functional, which, in its extension to excited states [50,51], includes a contribution from double excitations, thus, going beyond the single excitations introduced by TD-DFT routinely. We can observe how this method actually predicts $\Delta E_{ST} < 0$ values between -0.4 and -0.7 eV, roughly speaking that are, thus, too large with respect to the reference results. An additional concern arises from the inspection of individual $S_1 \leftarrow S_0$ and $T_1 \leftarrow S_0$ excitation energies, since the method seems to severely underestimate (by up to 1 eV) the former values progressively as a function of the N atoms introduced into the chemical structure.

The overestimation of the $T_1 \leftarrow S_0$ excitation energies is slightly attenuated with respect to the $S_1 \leftarrow S_0$ ones but again with values deviating too much with respect to the reference results. Shortly speaking, although this method is able to provide negative values for ΔE_{ST}, the results appear to be affected by a systematic error. The use of double-hybrid functionals has been recently and more systematically examined [52], with some of the models assessed being promising enough to display accurate individual excitation energies: we thus refer the reader to that study for further information and confirmation about the key role played by double excitations into the final values.

3.3. MC-PDFT Calculations

We will inspect the CASSCF results shown in Table 2 to first observe the effect of using both basis sets, def2-SVP and def2-TZVP, for this set of calculations. For the CASSCF(6,6) results, going from def2-SVP to def2-TZVP implies a slight increase of the $S_1 \leftarrow S_0$ and $T_1 \leftarrow S_0$ excitation energies, with the exception of the latter for the 7AP molecule, but asymmetrically, with the corresponding ΔE_{ST} values altered significantly. For the CASSCF(12,12) case, the variations for 7AP are also significant and deviate from the SCS-CC2/def2-TZVP reference values.

For MAP, the CASSCF(6,6)/def2-TZVP calculation already provided close results to reference SCS-CC2/def2-TZVP results, with $S_1 \leftarrow S_0$ and $T_1 \leftarrow S_0$ excitation energies differing by 0.15 and 0.10 eV, respectively, and thus leading to a negative ΔE_{ST} value of -0.18 compared to -0.22 eV as reference. However, a larger active space is not definitively giving any advantage here, as it was also found before [17], stabilizing too much the S_1 (T_1) state and leading consequently to an overly negative (positive) ΔE_{ST} value with the def2-SVP (def2-TZVP) basis set.

Table 1. Vertical excitation energies and associated ΔE_{ST} energy difference (all in eV) calculated with different methods.

Molecule	Method	$S_1 \leftarrow S_0$	$T_1 \leftarrow S_0$	ΔE_{ST}
MAP	ωB97	1.420	1.195	0.26
	ωB97X	1.385	1.167	0.22
	ωB97X-2	0.791	1.166	−0.38
4AP	ωB97	2.695	2.199	0.41
	ωB97X	2.531	2.150	0.38
	ωB97X-2	1.642	2.092	−0.45
5AP	ωB97	2.752	2.381	0.37
	ωB97X	2.687	2.335	0.35
	ωB97X-2	1.650	2.298	−0.65
7AP	ωB97	3.391	3.132	0.26
	ωB97X	3.310	3.059	0.26
	ωB97X-2	1.933	2.592	−0.66

For 4AP, the agreement is not so close to CASSCF(6,6), with both excitation energies largely overestimated (by 0.3–0.7 eV) independently of the basis set chosen. A larger active space, CASSCF(12,12), seems beneficial only with the def2-SVP basis set, which appears to indicate a not so balanced treatment of correlation effects in the absence of a dynamical correlation correction.

This overestimation was also found for 5AP and 7AP, particularly striking for the latter and again independently of the active space fixed. Overall, it seems that the CASSCF results do not suffice to lead to accurate and robust results by themselves, although negative ΔE_{ST} values are mostly obtained. Previous publications show how the need of dynamic correlation effects (i.e., NEVPT2) as a further step to obtain more accurate and robust results [17,18].

Table 2. Excited-state energies and associated ΔE_{ST} energy difference (all in eV) calculated with the CASSCF method.

Basis Set	Molecule	Method	$S_1 \leftarrow S_0$	$T_1 \leftarrow S_0$	ΔE_{ST}
def2-SVP	MAP	CASSCF(6,6)	1.218	1.436	−0.22
	4AP	CASSCF(6,6)	2.554	2.803	−0.25
	5AP	CASSCF(6,6)	2.686	2.916	−0.23
	7AP	CASSCF(6,6)	3.896	4.217	−0.32
	MAP	CASSCF(12,12)	0.145	0.696	−0.55
	4AP	CASSCF(12,12)	2.214	2.358	−0.14
	5AP	CASSCF(12,12)	2.581	2.519	0.06
	7AP	CASSCF(12,12)	2.752	3.210	−0.46
def2-TZVP	MAP	CASSCF(6,6)	1.256	1.427	−0.17
	4AP	CASSCF(6,6)	2.964	2.864	0.10
	5AP	CASSCF(6,6)	2.995	3.068	−0.07
	7AP	CASSCF(6,6)	5.237	4.437	0.80
	MAP	CASSCF(12,12)	0.179	0.722	−0.54
	4AP	CASSCF(12,12)	1.977	2.171	−0.19
	5AP	CASSCF(12,12)	2.762	2.688	0.07
	7AP	CASSCF(12,12)	4.334	4.637	−0.30

The use of the tPBE correlation functional together with Equation (7) is presented next in Table 3, again for both basis sets (def2-SVP and def2-TZVP) and both active spaces of the underlying CASSCF calculation. For MAP, the CASSCF(6,6) + tPBE results are considerably accurate with both basis sets, not only for the target ΔE_{ST} energy difference but also for the individual excitation energies. The use of the larger CASSCF(12,12) active space instead increases both excitation energies, especially the former, and reverse the sign of ΔE_{ST}.

For 4AP and 5AP, the CASSCF(12,12) + tPBE results are very accurate with the def2-TZVP basis set and with respect to the SCS-CC2 reference values. For 7AP, the CASSCF(12,12) + tPBE results are also relatively accurate, leading to a negative ΔE_{ST} value with both basis sets and correcting the overestimation of values found at the CASSCF(12,12) level with the def2-TZVP basis set. It thus appears that the addition of a (modified) correlation functional is qualitatively beneficial; however, more research is still needed to confirm the application of MC-PDFT to other chromophores and related systems.

Table 3. Excited-state energies and associated ΔE_{ST} energy difference (all in eV) calculated with MC-PDFT.

Basis Set	Molecule	Method	$S_1 \leftarrow S_0$	$T_1 \leftarrow S_0$	ΔE_{ST}
def2-SVP	MAP	CASSCF(6,6) + tPBE	1.168	1.284	−0.12
	4AP	CASSCF(6,6) + tPBE	2.135	1.871	0.26
	5AP	CASSCF(6,6) + tPBE	2.153	2.437	−0.28
	7AP	CASSCF(6,6) + tPBE	2.715	3.155	−0.44
	MAP	CASSCF(12,12) + tPBE	1.523	1.463	0.06
	4AP	CASSCF(12,12) + tPBE	2.181	2.713	−0.53
	5AP	CASSCF(12,12) + tPBE	2.720	2.889	−0.17
	7AP	CASSCF(12,12) + tPBE	2.849	3.373	−0.52
def2-TZVP	MAP	CASSCF(6,6) + tPBE	1.191	1.209	−0.02
	4AP	CASSCF(6,6) + tPBE	2.038	1.865	0.17
	5AP	CASSCF(6,6) + tPBE	1.140	2.769	−1.63
	7AP	CASSCF(6,6) + tPBE	4.641	3.325	1.32
	MAP	CASSCF(12,12) + tPBE	1.515	1.438	0.08
	4AP	CASSCF(12,12) + tPBE	2.420	2.572	−0.15
	5AP	CASSCF(12,12) + tPBE	2.304	2.579	−0.17
	7AP	CASSCF(12,12) + tPBE	1.906	2.148	−0.24

3.4. *Lie–Clementi (LC) and Colle–Salvetti (CS) Calculations*

These two functionals are here applied with the CASSCF(6,6) active space and the def2-SVP basis set to avoid a known (and long-standing) problem with these methods, i.e., the double counting of the dynamical correlation energy, which might be minimized using the smallest admissible active space in the underlying CASSCF calculations [53,54].

Table 4 presents these pioneering results, from which interesting features can be observed: (i) a negative ΔE_{ST} is provided in all cases, contrarily to some of the former methods and the isolated CASSCF(6,6) calculations of Table 2, which clearly underlines the major role played by dynamical correlation effects for describing at least quantitatively these states; (ii) the cost-effective Lie–Clementi functional overestimates the individual excitation energies, systematically increasing them with respect to the CASSCF(6,6) values; and (iii) the Colle–Salvetti functional, relying on the introduction of the $\rho_2(\mathbf{r})$ variable, provides closer values to reference results, again with the exception of 7AP for which these are still overestimated.

Given the variety of methods tested, Figure 2 reports the calculated ΔE_{ST} energy difference for a better comparison. Contrary to previous TD-DFT applications, and with very few exceptions, the figure clearly shows how it is possible to obtain negative values (i.e., inverted S_1 and T_1 excitation energies) with these methods thanks to the combination of non-dynamical and dynamical correlation effects.

Table 4. Excited-state energies and associated ΔE_{ST} energy difference (all in eV) calculated with non-standard correlation functionals.

Basis Set	Molecule	Method	$S_1 \leftarrow S_0$	$T_1 \leftarrow S_0$	ΔE_{ST}
def2-SVP	MAP	CASSCF(6,6) + LC	1.737	1.941	−0.20
		CASSCF(6,6) + CS	0.968	1.364	−0.40
	4AP	CASSCF(6,6) + LC	3.116	3.386	−0.27
		CASSCF(6,6) + CS	2.501	2.931	−0.43
	5AP	CASSCF(6,6) + LC	3.273	3.453	−0.18
		CASSCF(6,6) + CS	2.543	3.043	−0.50
	7AP	CASSCF(6,6) + LC	4.568	4.892	−0.32
		CASSCF(6,6) + CS	3.554	4.017	−0.46

Taking into account the use here of the def2-SVP basis set, we can next compare these results with the high-quality values available in literature: (i) For MAP, NEVPT2(6,6)/def2-SVP results [18] gave $S_1 \leftarrow S_0$ and $T_1 \leftarrow S_0$ excitation energies of 1.102 and 1.319 eV with the CASSCF(6,6) + CS results differing by only 0.13 and 0.05 eV, respectively. (ii) For 5AP, EOM-CCSD/cc-pVDZ results [13] are 2.251 and 2.329 eV for $S_1 \leftarrow S_0$ and $T_1 \leftarrow S_0$ excitation energies, respectively, with the CASSCF(6,6) + tPBE results of 2.153 and 2.437 eV differing by only 0.10 and 0.11 eV, respectively. (iii) For 7AP, DLPNO-NEVPT2(6,6)/def2-SVP results [13] led to $S_1 \leftarrow S_0$ and $T_1 \leftarrow S_0$ excitation energies of 2.552 and 2.906 eV with the CASSCF(12,12) + tPBE results being the closest ones but still differing by 0.30 and 0.47 eV, respectively.

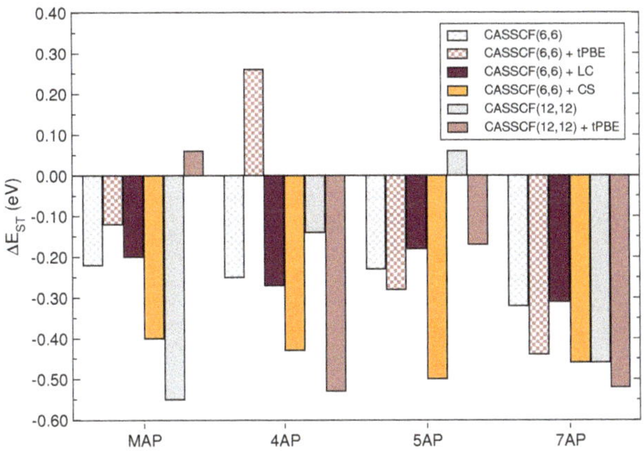

Figure 2. ΔE_{ST} values for the molecules selected (with the def2-SVP basis set).

4. Conclusions

The field of (TD-)DFT has so impressively advanced over the recent decades thanks to the two-fold and concurring efforts of continuously merging developments and applications. As a corollary, the latter would have not been possible without major advances from developments, often considered part of basic but completely needed Science. In this regard, new methods have historically been fostered by providing cost-effective yet accurate expressions and implementations for wide applications or, on the other hand, by tackling cutting-edge applications at the frontier of knowledge to move the field forward. In other words, inaccurate results are often needed to question why (TD-)DFT behaves as it does and how it can be rigorously and systematically improved.

Therefore, we attempted to continue building that interface between both worlds (developments and applications) by selecting a long-standing chemical problem of revisited interest: the energy inversion of the lowest spin-singlet and spin-triplet excited states of azaphenalene compounds intended to be used as organic emitters or photocatalysts. For that purpose, knowing from previous works than TD-DFT was not a reliable path to adequately address this issue, we applied methods employing not only the electronic density as the sole ingredient of an exchange-correlation functional but also other more involved magnitudes, e.g., the on-top pair density.

Overall, the use of the latter into the MC-PDFT scheme or as part of the explicit formulation of the Colle–Salvetti correlation functional offers an attractive way to overcome the limitations found for (TD-)DFT, although at a higher computational cost. However, we are also aware that further research is needed to benchmark these non-conventional methods as well as to reduce their computational cost and scaling with the system size, for which more challenging applications will be also welcome in the near future.

5. Concluding Remarks: A Personal Note

We would like to contribute with this article, as part of the Special Issue in honour of Professor Karlheinz Schwarz on the occasion of his 80th birthday, to celebrate the outstanding role played by Professor Schwarz in the field of Density Functional Theory (DFT) through his scientific career [55]. The authors met him as part of the International Scientific Committee of the "International Conference on Density-Functional Theory and its Applications", which is likely one of the longest-lived events to exist in the fields of theoretical and computational Chemistry and Physics after Paris (1995), Vienna (1997), Rome (1999), Madrid (2001), Brussels (2003), Geneva (2005), Amsterdam (2007), Lyon (2009), Athens (2011), Durham (2013), Debrecen (2015), Tällberg (2017), and Alicante (2019) editions.

Thanks to the strong activity of Heinz promoting DFT worldwide, we had the opportunity to enjoy, as local organizers of the last edition (see Figure 3), his compromise and illusion with this series of conferences. This work is, thus, our small recognition to his talented, vibrant, active, and kind figure, promoting DFT worldwide alongside an outstanding scientific career.

Figure 3. Group picture taken along the "18th International Conference on Density-Functional Theory and its Applications", celebrated (2019) in the campus of the University of Alicante. Professor Karlheinz Schwarz is standing on the first row (11th starting from the right to the left side).

Author Contributions: Conceptualization, J.-C.S.-G. and E.S.-F.; methodology, J.-C.S.-G. and E.S.-F.; software, J.-C.S.-G. and E.S.-F.; validation, J.-C.S.-G. and E.S.-F.; formal analysis, J.-C.S.-G. and E.S.-F.; investigation, J.-C.S.-G. and E.S.-F.; resources, J.-C.S.-G. and E.S.-F.; data curation, J.-C.S.-G. and E.S.-F.; writing, J.-C.S.-G. and E.S.-F.; funding acquisition, J.-C.S.-G. and E.S.-F. All authors have read and agreed to the published version of the manuscript.

Funding: This research was funded by "Ministerio de Ciencia e Innovación" of Spain grant number PID2019-106114GB-I00.

Acknowledgments: We kindly thank Chermette for the invitation to contribute to this Special Issue. We very much appreciate working together along the years with close colleagues, such as A.J. Pérez-Jiménez (U. Alicante, Spain), Y. Olivier (U. Namur, Belgium), E. Brémond (U. Paris, France), and C. Adamo (ChimieParis Tech, France), as well as with outstanding young researchers, such as G. Ricci (U. Namur, Belgium), for developments and applications of DFT to challenging problems, such as those treated here.

Conflicts of Interest: The authors declare no conflict of interest.

References

1. Liu, S.; Langenaeker, W. Hund's multiplicity rule: A unified interpretation. *Theor. Chem. Acc.* **2003**, *110*, 338–344. [CrossRef]
2. Köhler, A.; Beljonne, D. The singlet–triplet exchange energy in conjugated polymers. *Adv. Funct. Mater.* **2004**, *14*, 11–18. [CrossRef]
3. Gierschner, J.; Cornil, J.; Egelhaaf, H.J. Optical bandgaps of π-conjugated organic materials at the polymer limit: Experiment and theory. *Adv. Mater.* **2007**, *19*, 173–191. [CrossRef]
4. Köhler, A.; Bässler, H. Triplet states in organic semiconductors. *Mater. Sci. Eng. R Rep.* **2009**, *66*, 71–109. [CrossRef]
5. Becke, A.D. Singlet-triplet splittings from the virial theorem and single-particle excitation energies. *J. Chem. Phys.* **2018**, *148*, 044112. [CrossRef] [PubMed]
6. Becke, A.D. Communication: Optical gap in polyacetylene from a simple quantum chemistry exciton model. *J. Chem. Phys.* **2018**, *149*, 081102. [CrossRef]
7. Kollmar, H.; Staemmler, V. Violation of Hund's rule by spin polarization in molecules. *Theor. Chim. Acta* **1978**, *48*, 223–239. [CrossRef]
8. Olivier, Y.; Yurash, B.; Muccioli, L.; D'Avino, G.; Mikhnenko, O.; Sancho-García, J.C.; Adachi, C.; Nguyen, T.Q.; Beljonne, D. Nature of the singlet and triplet excitations mediating thermally activated delayed fluorescence. *Phys. Rev. Mater.* **2017**, *1*, 075602. [CrossRef]
9. Dhali, R.; Phan Huu, D.A.; Terenziani, F.; Sissa, C.; Painelli, A. Thermally activated delayed fluorescence: A critical assessment of environmental effects on the singlet–triplet energy gap. *J. Chem. Phys.* **2021**, *154*, 134112. [CrossRef]
10. Audebert, P.; Kroke, E.; Posern, C.; Lee, S.H. State of the Art in the Preparation and Properties of Molecular Monomeric s-Heptazines: Syntheses, Characteristics, and Functional Applications. *Chem. Rev.* **2021**, *121*, 2515–2544. [CrossRef]
11. Leupin, W.; Wirz, J. Low-lying electronically excited states of cycl[3.3.3]azine, a bridged 12π-perimeter. *J. Am. Chem. Soc.* **1980**, *102*, 6068–6075. [CrossRef]
12. Leupin, W.; Magde, D.; Persy, G.; Wirz, J. 1,4,7-Triazacycl[3.3.3]azine: Basicity, photoelectron spectrum, photophysical properties. *J. Am. Chem. Soc.* **1986**, *108*, 17–22. [CrossRef]
13. Pollice, R.; Friederich, P.; Lavigne, C.; dos Passos Gomes, G.; Aspuru-Guzik, A. Organic molecules with inverted gaps between first excited singlet and triplet states and appreciable fluorescence rates. *Matter* **2021**, *4*, 1654–1682. [CrossRef]
14. de Silva, P. Inverted singlet–triplet gaps and their relevance to thermally activated delayed fluorescence. *J. Phys. Chem. Lett.* **2019**, *10*, 5674–5679. [CrossRef]
15. Ehrmaier, J.; Rabe, E.J.; Pristash, S.R.; Corp, K.L.; Schlenker, C.W.; Sobolewski, A.L.; Domcke, W. Singlet–triplet inversion in heptazine and in polymeric carbon nitrides. *J. Phys. Chem. A* **2019**, *123*, 8099–8108. [CrossRef]
16. Sandoval-Salinas, M.E.; Carreras, A.; Casanova, D. Triangular graphene nanofragments: Open-shell character and doping. *Phys. Chem. Chem. Phys.* **2019**, *21*, 9069–9076. [CrossRef]
17. Ricci, G.; San-Fabián, E.; Olivier, Y.; Sancho-García, J.C. Singlet-triplet excited-state inversion in heptazine and related molecules: Assessment of TD-DFT and ab initio methods. *ChemPhysChem* **2021**, *22*, 553–560. [CrossRef] [PubMed]
18. Sanz-Rodrigo, J.; Ricci, G.; Olivier, Y.; Sancho-Garcia, J.C. Negative Singlet–Triplet Excitation Energy Gap in Triangle-Shaped Molecular Emitters for Efficient Triplet Harvesting. *J. Phys. Chem. A* **2021**, *125*, 513–522. [CrossRef] [PubMed]
19. Pios, S.; Huang, X.; Sobolewski, A.L.; Domcke, W. Triangular boron carbon nitrides: An unexplored family of chromophores with unique properties for photocatalysis and optoelectronics. *Phys. Chem. Chem. Phys.* **2021**, *23*, 12968–12975. [CrossRef] [PubMed]
20. Bhattacharyya, K. Can TDDFT render the electronic excited states ordering of Azine derivative? A closer investigation with DLPNO-STEOM-CCSD. *Chem. Phys. Lett.* **2021**, *779*, 138827. [CrossRef]
21. Sobolewski, A.L.; Domcke, W. Are Heptazine-Based Organic Light-Emitting Diode Chromophores Thermally Activated Delayed Fluorescence or Inverted Singlet–Triplet Systems? *J. Phys. Chem. Lett.* **2021**, *12*, 6852–6860. [CrossRef]
22. Li, J.; Nakagawa, T.; MacDonald, J.; Zhang, Q.; Nomura, H.; Miyazaki, H.; Adachi, C. Highly efficient organic light-emitting diode based on a hidden thermally activated delayed fluorescence channel in a heptazine derivative. *Adv. Mater.* **2013**, *25*, 3319–3323. [CrossRef]
23. Li, J.; Zhang, Q.; Nomura, H.; Miyazaki, H.; Adachi, C. Thermally activated delayed fluorescence from $^3n\pi^*$ to $^1n\pi^*$ up-conversion and its application to organic light-emitting diodes. *Appl. Phys. Lett.* **2014**, *105*, 98. [CrossRef]
24. Moscardó, F.; San-Fabián, E. Density-functional formalism and the two-body problem. *Phys. Rev. A* **1991**, *44*, 1549. [CrossRef]
25. Moscardó, F.; San-Fabián, E. A density functional for the correlation energy, deduced in the framework of the correlation factor approach. *Int. J. Quantum Chem.* **1991**, *40*, 23–32. [CrossRef]
26. Moscardó, F.; Pérez-Jiménez, Á.J. Self-consistent field calculations using two-body density functionals for correlation energy component: I. Atomic systems. *J. Comput. Chem.* **1998**, *19*, 1887–1898. [CrossRef]
27. Pastorczak, E.; Pernal, K. Electronic excited states from the adiabatic-connection formalism with complete active space wave functions. *J. Phys. Chem. Lett.* **2018**, *9*, 5534–5538. [CrossRef]
28. Pastorczak, E.; Hapka, M.; Veis, L.; Pernal, K. Capturing the dynamic correlation for arbitrary spin-symmetry CASSCF reference with adiabatic connection approaches: Insights into the electronic structure of the tetramethyleneethane diradical. *J. Phys. Chem. Lett.* **2019**, *10*, 4668–4674. [CrossRef]
29. Hoyer, C.E.; Ghosh, S.; Truhlar, D.G.; Gagliardi, L. Multiconfiguration pair-density functional theory is as accurate as CASPT2 for electronic excitation. *J. Phys. Chem. Lett.* **2016**, *7*, 586–591. [CrossRef] [PubMed]

30. Gagliardi, L.; Truhlar, D.G.; Li Manni, G.; Carlson, R.K.; Hoyer, C.E.; Bao, J.L. Multiconfiguration pair-density functional theory: A new way to treat strongly correlated systems. *Acc. Chem. Res.* **2017**, *50*, 66–73. [CrossRef]
31. Stoneburner, S.J.; Truhlar, D.G.; Gagliardi, L. Transition metal spin-state energetics by MC-PDFT with high local exchange. *J. Phys. Chem. A* **2020**, *124*, 1187–1195. [CrossRef] [PubMed]
32. Lykhin, A.O.; Truhlar, D.G.; Gagliardi, L. Dipole Moment Calculations Using Multiconfiguration Pair-Density Functional Theory and Hybrid Multiconfiguration Pair-Density Functional Theory. *J. Chem. Theory Comput.* **2021**, *17*, 7586–7601. [CrossRef]
33. Colle, R.; Salvetti, O. Approximate calculation of the correlation energy for the closed shells. *Theor. Chim. Acta* **1975**, *37*, 329–334. [CrossRef]
34. Colle, R.; Salvetti, O. Approximate calculation of the correlation energy for the closed and open shells. *Theor. Chim. Acta* **1979**, *53*, 55–63. [CrossRef]
35. Colle, R.; Salvetti, O. Generalization of the Colle–Salvetti correlation energy method to a many-determinant wave function. *J. Chem. Phys.* **1990**, *93*, 534–544. [CrossRef]
36. Lee, C.; Yang, W.; Parr, R.G. Development of the Colle–Salvetti correlation-energy formula into a functional of the electron density. *Phys. Rev. B* **1988**, *37*, 785. [CrossRef]
37. Lie, G.C.; Clementi, E. Study of the electronic structure of molecules. XXI. Correlation energy corrections as a functional of the Hartree-Fock density and its application to the hydrides of the second row atoms. *J. Chem. Phys.* **1974**, *60*, 1275–1287. [CrossRef]
38. Lie, G.C.; Clementi, E. Study of the electronic structure of molecules. XXII. Correlation energy corrections as a functional of the Hartree-Fock type density and its application to the homonuclear diatomic molecules of the second row atoms. *J. Chem. Phys.* **1974**, *60*, 1288–1296. [CrossRef]
39. Brandenburg, J.G.; Bannwarth, C.; Hansen, A.; Grimme, S. B97-3c: A revised low-cost variant of the B97-D density functional method. *J. Chem. Phys.* **2018**, *148*, 064104. [CrossRef]
40. Weigend, F.; Ahlrichs, R. Balanced basis sets of split valence, triple zeta valence and quadruple zeta valence quality for H to Rn: Design and assessment of accuracy. *Phys. Chem. Chem. Phys.* **2005**, *7*, 3297–3305. [CrossRef]
41. Weigend, F. Hartree–Fock exchange fitting basis sets for H to Rn. *J. Comput. Chem.* **2008**, *29*, 167–175. [CrossRef] [PubMed]
42. Chai, J.D.; Head-Gordon, M. Long-range corrected hybrid density functionals with damped atom–atom dispersion corrections. *Phys. Chem. Chem. Phys.* **2008**, *10*, 6615–6620. [CrossRef] [PubMed]
43. Chai, J.D.; Head-Gordon, M. Long-range corrected double-hybrid density functionals. *J. Chem. Phys.* **2009**, *131*, 174105. [CrossRef]
44. Neese, F. Software update: The ORCA program system, version 4.0. *Wiley Interdiscip. Rev. Comput. Mol. Sci.* **2018**, *8*, e1327. [CrossRef]
45. Barca, G.M.J.; Bertoni, C.; Carrington, L.; Datta, D.; De Silva, N.; Deustua, J.E.; Fedorov, D.G.; Gour, J.R.; Gunina, A.O.; Guidez, E.; et al. Recent developments in the general atomic and molecular electronic structure system. *J. Chem. Phys.* **2020**, *152*, 154102. [CrossRef]
46. Pérez-Jordá, J.M.; San-Fabián, E.; Moscardó, F. Spectroscopic constants of diatomic molecules computed correcting Hartree-Fock or general-valence-bond potential-energy curves with correlation-energy functionals. *Phys. Rev. A* **1992**, *45*, 4407. [CrossRef] [PubMed]
47. Moscardó, F.; Muñoz-Fraile, F.; Pérez-Jiménez, A.J.; Pérez-Jordá, J.M.; San-Fabián, E. Improvement of multiconfigurational wave functions and energies by correlation energy functionals. *J. Phys. Chem. A* **1998**, *102*, 10900–10902. [CrossRef]
48. Halpern, A.M.; Rossman, M.A.; Hosmane, R.S.; Leonard, N.J. Photophysics of the S_1 tautm. S_O transition in tri-s-triazine. *J. Phys. Chem.* **1984**, *88*, 4324–4326. [CrossRef]
49. Penfold, T.J. On predicting the excited-state properties of thermally activated delayed fluorescence emitters. *J. Phys. Chem. C* **2015**, *119*, 13535–13544. [CrossRef]
50. Grimme, S.; Neese, F. Double-hybrid density functional theory for excited electronic states of molecules. *J. Chem. Phys.* **2007**, *127*, 154116. [CrossRef]
51. Ottochian, A.; Morgillo, C.; Ciofini, I.; Frisch, M.J.; Scalmani, G.; Adamo, C. Double hybrids and time-dependent density functional theory: An implementation and benchmark on charge transfer excited states. *J. Comput. Chem.* **2020**, *41*, 1242–1251. [CrossRef] [PubMed]
52. Sancho-Garcia, J.C.; Bremond, E.; Ricci, G.; Pérez-Jiménez, Á.J.; Olivier, Y.; Adamo, C. Violation of Hund's Rule in Molecules: Predicting the Excited-State Energy Inversion by TD-DFT with Double-Hybrid Methods. *J. Chem. Phys.* **2021**, *156*. [CrossRef]
53. Sancho-García, J.; Moscardó, F. Usefulness of the Colle–Salvetti model for the treatment of the nondynamic correlation. *J. Chem. Phys.* **2003**, *118*, 1054–1058. [CrossRef]
54. Moscardó, F.; San-Fabián, E.; Pastor-Abia, L. The Colle–Salvetti wavefunction revisited: A comparison between three approaches for obtaining the correlation energy. *Theor. Chem. Acc.* **2006**, *115*, 334–342. [CrossRef]
55. Blaha, P.; Schwarz, K.; Tran, F.; Laskowski, R.; Madsen, G.K.; Marks, L.D. WIEN2k: An APW+ lo program for calculating the properties of solids. *J. Chem. Phys.* **2020**, *152*, 074101. [CrossRef]

computation

MDPI

Article

Density Functional Theory of Coulombic Excited States Based on Nodal Variational Principle †

Ágnes Nagy

Department of Theoretical Physics, University of Debrecen, H-4002 Debrecen, Hungary; anagy@phys.unideb.hu

† Dedicated to Professor Karlheinz Schwarz on the occasion of his 80th birthday.

Abstract: The density functional theory developed earlier for Coulombic excited states is reconsidered using the nodal variational principle. It is much easier to solve the Kohn–Sham equations, because only the correct number of nodes of the orbitals should be insured instead of the orthogonality.

Keywords: density functional theory; Coulomb systems; excited states; nodal variational principle

Citation: Nagy, Á. Density Functional Theory of Coulombic Excited States Based on Nodal Variational Principle. *Computation* **2021**, 9, 93. https://doi.org/10.3390/computation9080093

Academic Editor: Henry Chermette

Received: 27 July 2021
Accepted: 19 August 2021
Published: 23 August 2021

1. Introduction

The density functional theory (DFT) [1,2] has been originally worked out for the ground state. It has rigorously been extended to excited states by Theophilou [3] and later by Gross, Oliveira, and Kohn [4–6]. For further extensions and applications of these subspace and ensemble theories, see reference [7]. Subsequently, theories for individual excited states were presented [8–15]. Several works on excited states have been done within the local potential framework [16–29]. Recently, a comprehensive theory for Coulombic excited states has been put forward in a series of papers [30–32]. It takes advantage of the fact that the Coulomb density determines not only its Hamiltonian but the degree of excitation as well and consequently, there is a universal functional valid for any excited state. In addition, excited state Kohn–Sham (KS) equations similar to the ground-state KS equations can be derived.

Recently, Zahariev, Gordon, and Levy [33] have presented a nodal variational principle for excited states. They have proved that the minimum of the energy expectation value of trial wave functions that are analytically well behaved and have nodes of the exact wave function is the exact eigenvalue. This minimum is achieved at the exact eigenfunction.

In this paper, the Coulombic excited state theory is reconsidered utilizing the nodal variational principle. Certainly, the functionals are the same as in the original theory, but it is much easier to solve the Kohn–Sham equations, because only the correct number of nodes of the orbitals should be insured instead of the orthogonality. It is especially important in case of highly excited orbitals.

The paper is organized as follows. In Section 2, the DFT for Coulombic excited states [30–32] is reworked. Section 3 is dedicated to the discussion.

2. Coulombic Excited State Theory Using Nodal Variation Principle

The theory is valid for Coulomb external potential v^{Coul}. The Hamiltonian has the form

$$\hat{H} = \hat{T} + \hat{V}_{ee} + \sum_{i=1}^{N} v^{Coul}(\mathbf{r}_i) \,, \tag{1}$$

where $\hat{T}$ and $\hat{V}_{ee}$ are the kinetic energy and the electron–electron energy operators. N is the number of electrons and

$$v^{Coul}(\mathbf{r}) = -\sum_{\beta=1}^{M} \frac{Z_\beta}{r_\beta}. \tag{2}$$

M is the number of nuclei and $r_\beta = |\mathbf{r} - \mathbf{R}_\beta|$. $\mathbf{R}_\beta$ and Z_β denote the position and the charge of the nucleus β. Kato's theorem [34–40]

$$\left.\frac{\partial \bar{n}_\beta(r_\beta)}{\partial r_\beta}\right|_{r_\beta=0} = -2Z_\beta n(\mathbf{r} = \mathbf{R}_\beta) \tag{3}$$

is valid both for the ground and any excited state. It has the consequence that the cusps of the density n exhibit the atomic numbers and the positions of the nuclei. In addition, N is given by the the integral of n. Hence, n specifies all parameters of the Coulomb potential (2), thus determines the external potential, the Hamiltonian (1), and all properties of the Coulomb system. Furthermore, n cannot be the density for any other Coulomb external potential, that is, two different excited states cannot have the same electron density [30]. Therefore, we might think that the expression

$$F^{Coul}[n] = E[n] - \int n(\mathbf{r}) v^{Coul}[n;\mathbf{r}] d\mathbf{r} \tag{4}$$

would be the appropriate functional for Coulombic densities. However, it is not known how to decide if a density is Coulombic or not. Therefore, instead of (4) F is defined in another way: it is defined for all electron densities not only for Coulombic densities.

As a first step consider a bifunctional

$$F[n, n^{Coul}] = \min_{\substack{\Psi \to n \\ \{\langle \Psi | \Psi_l^{Coul}[n^{Coul}] \rangle = 0\}_{l=1}^{k-1}}} \langle \Psi | \hat{T} + \hat{V}_{ee} | \Psi \rangle \ , \tag{5}$$

where the minimum is searched over the wave functions that provide the excited state density n and is orthogonal to the first $k-1$ eigenfunctions of the Coulomb system of n^{Coul}.

Using the nodal variation principle instead of Equation (5) we can write

$$F[n, n^{Coul}] = \min_{\substack{\Psi \to n \\ \{\Psi \text{ has the nodes of the exact wave function}\}}} \langle \Psi | \hat{T} + \hat{V}_{ee} | \Psi \rangle \ . \tag{6}$$

It is assumed that a Coulomb density close to n exists.

$$F_\epsilon^{Coul}[n] = \min_{n^{Coul}} F[n, n^{Coul}]; \quad ||n^{Coul} - n|| \le \epsilon. \tag{7}$$

The smallest F is taken, if there are more than one Coulomb density at the same distance from n:

$$F^{Coul}[n] = F_{\epsilon_{min}}^{Coul}[n]. \tag{8}$$

To measure the distance a Sobolev-type norm is applied:

$$d(n^{Coul}, n) \equiv \int \left| \sqrt{n^{Coul}(\mathbf{r}) - n(\mathbf{r})} \right|^2 d\mathbf{r} + \int \left| \nabla \sqrt{n^{Coul}(\mathbf{r}) - n(\mathbf{r})} \right|^2 d\mathbf{r}. \tag{9}$$

The Euler equation is obtained by functional derivation

$$v^{Coul}([n], \mathbf{r}) = -\frac{\delta F^{Coul}[n]}{\delta n(\mathbf{r})} \tag{10}$$

up to a constant.

It is worth emphasizing that the theory above is based on the following statements:

(a) The cusps and the asymptotic decay of the Coulombic density determine the external potential and the ionization potential;

(b) It is supposed that bifunctional $F[n, n^{Coul}]$ (Equation (5) or (6)) exists, where n^{Coul} is close to n. Further, the existence of $F^{Coul}[n]$ (defined by the Equations (7) and (8)) is assumed;

(c) Equation (6) is based on the assumption that the nodes of the exact excited state wave functions are known;

(d) It is assumed that the functional derivative of $F^{Coul}[n]$ exists. It is needed to derive the Euler Equation (10).

Consider now the Kohn–Sham (KS) system. In our original definition the non-interacting kinetic energy bifunctional was written

$$T_s^{Coul}[n, n^{Coul}] = \min_{\substack{\Phi \to n \\ \{\langle \Phi | \Phi_l[n^{Coul}] \rangle = 0\}_{l=1}^{k-1} \\ ||n_1^{Coul} - n_1^0|| \leq \delta}} \langle \Phi | \hat{T} | \Phi \rangle, \tag{11}$$

where the search is over the wave functions Φ having the excited state density n and orthogonal to the first $l-1$ eigen functions of the non-interacting system. The excited state density is the same in the real and the KS systems. If there are more than one KS system with the same density n^{Coul}, the one closest to the true ground-state density n_1^{Coul} is taken. Instead of Equation (11) we can write

$$T_s^{Coul}[n, n^{Coul}] = \min_{\substack{\Phi \to n \\ \{\Phi \text{ has the nodes of the exact wave function}\} \\ ||n_1^{Coul} - n_1^0|| \leq \delta}} \langle \Phi | \hat{T} | \Phi \rangle \tag{12}$$

using the nodal variation principle. The existence of a unique Coulomb density close to the non-Coulomb density n is assumed:

$$T_{s,\epsilon}^{Coul}[n] = \min_{n^{Coul}} T_s[n, n^{Coul}]; \quad ||n^{Coul} - n|| \leq \epsilon. \tag{13}$$

It is supposed that there is at least one Coulomb density closer to n than ϵ, provided that ϵ is large enough. The minimum specifies the final form:

$$T_s^{Coul}[n] = T_{s,\epsilon_{min}}^{Coul}[n]. \tag{14}$$

The functional derivation yields an Euler equation, within an additive constant,

$$w^{Coul}([n], \mathbf{r}) = -\frac{\delta T_s^{Coul}[n]}{\delta n(\mathbf{r})}. \tag{15}$$

The KS theory presented above is based on the following statements:

(a) The existence of the non-interacting kinetic energy bifunctional $T_s^{Coul}[n, n^{Coul}]$ (Equation (11) or (12)) with n^{Coul} close to n is assumed. Further, it is presumed that $T_s^{Coul}[n]$ constructed by Equations (13) and (14) exists;

(b) Equation (12) is based on the assumption that the nodes of the non-interacting excited state wave functions are known;

(c) It is supposed that the functional derivative $T_s^{Coul}[n]$ exists and the Euler Equation (15) can be derived.

It is convenient to partition $F^{Coul}[n]$ as

$$F^{Coul}[n] = T_s^{Coul}[n] + J^{Coul}[n] + E_{xc}^{Coul}[n], \tag{16}$$

where $J^{Coul}[n]$ and $E_{xc}^{Coul}[n]$ are the classical Coulomb and exchange-correlation energies. Equations (10), (15) and (16) lead to the KS potential

$$w^{Coul}([n], \mathbf{r}) = v^{Coul}([n], \mathbf{r}) + v_J^{Coul}([n], \mathbf{r}) + v_{xc}^{Coul}([n], \mathbf{r}) \tag{17}$$

as the sum of the external, the classical Coulomb and the exchange-correlation potentials. The density has the form

$$n = \sum_{i=1}^{K} \lambda_i |\phi_i|^2, \tag{18}$$

where the KS orbitals ϕ_i are solutions of the KS equations

$$\left[-\frac{1}{2}\nabla^2 + w^{Coul}([n], \mathbf{r}) \right] \phi_i = \varepsilon_i \phi_i. \tag{19}$$

The occupation numbers λ_i are 0, 1, or 2 for a non-degenerate system. K denotes the orbital having the highest orbital energy with non-zero occupation number.

3. Discussion

In the present version of the Coulombic excited state theory, the variation is done over the trial wave functions having the nodes of the exact wave functions both in the interacting and the non-interacting systems. That is, the sole difference between the original and the present forms of the theory is using Equations (6) and (12), instead of Equations (5) and (11). Despite this difference, the functionals are the same as in both versions. Generally, the nodes are not known. The wave functions are not known either. In DFT we define functional $F[n]$ via the wave function, but we do not actually use this definition in calculations. Only, F as a functional of n is applied.

On the other hand, in DFT the exact functionals are not known and approximate functionals are applied in calculations. Additionally, in actual calculations the KS Equations (19) are solved. The nodal variational principle leads to a huge simplification, inducing much easier calculations. It is the consequence of the fact that the variational problem reduces to the solution of the KS equations. The orbitals, that is, one-particle functions have to be obtained. If the electron configuration of the state is known, we have to solve the KS equations insuring either the orthogonality of orbitals or the correct number of nodes of the orbitals. The latter is simpler as it is explained in the example below. Certainly, we have to know the correct number of nodes of the orbitals.

The nodal behavior of eigenfunctions were discussed in several papers (see, e.g., [41–44]). Still the number of nodal surfaces is rarely counted in calculations. Hatano and coworkers [43,44] developed a computer program to count the number of nodal regions and applied it in molecular orbital calculations.

Recently, the original Coulombic excited state theory [30–32] has been discussed [7]. The localized Hartree–Fock (LHF) [45] and the Krieger, Li, and Iafrate (KLI) [46] methods combined with correlation have been generalized for excited states. In addition, several highly excited states of Li and Na atoms have been studied.

The radial KS equations can be solved using Numerov's algorithm [47] searching eigenvalues with the correct number of nodes. This method was used by Herman and Skillman in their Hartree–Fock–Slater computer code [48]. We do not have to check the orthogonality of the orbitals during the calculations, only the number of nodes has to be counted. The correct number of nodes is enough to insure the orthogonality. It is especially beneficial in studying higher excited states. In [7], several highly excited states of Li and Na atoms have been studied. Calculations have been performed with KLI and KLI plus (local Wigner) correlation (see details in [7]). Take, for example, the configuration $1s^2 5s$. The orbital ϕ_{5s} should be orthogonal to all the orbitals below, that is, ϕ_{1s}, ϕ_{2s}, ϕ_{3s}, and ϕ_{4s}. The orbitals ϕ_{2s}, ϕ_{3s}, and ϕ_{4s} have zero occupation numbers, do not contribute to the density, so we do not have to calculate them. It is enough to calculate the orbital ϕ_{5s} and the correct number of nodes insures the orthogonality to all the orbitals below. We emphasize that as the configuration ($1s^2 5s$) is known, we know the exact number of nodes of the radial orbitals. The radial orbital ϕ_{ks} has $k-1$ nodes. (Because of the spherical symmetry of the system, the radial KS equations should be solved.) We can easily check numerically that the orbitals with the correct number of nodes are really orthogonal. We calculated the orbitals ϕ_{2s}, ϕ_{3s}, and ϕ_{4s}, and the integrals $\int \phi_{k_1}\phi_{k_2} d\mathbf{r}$, where k_1 and k_2 can be $1s, \ldots, 5s$. We found that the absolute value of the integral was always less than 10^{-6} for $k_1 \neq k_2$.

In summary, the Coulombic excited state theory has been re-examined based on the nodal variational principle. The functionals are the same as in the original theory, but the solution of the Kohn–Sham equation is much easier as only the correct number of nodes of the orbitals should be insured instead of the orthogonality.

Funding: This research was supported by the National Research, Development and Innovation Fund of Hungary, financed under 123988 funding scheme.

Institutional Review Board Statement: Not applicable.

Informed Consent Statement: Not applicable.

Data Availability Statement: Data sharing is not applicable to this article.

Conflicts of Interest: The author declares no conflict of interest.

References

1. Hohenberg, P.; Kohn, W. Inhomogeneous Electron Gas. *Phys. Rev. J. Arch.* **1964**, *136*, B864. [CrossRef]
2. Kohn, W.; Sham, L.J. Self-Consistent Equations Including Exchange and Correlation Effects. *Phys. Rev. J. Arch.* **1965**, *140*, A1133. [CrossRef]
3. Theophilou, A.K. The energy density functional formalism for excited states. *J. Phys. C Solid State Phys.* **1979**, *12*, 5419–5430. [CrossRef]
4. Gross, E.K.U.; Oliveira, L.N.; Kohn, W. Rayleigh-Ritz variational principle for ensembles of fractionally occupied states. *Phys. Rev. A* **1988**, *37*, 2805–2808. [CrossRef] [PubMed]
5. Gross, E.K.U.; Oliveira, L.N.; Kohn, W. Density-functional theory for ensembles of fractionally occupied states. I. Basic formalism. *Phys. Rev. A* **1988**, *37*, 2809–2820. [CrossRef]
6. Oliveira, L.N.; Gross, E.K.U.; Kohn, W. Density-functional theory for ensembles of fractionally occupied states. II. Application to the He atom. *Phys. Rev. A* **1988**, *37*, 2821–2833. [CrossRef]
7. Nagy, Á. Density Functional Theory of Highly Excited States of Coulomb Systems. *Computation* **2021**, *9*, 73. computation9060073. [CrossRef]
8. Nagy, Á. Excited states in density functional theory. *Int. J. Quantum Chem.* **1998**, *70*, 681–691.:4/5<681::AID-QUA14>3.0.CO;2-5. [CrossRef]
9. Nagy, Á. *Electron Correlations and Materials Properties*; Gonis, A., Kioussis, N., Ciftan, M., Eds.; Kluwer: New York, NY, USA, 1999; pp. 451–462.
10. Nagy, Á. Theories for excited states. *Adv. Quant. Chem.* **2003**, *42*, 363–381. [CrossRef]
11. Levy, M.; Nagy, Á. Variational Density-Functional Theory for an Individual Excited State. *Phys. Rev. Lett.* **1999**, *83*, 4361–4364. [CrossRef]
12. Nagy, Á.; Levy, M. Variational density-functional theory for degenerate excited states. *Phys. Rev. A* **2001**, *63*, 052502. [CrossRef]
13. Görling, A. Density-functional theory beyond the Hohenberg-Kohn theorem. *Phys. Rev. A* **1999**, *59*, 3359–3374. doi:10.1103/PhysRevA.59.3359. [CrossRef]
14. Görling, A. Proper Treatment of Symmetries and Excited States in a Computationally Tractable Kohn-Sham Method. *Phys. Rev. Lett.* **2000**, *85*, 4229–4232. [CrossRef]
15. Sahni, V.; Massa, L.; Singh, R.; Slamet, M. Quantal Density Functional Theory of Excited States. *Phys. Rev. Lett.* **2001**, *87*, 113002. [CrossRef] [PubMed]
16. Sen, K.D. An accurate local exchange potential for atomic one- and two-electron excited states. *Chem. Phys. Lett.* **1992**, *188*, 510–512. [CrossRef]
17. Singh, R.; Deb, B.M. Excited-state density functional calculations on the helium isoelectronic sequence. *Proc. Indian Acad. Sci. Chem. Sci.* **1994**, *106*, 1321–1328.
18. Singh, R.; Deb, B.M. Density functional calculation of complex atomic spectra. *J. Mol. Struct. Theochem* **1996**, *361*, 33–39. [CrossRef]
19. Singh, R.; Deb, B.M. Density functional calculation for doubly excited autoionizing states of helium atom. *J. Chem. Phys.* **1998**, *104*, 5892–5897. [CrossRef]
20. Roy, A.K.; Deb, B.M. Atomic inner-shell transitions: A density functional approach. *Phys. Lett. A* **1997**, **234**, 465–471. [CrossRef]
21. Roy, A.K.; Deb, B.M. Density functional calculations on neon satellites. *Chem. Phys. Lett.* **1998**, *292*, 461–466. [CrossRef]
22. Roy, A.K.; Singh, R.; Deb, B.M. Density-functional calculations for doubly excited states of He, and Li^+, Be^{2+} and Be^{3+} ($^{1,3}S^e$, $^3P^0$, $^{1,3}D^e$, $^{1,3}P^0$,$^1G^e$. *J. Phys. B* **1997**, *30*, 4763–4782. [CrossRef]
23. Roy, A.K.; Singh, R.; Deb, B.M. Density functional calculations on triply excited states of lithium isoelectronic sequence. *Int. J. Quaant. Chem.* **1997**, *65*, 317–332. [CrossRef]

24. Singh, R.; Roy, A.K. Deb, B.M. Density functional calculations on low-lying singly excited states of open-shell atoms. *Chem. Phys. Lett.* **1998**, *296*, 530–536. [CrossRef]
25. Slamet, M.; Sahni, V. Quantal density functional theory of excited states: Application to an exactly solvable model. *Int. J. Quant. Chem.* **2001**, *85*, 436–448. [CrossRef]
26. Sahni, V.; Pan, X.Y. Quantal Density Functional Theory of Degenerate States. *Phys. Rev. Lett.* **2003**, *90*, 123001. [CrossRef]
27. Slamet, M.; Singh, R; Massa, L.; Sahni, V. Quantal density-functional theory of excited states: The state arbitrariness of the model noninteracting system. *Phys. Rev. A* **2003**, *68*, 042504. [CrossRef]
28. Sahni, V.; Slamet, M.; Pan, X.Y. Local effective potential theory: Nonuniqueness of potential and wave function. *J. Chem. Phys.* **2007**, *126*, 204106. [CrossRef]
29. Li, Y.Q.; Pan, X.Y.; Li, B.; Sahni, V. Demonstration of the Gunnarsson-Lundqvist theorem and the multiplicity of potentials for excited states *Phys. Rev. A* **2012**, *85*, 032517. [CrossRef]
30. Ayers, P.W.; Levy, M.; Nagy, Á. Time-independent density-functional theory for excited states of Coulomb systems. *Phys. Rev. A* **2012**, *85*, 042518. [CrossRef]
31. Ayers, P.W.; Levy, M.; Nagy, Á. Kohn-Sham theory for excited states of Coulomb systems. *J. Chem. Phys.* **2015**, *143*, 191101. [CrossRef]
32. Ayers, P.W.; Levy, M.; Nagy, Á. Time-independent density functional theory for degenerate excited states of Coulomb systems. *Theor. Chim. Account.* **2018**, *137*, 152. [CrossRef]
33. Zahariev, F.; Gordon, M.S.; Levy, M. Nodal variational principle for excited states. *Phys. Rev. A* **2018**, *98*, 012144. [CrossRef]
34. Kato, T. On the eigenfunctions of many particle systems in quantum mechanics. *Commun. Pure Appl. Math.* **1957**, *10*, 151–177. [CrossRef]
35. Steiner, E. Charge Densities in Atoms. *J. Chem. Phys.* **1963**, *39*, 2365–2366. [CrossRef]
36. March, N.H. *Self-Consistent Fields in Atoms*; Pergamon Press: Oxford, UK, 1975.
37. Nagy, Á.; Sen, K.D. Exact results on the curvature of the electron density at the cusp in certain highly excited states of atoms. *Chem. Phys. Lett.* **2000**, *332*, 154–158. [CrossRef]
38. Ayers, P.W. Density per particle as a descriptor of Coulombic systems. *Proc. Natl. Acad. Sci. USA* **2000**, *97*, 1959–1964. [CrossRef]
39. Nagy, Á.; Sen, K.D. Higher-order cusp of the density in certain highly excited states of atoms and molecules. *J. Phys. B* **2000**, *33*, 1745–1752. [CrossRef]
40. Nagy, Á.; Sen, K.D. Ground- and excited-state cusp conditions for the electron density. *J. Chem. Phys.* **2001**, *115*, 6300–6308. [CrossRef]
41. Wilson, B.E. Symmetry, nodal surfaces, and energy ordering of molecular orbitals. *J. Chem. Phys.* **1975**, *63*, 4870–4879. [CrossRef]
42. Korsch, H.J. On the nodal behaviour of eigenfunctions. *Phys. Lett. A* **1983**, *97*, 77–80. [CrossRef]
43. Hatano, Y.; Yamamoto, S.; Tatewaki, H. Characterization of molecular orbitals by counting nodal regions. *J. Comp. Chem.* **2005**, *26*, 325–333. [CrossRef] [PubMed]
44. Takeda, N.; Hatano, Y.; Yamamoto, S; Tatewaki, H. Counting nodal surfaces in molecular orbitals: Elimination of artificial nodes. *Comp. Theor. Chem.* **2014**, *1045*, 99–112. [CrossRef]
45. Della-Sala, F.; Görling, A. Efficient localized Hartree—Fock methods as effective exact-exchange Kohn—Sham methods for molecules. *J. Chem. Phys.* **2001**, *115*, 5718–5732. [CrossRef]
46. Krieger, J.B.; Li, Y.; Iafrate, G.J. Derivation and application of an accurate Kohn-Sham potential with integer discontinuity. *Phys. Lett. A* **1990**, *146*, 256–260. [CrossRef]
47. Noumerov, B.V. A method of extrapolation of perturbations. *Mon. Not. R. Astron. Soc.* **1924**, *84*, 592–592. [CrossRef]
48. Herman, F.; Skillman, S. *Atomic Structure Calculations*; Prentice-Hall: Englewood Cliffs, NJ, USA; New York, NY, USA, 1963.

computation

MDPI

Article

Plasma Confined Ground and Excited State Helium Atom: A Comparison Theorem Study Using Variational Monte Carlo and Lagrange Mesh Method

Salah B. Doma [1], Mahmoud A. Salem [2] and Kalidas D. Sen [3,*]

1 Faculty of Science, Alexandria University, Alexandria 21544, Egypt; sbdoma@alexu.edu.eg
2 Department of Basic and Applied Sciences, Arab Academy for Science, Technology and Maritime Transport, Alexandria 21544, Egypt; mahmoud_ms2008@yahoo.com
3 School of Chemistry, University of Hyderabad, Hyderabad 500046, India
* Correspondence: kds77@uohyd.ac.in

Abstract: The energy eigenvalues of the ground state helium atom and lowest two excited states corresponding to the configurations 1s2s embedded in the plasma environment using Hulthén, Debye–Hückel and exponential cosine screened Coulomb model potentials are investigated within the variational Monte Carlo method, starting with the ultracompact trial wave functions in the form of generalized Hylleraas–Kinoshita functions and Guevara–Harris–Turbiner functions. The Lagrange mesh method calculations of energy are reported for the He atom in the ground and excited ^1S and ^3S states, which are in excellent agreement with the variational Monte Carlo results. Interesting relative ordering of eigenvalues are reported corresponding to the different screened Coulomb potentials in the He ground and excited electronic states, which are rationalized in terms of the comparison theorem of quantum mechanics.

Keywords: helium atom; screened Coulomb potential; variational Monte Carlo method; Lagrange mesh method; comparison theorem

Citation: Doma, S.B.; Salem, M.A.; Sen, K.D. Plasma Confined Ground and Excited State Helium Atom: A Comparison Theorem Study Using Variational Monte Carlo and Lagrange Mesh Method. *Computation* **2021**, *9*, 138. https://doi.org/10.3390/computation9120138

Academic Editor: Henry Chermette

Received: 23 November 2021
Accepted: 7 December 2021
Published: 10 December 2021

1. Introduction

The theoretical studies of atomic systems in dense plasmas at different temperatures play a very important role in some physical situations and have gained considerable interest in recent years [1–8]. The dilute plasma environment is represented by the screened Coulomb potentials given by the Debye–Hückel model (DHM) or screened Coulomb potential (SCP) [9], which provides a suitable treatment of nonideality in plasma via the screening effect under the low-density and high-temperature conditions. A closely related Hulthén potential is also used as a model potential for the dilute plasma environment in which the atoms are embedded. On the other hand, the dense quantum plasmas environment is represented by using the modified Debye–Hückel model (MDHM) [10] or exponential cosine screened Coulomb potential (ECSCP). Due to its oscillatory nature, the MDHM potential represents a stronger screening effect than the DHM potential.

Considerable attention has been given to the screened Coulomb potentials and exponential cosine screened Coulomb potential in field theory, nuclear, and plasma physics [11–17]. Accurate B-spline configuration interaction (BSCI) method was recently employed to study the spectral/structural data of the helium atom with exponential cosine screened Coulomb potentials [18]. Roy [19] discussed the critical parameter for the spherically confined H atom embedded within a diverse set of screened Coulomb potentials. Ghoshal and Ho [20] investigated the two-electron system in the field of generalized screened potential within the framework of highly correlated and extensive wave functions in Ritz's variational principle, where they were able to determine accurate ground state energies and wave functions of the two-electron system for different values of the screening parameter. Nasser, Zeama, and

Abdel-Hady [21] made a comparative study of the atomic Rényi and Shannon entropies with different wave function within the ECSCP for the $1s^2$-state of the helium isoelectronic series in the Hylleraas-space with few variational parameters. Several other interesting studies on the few electron atoms embedded inside the different plasma potentials have been presented in the literature [22–24]. Very recently a comprehensive compilation of accurate energy values and other structural parameters for the He-like atoms has been published [25,26] using the Hylleraas wave functions.

The purpose of this paper is twofold. Firstly, we report the energy values corresponding to the ground state and two low-lying excited electronic states of the He-like atoms embedded in three different plasma environments using the variational Monte Carlo (VMC) method [27–31] and the Lagrange mesh method (LMM) [32]. The second purpose of this work concerns with the application of the comparison theorem of quantum mechanics. According to the comparison theorem, if a set of spherical potentials $V_1(r)$, $V_2(r)$, $V_3(r)$ satisfy the condition $V_1(r) < V_2(r) < V_3(r)$ at all radial distances, then their eigenvalues obey $E_1 < E_2 < E_3$ for all $n\ell$-states. For nonrelativistic Hamiltonians bounded from below, this theorem follows directly from the variational characterization of the eigenspectrum [33]. Generalized comparison theorems have been proposed [34–38] that allow the two potential curves to cross over in a controlled fashion while maintaining a definite ordering of the respective eigenvalues. Refined comparison theorems applicable to the relativistic Dirac Hamiltonian, which is not bounded from below, have also been established [39–41]. We refer to the works on the generalized comparison theorem [42] and the refined comparison theorem [43], which covers the current research trends in this area. In this work, we used the comparison theorem in order to rationalize the relative ordering of eigenspectra of the He atom under a set of different plasma screened Coulomb potentials for the ground and a few low-lying excited states. To the best of our knowledge, a comprehensive numerical test of the comparison theorem using the VMC and the LMM computations including the excited states, as reported in this work, has not been attempted earlier.

The outline of this paper follows. In Section 2, we define the three different screened Coulomb model potentials that are introduced above. In Section 3, an outline of the variational Monte Carlo (VMC) method employed in this work is presented. This is followed by the computational details and the choice of the trial wave functions for the low-lying excited states of He, described in Sections 4 and 5, respectively. A brief description of the Lagrange mesh method employed in this work is presented in Section 6. Our results are presented and discussed in Section 7. Finally, the main conclusions of this work are listed in Section 8.

2. Plasma Model Potentials

The collective effects of correlated many-particle interactions lead to screened Coulomb interactions in hot dense plasma conditions, which are commonly represented by the DHM or SCP and given by

$$V_{DH}(r) = -\frac{Ze^2}{r}exp(-\mu r), \tag{1}$$

where $\mu = \frac{1}{\lambda_D}$ represents the Debye screening parameter that determines the electronic interaction in the Debye plasma. It depends on the temperature and density of the plasma in the following form [44]:

$$\mu = \frac{1}{\lambda_D} = \sqrt{4\pi e^2 N_e / KT_e}, \tag{2}$$

where λ_D is called Debye screening length, K is the Boltzmann constant, T_e is the electron temperature, e is the electronic charge, Z is the atomic number, and N_e is the plasma–electron density. The Hulthén [45] potential is given by

$$V_{Hu\_\mu}(r) = -Ze^2\frac{\mu e^{-\mu r}}{1 - e^{-\mu r}}\,. \tag{3}$$

A useful form of Hulthén potential in which the screening parameter μ in Equation (3) is simply multiplied by a factor of 2 can be defined as

$$V_{Hu\_2\mu}(r) = -Ze^2 \frac{2\mu e^{-2\mu r}}{1 - e^{-2\mu r}}. \tag{4}$$

It was shown that the study of effective screened potential in dense quantum plasmas can be represented by using MDHM [10] or ECSCP, which is given by

$$V_{MDH}(r) = -\frac{Ze^2}{r} exp(-\mu r)\cos(\mu r). \tag{5}$$

Usually, in quantum plasmas, μ is related to the quantum wave number of the electron, which is related to the electron plasma frequency. Furthermore, the definitions of μ in the two model potentials are different. In the present paper, we are considering μ as a parameter so that the physical difference of μ between these model potentials [14,19,46] is not discussed.

3. Variational Monte Carlo Method

Quantum Monte Carlo methods have already been used for quantum mechanical systems. There are several quantum Monte Carlo techniques such as VMC, diffusion Monte Carlo and Green's function Monte Carlo methods. In this paper, we concentrate on the VMC method, which is used to approximate the eigenstate of the Hamiltonian $\hat{H}$ of a quantum mechanical system by some trial wave function $\psi_T(R)$ whose form is chosen from the analysis of the quantum mechanical system under study. Therefore, the expectation value of the Hamiltonian $\hat{H}$ is written as [46]

$$\hat{H} = E_{VMC} = \frac{\int \psi_T^*(R)\hat{H}\psi_T(R)dR}{\int \psi_T^*(R)\psi_T(R)dR} = \frac{\int dR\psi_T^2(R)E_L(R)}{\int dR\psi_T^2(R)} = \int dR\, \rho(R)E_L(R) \tag{6}$$

where $E_L(R) = (H\psi_T(R))/\psi_T(R)$ is the local energy depending on the $3N$ coordinates R of the N electrons, and $\rho(R) = \frac{\psi_T^2(R)}{\int dR\psi_T^2(R)}$ is the normalized probability density. The variational energy can be calculated as the average value of $E_L(R)$ on a sample of M points R_k, sampled from the probability density $\rho(R)$ as follows:

$$E_{VMC} \approx E_L = \frac{1}{M}\sum_{k=1}^{M} E_L(R_k). \tag{7}$$

In practice, the points R_k are sampled using the Metropolis–Hastings algorithm [27,28].

When evaluating the energy of the system it is important to calculate the standard deviation of this energy, given by [47]

$$\sigma = \sqrt{\frac{\langle E_L^2\rangle - \langle E_L\rangle^2}{N(M-1)}}.$$

Since E_L will be exact when an exact trial wave function is used, then the standard deviation of the local energy will be zero for this case. Thus, in the Monte Carlo method, the minimum of E_L should coincide with a minimum in the standard deviation.

4. Theoretical Details

The nonrelativistic Hamiltonian in Hylleraas coordinates [47] for the two electron systems, under effective SCP in dense plasmas is given, in atomic units, by

$$H_1 = -\frac{1}{2}\sum_i^2 \nabla_i^2 - 2\left[\frac{exp(-\mu r_1)}{r_1} + \frac{exp(-\mu r_2)}{r_2}\right] + \frac{exp(-\mu r_{12})}{r_{12}} \tag{8}$$

where r_1 and r_2 are the radius vectors of the twoelectrons relative to the nucleus, and $r_{12} = |r_1 - r_2|$ is their relative distance.

Moreover, the nonrelativistic Hamiltonian in the effective ECSCP is given by

$$H_2 = -\frac{1}{2}\sum_{i}^{2}\nabla_i^2 - 2\left[\frac{exp(-\mu r_1)}{r_1}\cos(\mu r_1) + \frac{exp(-\mu r_2)}{r_2}\cos(\mu r_2)\right] + \frac{exp(-\mu r_{12})}{r_{12}}\cos(\mu r_{12}) \quad (9)$$

The ground state of the helium atom is a spin singlet two-electron atom. Our calculations for this two-electron system depend on using an ultracompact accurate symmetric function, a nontrivial seven-parameter function, which is constructed by Turbiner et al. [48] as follows:

$$\Psi = (1 + P_{12})[\phi(r_1, r_2)\chi_A] \quad (10)$$

with space wave function:

$$\phi(r_1, r_2) = (1 - ar_1 + br_{12})e^{-\alpha_1 z r_1 - \beta_1 z r_2 + \gamma r_{12}\frac{1+cr_{12}}{1+dr_{12}}} \quad (11)$$

where $a, b, c, d, \alpha_1, \beta_1, \gamma$ are z-dependent parameters and P_{12} is a permutation operator. This function leads for helium atom ($Z = 2$) to a certain improvement of the variational energy and the electron–nuclear cusp and at the same time, the electron–electron cusp. The function χ_A represents the antisymmetric spin wave function with (α) spin up and (β) spin down as follows:

$$\chi_A = \alpha(1)\beta(2) - \alpha(2)\beta(1) \quad (12)$$

This function allows us to obtain the same relative accuracy in both cusp parameters and electronic correlation energy. The function appears as a uniform, locally accurate approximation of the exact ground state eigenfunction. It provides the same relative accuracies in energies and several expectation values together with both cusp parameters.

5. Trial Wave Functions for the Low-Lying Excited States of the Helium Atom

The study of the low-lying excited states of the helium atom has received considerable attention in theoretical investigations. Therefore, for the lowest two excited states, corresponding to the configurations 1s2s, we used the following trial wave functions:

1. For the lowest ortho (space-antisymmetric) state 2^3S, corresponding to the configuration 1s2s, we considered the following simple trial wave function

$$\Psi_{2^3S}(r_1, r_2) = N[(\psi_{1s}(r_1)\psi_{2s}(r_2) - \psi_{1s}(r_2)\psi_{2s}(r_1))\chi_s]f(r_{12}). \quad (13)$$

2. The state 2^1S is a para (space-symmetric) state corresponding to the configuration 1s2s and its trial wave function is, then, taken of the form

$$\Psi_{2^1S}(r_1, r_2) = N[(\psi_{1s}(r_1)\psi_{2s}(r_2) + \psi_{1s}(r_2)\psi_{2s}(r_1))\chi_A]f(r_{12}). \quad (14)$$

In these equations, z_0 and z_i are variational parameters and N is the normalization constant. For spin functions, χ_A represents the singlet antisymmetric spin wave function with (α) spin up and (β) spin down as described in Equation (12).

The function $f(r_{12})$ is the Jastrow correlation function given by [49]

$$f(r_{12}) = e^{\frac{r_{12}}{\alpha_2(1+\beta_2 r_{12})}}, \quad (15)$$

where α_2 and β_2 are variational parameters.

For the relationship of the electron–electron interaction, one obtains the cusp conditions

$$\left.\begin{array}{l} \left.\frac{1}{\Psi}\frac{\partial\Psi}{\partial \mathrm{r_{ij}}}\right|_{\mathrm{r_{ij}}=0} = \frac{1}{2} \text{ for unlike spins} \\ \left.\frac{1}{\Psi}\frac{\partial\Psi}{\partial \mathrm{r_{ij}}}\right|_{\mathrm{r_{ij}}=0} = \frac{1}{4} \text{ for like spins} \end{array}\right\}.$$

6. Lagrange Mesh Method

The Lagrange mesh method (LMM) [50–52] is a numerical procedure wherein the Schrödinger equation is placed into a nonuniform heterogeneous lattice defined by zeroes of classical orthogonal polynomials, using a basis of Laguerre functions and the associated Gauss quadratures. An exhaustive compilation of accurate energies using the LMM for the He-like atoms can be found in [53]. The wave function is expressed in terms of the perimetric coordinates [54,55]. We employed the lattice parameters [56] $N_x = N_y = 50$, $N_z = 40$, and the scaling parameters $h_x = h_y = 0.3$, $h_z = 0.4$. We used the suitably modified PERILAG code [50] in order to implement the screened Coulomb potentials given by Equations (1), (3) and (5). This code was recently employed [57,58] to carry out accurate calculations of energy for the ground and excited electronic states of He-like atoms embedded inside different plasma model potentials.

7. Results and Discussion

The numerical method used in our calculations, the VMC method, is based on a combination of the well-known variational method and the Monte Carlo technique of calculating the multidimensional integrals. By a suitable choice of the trial wave function, it is then possible to obtain minimum energy eigenvalues in agreement with the exact values for the ground as well as the excited states of the given atom. Accordingly, we investigated the effect of the plasma environment by using the SCP and the ECSCP models on the energy eigenvalues of the helium atom. The calculations are performed using a set of 10^8 Monte Carlo integration points to assess the accuracy with standard deviation of the order 10^{-5}. All our results are obtained in atomic units, i.e., $(\hbar = e = m_e = 1)$. For the value of the ground state energy of the He atom that corresponds to Debye screening length $\lambda_D = \infty$ with screening parameter $\mu = 0.0$ and expresses the case of pure Coulomb potential, we obtained the value -2.902662 a.u., which nearly coincides with the value -2.9027 reported in [48].

Table 1 shows the ground state energies of the helium atom under effective SCP in dense quantum plasma with the He^+ threshold energies and ionization potential $[E_{He^+} - E_{He}]$ of He. The results show good agreement with the most accurate previous results, where the z parameter equals 2, and it is equivalent to the atomic nuclear charge for screening parameter $\mu < 0.5$ ($\lambda_D > 2$, Debye screening length). For $\mu > 0.5$, the parameter z starts to decrease slightly around the value 2; at $\mu = 0.5, z \approx 1.9$ and at $\mu = 1$, $z \approx 1.78$.

In Table 2, we present the ground state energies of the helium atom under effective ECSCP in dense quantum plasma. The He^+ threshold energies and ionization potential $[E_{He^+} - E_{He}]$ of He are also given.

In Table 3, we present the results of our calculations of the ground state energies of the helium atom under Hulthén potential in dense quantum plasma with the He^+ threshold energies and ionization potential $[E_{He^+} - E_{He}]$ of He. For the one electron atoms, the energy ordering of $E^{\mu}_{Hulthen} < E^{\mu}_{SCP} < E^{\mu}_{ECSCP}$, has been a well-known consequence of the comparison theorem [34]. More recently, based on a detailed mathematical analysis, a similar ordering has been conjectured for the He-like atoms [57]. For the ground state He, the proposed conjecture has been validated numerically [25,56]. The energy data presented in Tables 1–3 is employed in Figure 1, where we display the variation of E^{μ}_{SCP}, E^{μ}_{ECSCP}, and $E^{\mu}_{Hulthen}$ as a function of μ. The adherence to the energy ordering $E^{\mu}_{Hulthen} < E^{\mu}_{SCP} < E^{\mu}_{ECSCP}$ is numerically validated for the He atom in its ground state.

Another interesting ordering of energy levels for the one electron atoms is given by $E^{\mu}_{Hulthen} < E^{\mu}_{SCP} < E^{2\mu}_{Hulthen} < E^{\mu}_{ECSCP}$, where $E^{2\mu}_{Hulthen}$ denotes the energy of the Hulthén potential at the screening parameter of 2μ, as defined in Equation (4). As a natural extension of the conjecture [56], this energy ordering is now tested numerically for the He atom in the ground and a few excited states.

Table 1. Ground state energies of the helium atom under effective SCP in dense quantum plasma. The He^+ threshold energies and ionization potential $[E_{He^+} - E_{He}]$ of He are also given.

μ	$-E_{He}$	$-E_{He^+}$	$-[E_{He^+} - E_{He}]$
0	2.902662	2.000000	0.902662
	2.903724[4]	2.000000[4]	0.903724[4]
	2.90337[16]	-	-
0.01	2.872771	1.980070	0.892701
	2.873839[4]	1.980075[4]	0.893764[4]
0.02	2.843117	1.960268	0.882849
	2.844181[4]	1.960298[4]	0.883883[4]
0.04	2.784473	1.920408	0.864065
0.05	2.755475	1.901848	0.853627
	2.756549[4]	1.901845[4]	0.854704[4]
0.1	2.613758	1.807262	0.806496
	2.614853[4]	1.807266[4]	0.807587[4]
	2.61451[16]	-	-
0.2	2.345674	1.628183	0.717491
	2.347006[4]	1.628232[4]	0.718774[4]
	2.34666[16]	-	-
0.4	1.864564	1.306890	0.557674
	1.868451[4]	1.307234[4]	0.561217[4]
0.5	1.652445	1.162981	0.489464
	1.655401[4]	1.163678[4]	0.491723[4]
	1.65504[16]	-	-
1	0.803519	0.585547	0.217972
	0.818214[4]	0.592468[4]	0.225746[4]
	0.81704[16]	-	-

Table 2. Ground state energies of the helium atom under effective ECSCP in dense quantum plasma. The He^+ threshold energies and ionization potential $[E_{He^+} - E_{He}]$ of He are also given.

μ	$-E_{He}$	$-E_{He^+}$	$-[E_{He^+} - E_{He}]$
0	2.902662	2.000000	0.902662
	2.903724[5]	2.000000[5]	0.903724[5]
	2.90337[16]	-	-
0.01	2.872533	1.979987	0.892546
	2.873725[5]	1.979988[5]	0.893737[5]
0.02	2.842163	1.959989	0.882174
	2.843730[5]	1.959991[5]	0.883739[5]
0.04	2.780706	1.919235	0.861471
0.05	2.749620	1.900052	0.849568
	2.753816[5]	1.900048[5]	0.853768[5]
0.1	2.590905	1.800491	0.790414
	2.604436[5]	1.800457[5]	0.803978[5]
	2.60409[16]	-	-
0.2	2.259199	1.603504	0.655695
	2.309111[5]	1.603527[5]	0.705587[5]
0.4	1.770741	1.225043	0.545698
0.5	1.444394	1.046606	0.397788
	1.476958[5]	1.047060[5]	0.429898[5]
	1.47653[16]	-	-
1	0.402097	0.296429	0.105668
	0.405261[5]	0.310714[5]	0.094547[5]

In Table 4, we list the VMC estimates of $E^{\mu}_{Hulthen}$, E^{μ}_{SCP}, $E^{2\mu}_{Hulthen}$, and E^{μ}_{ECSCP} for $\mu = 0 - 1.0$ corresponding to the ground state He. The LMM estimates are given below the VMC estimates in each case. The two sets of values are found to be in good agreement with each other. The LMC estimates are uniformly below the VMC results. The latter values can in principle be improved further following a more accurate choice of the trial wave function. In Table 4, we present the results of our calculations of ground state energy for He corresponding to the Hulthen (μ), SCP, Hulthen (2μ), and ECSCP potentials, given by Equations (1), (2), (4) and (5), respectively, over a representative set of μ values. In each case, we include the estimates obtained from the LMM and the VMC calculations. A numerical validation of the energy ordering $E^{\mu}_{Hulthen} < E^{\mu}_{SCP} < E^{2\mu}_{Hulthen} < E^{\mu}_{ECSCP}$ for the ground state He atom is evident from the data in Table 4. It is clear from the present calculations that the inclusion of $E^{2\mu}_{Hulthen}$ makes the bounds to E^{μ}_{SCP} tighter than given by the ordering without $E^{2\mu}_{Hulthen}$.

Table 3. Ground state energies of the helium atom under Hulthén potential in dense quantum plasma. The He^+ threshold energies and ionization potential $[E_{He^+} - E_{He}]$ of He are also given.

μ	$-E_{He}$	$-E_{He^+}$	$-[E_{He^+} - E_{He}]$
0	2.902662	2.000000	0.902662
0.01	2.887679	1.989821	0.897858
0.02	2.872731	1.979252	0.893479
0.04	2.842973	1.960275	0.882698
0.05	2.828137	1.949522	0.878615
0.1	2.754565	1.900474	0.854091
0.2	2.610298	1.804300	0.805998
0.4	2.333143	1.619535	0.713608
0.5	2.200208	1.530918	0.669290
1	1.592340	1.124164	0.468176

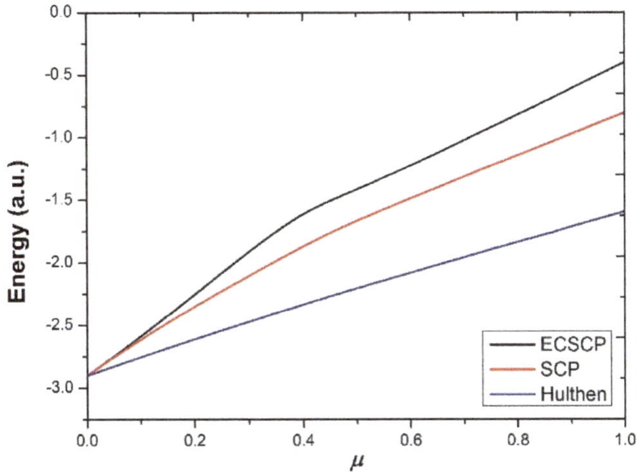

Figure 1. Groundstate energy of the helium atom in SCP, ECSCP, and Hulthén potentials for different values of the screening parameter μ.

In Tables 5 and 6, we present the results of our calculations of $E^{\mu}_{Hulthen}$, E^{μ}_{SCP}, E^{μ}_{ECSCP}, and $E^{2\mu}_{Hulthen}$ corresponding to the (1s2s) excited states of ^1S and ^3S. As observed in the case of the ground state He in Table 4, the VMC and the LMM estimates are in good agreement with each other and the comparison theorembased ordering $E^{\mu}_{Hulthen} < E^{\mu}_{SCP} < E^{2\mu}_{Hulthen} < E^{\mu}_{ECSCP}$ is also obeyed in the excited states. To the best of our knowledge, Tables 4–6 present for the first time numerical validation of the conjecture $E^{\mu}_{Hulthen} < E^{\mu}_{SCP} < E^{2\mu}_{Hulthen} < E^{\mu}_{ECSCP}$ for the ground and excited states of the He atom.

Table 4. Comparison between (a) LMM and (b) VMC for the ground state energies of the helium atom under $E^{\mu}_{Hulthen}$, E^{μ}_{SCP}, $E^{2\mu}_{Hulthen}$, and E^{μ}_{ECSCP}.

μ	Hulthen (μ)	SCP	Hulthén (2μ)	ECSCP
0	−2.903724377 (a) −2.902662 (b)	−2.903724377 (a) −2.902662 (b)		−2.903724377 (a) −2.902662 (b)
0.01	−2.888743509 (a) −2.887679 (b)	−2.873838795 (a) −2.872771 (b)	−2.873800905 (a) −2.872731 (b)	−2.873725125 (a) −2.872533 (b)
0.02	−2.873800905 (a) −2.872731 (b)	−2.844180576 (a) −2.843117 (b)	−2.84403049 (a) −2.842973 (b)	−2.843730329 (a) −2.842163 (b)
0.04	−2.84403049 (a) −2.842973 (b)	−2.785537653 (a) −2.784473 (b)		−2.783771455 (a) −2.780706 (b)
0.05	−2.829202681 (a) −2.828137 (b)	−2.756548811 (a) −2.755475 (b)	−2.755637701 (a) −2.754565 (b)	−2.753815807 (a) −2.749620 (b)
0.1	−2.755637701 (a) −2.754565 (b)	−2.614852947 (a) −2.613758 (b)	−2.611379351 (a) −2.610298 (b)	−2.604435567 (a) −2.590905 (b)
0.2	−2.611379351 (a) −2.610298 (b)	−2.347006184 (a) −2.345674 (b)	−2.334370372 (a) −2.333143 (b)	−2.309114171 (a) −2.259199 (b)
0.4	−2.334370372 (a) −2.333143 (b)	−1.868450546 (a) −1.864564 (b)		−1.742851883 (a) −1.770741 (b)
0.5	−2.201638137 (a) −2.200208 (b)	−1.655401315 (a) −1.652445 (b)	−1.596227498 (a) −1.592340 (b)	−1.47695782 (a) −1.444394 (b)
1	−1.596227498 (a) −1.592340 (b)	−0.818214183 (a) −0.803519 (b)		−0.405261234 (a) −0.402097 (b)

Table 5. Comparison between (a) LMM and (b) VMC for (1s2s) excited states of ^1S energies of the helium atom under $E^{\mu}_{Hulthen}$, E^{μ}_{SCP}, $E^{2\mu}_{Hulthen}$, and E^{μ}_{ECSCP}.

μ	Hulthen (μ)	SCP	Hulthén (2μ)	ECSCP
0	−2.14596983 (a) −2.145788 (b)	−2.14596983 (a) −2.145788 (b)		−2.14596983 (a) −2.145788 (b)
0.01	−2.131029234 (a) −2.129545 (b)	−2.116300015 (a) −2.115372 (b)	−2.116194799 (a) −2.114716 (b)	−2.11598439 (a) −2.113992 (b)
0.02	−2.116194799 (a) −2.114716 (b)	−2.08725862 (a) −2.086387 (b)	−2.086857072 (a) −2.085402 (b)	−2.08605418 (a) −2.085065 (b)
0.03	−2.101470743 (a) −2.100005 (b)	−2.058823511 (a) −2.058013 (b)	−2.05796091 (a) −2.057528 (b)	−2.05623611 (a) −2.055263 (b)
0.04	−2.086857072 (a) −2.085402 (b)	−2.030971244 (a) −2.028217 (b)	−2.029506384 (a) −2.028095 (b)	−2.026576457 (a) −2.025621 (b)
0.05	−2.072353792 (a) −2.070909 (b)	−2.003680728 (a) −2.000982 (b)	−2.001493594 (a) −2.000105 (b)	−1.99711615 (a) −1.994182 (b)
0.06	−2.05796091 (a) −2.056528 (b)	−1.976932817 (a) −1.974287 (b)		−1.967891767 (a) −1.964984 (b)
0.08	−2.029506384 (a) −2.028095 (b)	−1.924996108 (a) −1.922452 (b)		−1.910280654 (a) −1.907431 (b)
0.1	−2.001493594 (a) −2.000105 (b)	−1.875036337 (a) −1.872588 (b)	−1.868060722 (a) −1.866782 (b)	−1.853980921 (a) −1.851209 (b)
0.2	−1.868060722 (a) −1.866782 (b)	−1.651488956 (a) −1.649457 (b)	−1.634414579 (a) −1.633359 (b)	−1.601389925 (a) −1.599132 (b)
0.4	−1.634414579 (a) −1.633359 (b)	−1.304102639 (a) −1.302586 (b)	−1.275179743 (a) −1.274386 (b)	−1.219166537 (a) −1.217393 (b)
0.5	−1.534021688 (a) −1.533064 (b)	−1.158970629 (a) −1.157579 (b)	−1.119597793 (a) −1.118824 (b)	−1.041082207 (a) −1.039325 (b)
0.8	−1.275179743 (a) −1.274386 (b)	−0.787833216 (a) −0.786741 (b)		−0.563849944 (a) −0.562433 (b)
1	−1.119597793 (a) −1.118824 (b)	−0.586621163 (a) −0.585698 (b)		−0.304567693 (a) −0.303396 (b)

Table 6. Comparison between (a) LMM and (b) VMC for (1s2s) excited states of ^3S energies of the helium atom under $E^{\mu}_{Hulthen}$, E^{μ}_{SCP}, $E^{2\mu}_{Hulthen}$, and E^{μ}_{ECSCP}.

μ	Hulthen (μ)	SCP	Hulthén (2μ)	ECSCP
0	−2.175228899 (a) −2.168892 (b)	−2.175228899 (a) −2.168892 (b)		−2.175228899 (a) −2.168892 (b)
0.01	−2.160277331 (a) −2.158999 (b)	−2.145513403 (a) −2.144725 (b)	−2.14542119 (a) −2.143899 (b)	−2.145236778 (a) −2.142392 (b)
0.02	−2.14542119 (a) −2.143899 (b)	−2.11635153 (a) −2.115424 (b)	−2.115996636 (a) −2.114533 (b)	−2.115287001 (a) −2.112255 (b)
0.03	−2.130660957 (a) −2.129201 (b)	−2.08772446 (a) −2.086901 (b)	−2.086955753 (a) −2.085513	−2.085418757 (a) −2.082449 (b)
0.04	−2.115996636 (a) −2.114533 (b)	−2.059614751 (a) −2.058829 (b)	−2.058298598 (a) −2.056876 (b)	−2.05566674 (a) −2.052696 (b)
0.05	−2.101428233 (a) −2.099977 (b)	−2.032006504 (a) −2.029274 (b)	−2.030025255 (a) −2.028622 (b)	−2.026062181 (a) −2.023112 (b)
0.06	−2.086955753 (a) −2.085513 (b)	−2.004885122 (a) −2.002199 (b)		−1.996633374 (a) −1.993701 (b)
0.08	−2.058298598 (a) −2.056876 (b)	−1.952049995 (a) −1.949456 (b)		−1.938404392 (a) −1.935519 (b)
0.1	−2.030025255 (a) −2.028622 (b)	−1.901012328 (a) −1.898506 (b)	−1.894420022 (a) −1.893113 (b)	−1.88116494 (a) −1.878328 (b)
0.2	−1.894420022 (a) −1.893113 (b)	−1.670095721 (a) −1.667977 (b)	−1.652094063 (a) −1.650979 (b)	−1.615667068 (a) −1.613269 (b)
0.4	−1.652094063 (a) −1.650979 (b)	−1.307190384 (a) −1.305633 (b)	−1.276412172 (a) −1.275605 (b)	−1.219400057 (a) −1.217612 (b)
0.5	−1.54542779 (a) −1.544411 (b)	−1.160139744 (a) −1.158742 (b)	−1.120174566 (a) −1.119406 (b)	−1.041273569 (a) −1.039521 (b)
0.8	−1.276412172 (a) −1.275605 (b)	−0.7882615 (a) −0.7871699 (b)		−0.564149249 (a) −0.5627306 (b)
1	−1.120174566 (a) −1.119406 (b)	−0.587027809 (a) −0.5861029 (b)		−0.305119077 (a) −0.3039526 (b)

From the Hamiltonian form of the two-body interactions, the overall potential strength decreases when going from pure Coulomb potentials to SCP and to ECSCP. Physically, we expect for screened potentials that the energy levels increase as μ increases. Furthermore, due to stronger screening effects, for a given μ, the ECSCP values should lie above the corresponding SCP data. Figure 1 represents this situation.

8. Conclusions

In our opinion, the present study is a useful contribution to understanding the ground and a few low-lying excited states of two electron atoms under the influence of three commonly used model potentials describing the plasma environment. The computations are carried out using accurate numerical algorithms based on the VMC and the Lagrange mesh methods. A comprehensive set of numerical results including the ground state of Helium ($1s^2$) is presented, which describe the screening of charges in a plasma where both positive and negative charges are present, and where their motion is thermal. Furthermore, we carried out an investigation to determine the effect of Debye plasma and dense quantum plasmas on the low-lying excited states of helium atom using trial wave functions for the lowest two excited states, corresponding to the configuration 1s2s. The energy ordering of $E^{\mu}_{Hulthen}$, E^{μ}_{SCP}, E^{μ}_{ECSCP}, and $E^{2\mu}_{Hulthen}$ derived from the comparison theorem of quantum mechanics for the hydrogen-like atoms is successfully tested numerically for the He atom in the ground and a few low-lying excited states, which vindicates the proposed conjecture [57] for the first time for the ground as well as the excited states, thus implying the general validity of the comparison theorem in the presence of electron repulsion interaction.

Author Contributions: S.B.D. and K.D.S.: visualization, supervision, investigation, computations, formal analysis, writing—original draft, review and editing, software, visualization, supervision. M.A.S.: writing—review and editing, formal analysis, computations. All authors have read and agreed to the published version of the manuscript.

Funding: This research received no external funding.

Institutional Review Board Statement: Not applicable.

Informed Consent Statement: Not applicable.

Data Availability Statement: No supporting data reported.

Acknowledgments: K.D.S. thanks the Indian National Science Academy, New Delhi, for the award of a senior scientist status. He is also grateful to Daniel Baye for a copy of the PERILAG code. We thank the anonymous reviewers for their helpful suggestions.

Conflicts of Interest: The authors declare no conflict of interest.

References

1. Ghoshal, A.; Ho, Y.K. An Investigation on the $He^-(1s2s^2\,^2S)$ Resonance in Debye Plasmas. *Atoms* **2017**, *5*, 2. [CrossRef]
2. Ning, Y.; Yan, Z.C.; Ho, Y.K. Quantum entanglement for helium atom in the Debye plasmas. *Phys. Plasmas* **2015**, *22*, 013302. [CrossRef]
3. Ning, Y.; Yan, Z.-C.; Ho, Y.K. Natural and Unnatural Parity Resonance States in the Positron-Hydrogen System with Screened Coulomb Interactions. *Atoms* **2015**, *4*, 3. [CrossRef]
4. Kar, S.; Ho, Y.K. Bound states of helium atom in dense plasmas. *Int. J. Quantum Chem.* **2005**, *106*, 814–822. [CrossRef]
5. Ghoshal, A.; Ho, Y.K. Ground states of helium in exponential-cosine-screened Coulomb potentials. *J. Phys. B At. Mol. Opt. Phys.* **2009**, *42*. [CrossRef]
6. Kar, S.; Ho, Y.K.; Jiang, Z. Dispersion coefficients for interactions between hydrogen and helium atoms with Coulomb and screened Coulomb potentials. *J. Phys. B At. Mol. Opt. Phys.* **2011**, *44*. [CrossRef]
7. Kar, S.; Ho, Y. Doubly-excited $2s^{21}Se$ resonance state of helium embedded in Debye plasmas. *Chem. Phys. Lett.* **2005**, *402*, 544–548. [CrossRef]
8. Janev, R.; Zhang, S.-B.; Wang, J. Review of quantum collision dynamics in Debye plasmas. *Matter Radiat. Extrem.* **2016**, *1*, 237–248. [CrossRef]
9. Margenau, H.; Lewis, M. Structure of Spectral Lines from Plasmas. *Rev. Mod. Phys.* **1959**, *31*, 569–615. [CrossRef]
10. Shukla, P.; Eliasson, B. Screening and wake potentials of a test charge in quantum plasmas. *Phys. Lett. A* **2008**, *372*, 2897–2899. [CrossRef]
11. Lin, C.Y.; Ho, Y.K. Effects of screened Coulomb (Yukawa) and exponential-cosine-screened Coulomb potentials on photoionization of H and He+. *Eur. Phys. J. D* **2010**, *57*, 21–26. [CrossRef]
12. Soylu, A. Plasma screening effects on the energies of hydrogen atom. *Phys. Plasmas* **2012**, *19*, 072701. [CrossRef]
13. Chang, T.N.; Fang, T.K. Atomic photoionization in a changing plasma environment. *Phys. Rev. A* **2013**, *88*. [CrossRef]
14. Ghoshal, A.; Ho, Y.K. Doubly excited resonance states of helium in exponential cosine-screened Coulomb potentials. *Phys. Rev. A* **2009**, *79*, 062514. [CrossRef]
15. Ghoshal, A.; Ho, Y.K. Properties of hydrogen molecular ion with static screened coulomb and exponential cosine screened coulomb potentials. *Int. J. Quantum Chem.* **2011**, *111*, 4288–4295. [CrossRef]
16. Ancarani, L.U.; Rodriguez, K.V. Correlated expansions of n1S and n3S states for two-electron atoms in exponential cosine screened potentials. *Phys. Rev. A* **2014**, *89*, 012507. [CrossRef]
17. Fang, T.K.; Wu, C.S.; Gao, X.; Chang, T.N. Redshift of the Heα emission line of He-like ions under a plasma environment. *Phys. Rev. A* **2017**, *96*, 052502. [CrossRef]
18. Lin, Y.-C.; Lin, C.-Y.; Ho, Y.K. Spectral/structural data of helium atoms with exponential-cosine-screened coulomb potentials. *Int. J. Quantum Chem.* **2015**, *115*, 830–836. [CrossRef]
19. Roy, A.K. Critical parameters and spherical confinement of H atom in screened Coulomb potential. *Int. J. Quantum Chem.* **2016**, *116*, 953–960. [CrossRef]
20. Ghoshal, A.; Ho, Y.K. Two-Electron System in The Field of Generalized Screened PotentiaL. *Mod. Phys. Lett. B* **2011**, *25*, 1619–1629. [CrossRef]
21. Nasser, I.; Zeama, M.; Abdel-Hady, A. The Rényi entropy, a comparative study for He-like atoms using the exponential-cosine screened Coulomb potential. *Results Phys.* **2017**, *7*, 3892–3900. [CrossRef]
22. Nasyrov, V.V. Radial correlation of the helium atom in the ground state. *J. Exp. Theor. Phys.* **2017**, *125*, 369–371. [CrossRef]
23. Bressanini, D.; Morosi, G. A compact boundary-condition- determined wavefunction for two-electron atomic systems. *J. Phys. B At. Mol. Opt. Phys.* **2008**, *41*. [CrossRef]
24. Rodriguez, K.V.; Gasaneo, G.; Mitnik, D.M. Accurate and simple wavefunctions for the helium isoelectronic sequence with correct cusp conditions. *J. Phys. B At. Mol. Opt. Phys.* **2007**, *40*, 3923–3939. [CrossRef]

25. Zhang, Y.Z.; Jiao, L.G.; Liu, F.; Liu, A.H.; Ho, Y.K. Energy levels of ground and singly excited states of two-electron atoms in dense quantum plasmas. *At. Data Nucl. Data Tables* **2021**, *140*, 101420. [CrossRef]
26. Wang, X.; Jiang, Z.; Kar, S.; Ho, Y.K. The ^{1,3}P0 states of exponential cosine-screened helium like atoms. *At. Data and Nucl. Data Tables* **2021**, in press.
27. Papoulis, A. *Probability, Random Variables, and Stochastic Processes*; McGraw-Hill: New York, NY, USA, 1965.
28. Metropolis, N.; Rosenbluth, A.W.; Rosenbluth, M.N.; Teller, A.H.; Teller, E. Equation of State Calculations by Fast Computing Machines. *J. Chem. Phys.* **1953**, *21*, 1087–1092. [CrossRef]
29. Doma, S.B.; Abu-Shady, M.; El-Gammal, F.N.; Amer, A.A. Ground states of the hydrogen molecule and its molecular ion in the presence of a magnetic field using the variational Monte Carlo method. *Mol. Phys.* **2016**, *114*, 1787–1793. [CrossRef]
30. Doma, S.; Shaker, M.O.; Farag, A.M.; El-Gammal, F. Ground States of Helium Atom and Hydrogen Negative Ion in the Presence of Magnetic Field Using Variational Monte Carlo Technique. *Acta Phys. Pol. A* **2014**, *126*, 700–705. [CrossRef]
31. Doma, S.B.; Shaker, M.O.; Farag, A.M.; El-Gammal, F.N. Variational Monte Carlo calculations of lithium atom in strong magnetic field. *J. Exp. Theor. Phys.* **2017**, *124*, 1–9. [CrossRef]
32. Baye, D. The Lagrange-mesh method. *Phys. Rep.* **2014**, *565*, 1–107. [CrossRef]
33. Thirring, W. *A Course in Mathematical Physics 3: Quantum Mechanics of Atoms and Molecules*; Springer: New York, NY, USA, 1981.
34. Wang, X.R. General theorem on the Schrödinger equation. *Phys. Rev. A* **1992**, *46*, 7295–7296. [CrossRef] [PubMed]
35. Hall, R. Refining the comparison theorem of quantum mechanics. *J. Phys. A Math. Gen.* **1992**, *25*, 4459–4469. [CrossRef]
36. Santos, E.S.; Rivelino, R.; De Montigny, M.; De Melo, G.R. A spin Hamiltonian for non-relativistic electrons and their interaction with an external field. *J. Phys. A Math. Theor.* **2010**, *43*. [CrossRef]
37. Ragusa, M.A.; Tachikawa, A. On continuity of minimizers for certain quadratic growth functionals. *J. Math. Soc. Jpn.* **2005**, *57*. [CrossRef]
38. Fassari, S. On the bound states of non-relativistic Krönig-Penney Hamiltonians with short range impurities. *Helv. Phys. Acta* **1990**, *63*, 849–883.
39. Hall, R.L.; Katatbeh, Q.D. Generalized comparison theorems in quantum mechanics. *J. Phys. A Math. Gen.* **2002**, *35*, 8727–8742. [CrossRef]
40. Chen, G. Spectral comparison theorem for theN-dimensional Dirac equation. *Phys. Rev. A* **2005**, *72*, 044102. [CrossRef]
41. Hall, R.L. Special Comparison Theorem for the Dirac Equation. *Phys. Rev. Lett.* **2008**, *101*, 090401. [CrossRef] [PubMed]
42. Hall, R.L.; Yeşiltaş, Ö. Comparison theorems for the Dirac equation with spin-symmetric and pseudo-spin-symmetric interactions. *J. Phys. A Math. Theor.* **2010**, *43*. [CrossRef]
43. Semay, C. General comparison theorem for eigenvalues of a certain class of Hamiltonians. *Phys. Rev. A* **2011**, *83*. [CrossRef]
44. Ichimaru, S.; Iyetomi, H.; Tanaka, S. Statistical physics of dense plasmas: Thermodynamics, transport coefficients and dynamic correlations. *Phys. Rep.* **1987**, *149*, 91–205. [CrossRef]
45. Hulthén, L. On the characteristic solutions of the Schrödinger deuteron equation. *Ark. Met. Astron. Fys. A* **1942**, *28*, 5.
46. Pottorf, S.; Pudzerd, A.; Choud, M.Y.; Hasbun, J.E. The simple harmonic oscillator ground state using a variational Monte Carlo method. *Eur. J. Phys.* **1999**, *20*, 205–212. [CrossRef]
47. Ruiz, M.B. Hylleraas method for many-electron atoms. I. The Hamiltonian. *Int. J. Quantum Chem.* **2004**, *101*, 246–260. [CrossRef]
48. Turbiner, A.V.; del Valle, J.C.; Vieyra, J.C.L.; Nader, D.J. Ultra-compact accurate wave functions for He-like and Li-like iso-electronic sequences and variational calculus: I. Ground state. *Int. J. Quantum Chem.* **2021**, *121*, e26586. [CrossRef]
49. Filippi, C.; Umrigar, C.J. Multiconfiguration wave functions for quantum Monte Carlo calculations of first-row diatomic molecules. *J. Chem. Phys.* **1996**, *105*, 213–226. [CrossRef]
50. Hesse, M.; Baye, D. Lagrange-mesh calculations of three-body atoms and molecules. *J. Phys. B At. Mol. Opt. Phys.* **1999**, *32*, 5605–5617. [CrossRef]
51. Baye, D.; Sen, K.D. Confined hydrogen atom by the Lagrange-mesh method: Energies, mean radii, and dynamic polarizabilities. *Phys. Rev. E* **2008**, *78*, 026701. [CrossRef]
52. Montgomery, J.H.E.; Sen, K.D.; Katriel, J. Critical screening in the one- and two-electron Yukawa atoms. *Phys. Rev. A* **2018**, *97*, 022503. [CrossRef]
53. Turbiner, A.V.; Vieyra, J.C.L.; Olivares-Pilón, H. Few-electron atomic ions in non-relativistic QED: The ground state. *Ann. Phys.* **2019**, *409*, 167908. [CrossRef]
54. Coolidge, A.S.; James, H.M. On the Convergence of the Hylleraas Variational Method. *Phys. Rev.* **1937**, *51*, 855–859. [CrossRef]
55. Pekeris, C.L. Ground State of Two-Electron Atoms. *Phys. Rev.* **1958**, *112*, 1649–1658. [CrossRef]
56. Bollhöfer, M.; Notay, Y. JADAMILU: A software code for computing selected eigenvalues of large sparse symmetric matrices. *Comput. Phys. Commun.* **2007**, *177*, 951–964. [CrossRef]
57. Sen, K.; Katrielb, J.; Montgomery, H. A comparative study of two-electron systems with screened Coulomb potentials. *Ann. Phys.* **2018**, *397*, 192–212. [CrossRef]
58. Adan, M.; Sanchez, M.; Martinez-Flores, C.; Vargas, R.; Cabera-Trujillo, R.; Garza, J.; Sen, K.D. Ionization of many-electron atoms by the action of two plasma models. *Phys. Rev. E* **2021**, *103*, 043202.

computation

Article

Impact of Charge-Resonance Excitations on CT-Mediated J-Type Aggregation in Singlet and Triplet Exciton States of Perylene Di-Imide Aggregates: A TDDFT Investigation

Yasi Dai [1], Maria Zubiria-Ulacia [2,3], David Casanova [2,4] and Fabrizia Negri [1,5,*]

[1] Dipartimento di Chimica "Giacomo Ciamician", Università di Bologna, 40126 Bologna, Italy; yasi.dai2@unibo.it
[2] Donostia International Physics Center (DIPC), 20018 Donostia, Euskadi, Spain; maria.zubiria@dipc.org (M.Z.-U.); david.casanova@ehu.eus (D.C.)
[3] Universidad del País Vasco/Euskal Herriko Unibertsitatea, Manuel Lardizabal 3, 20018 Donostia, Euskadi, Spain
[4] IKERBASQUE, Basque Foundation for Science, 48009 Bilbao, Euskadi, Spain
[5] INSTM, UdR Bologna, 40126 Bologna, Italy
* Correspondence: fabrizia.negri@unibo.it

Abstract: The modulation of intermolecular interactions upon aggregation induces changes in excited state properties of organic molecules that can be detrimental for some optoelectronic applications but can be exploited for others. The time-dependent density functional theory (TDDFT) is a cost-effective approach to determining the exciton states of molecular aggregates, and it has been shown to provide reliable results when coupled with the appropriate choice of the functional. Here we apply a general procedure to analyze the aggregates' exciton states derived from TDDFT calculations in terms of diabatic states chosen to coincide with local (LE) and charge-transfer (CT) excitations within a restricted orbital space. We apply the approach to study energy profiles, interstate couplings, and the charge-transfer character of singlet and triplet exciton states of perylene di-imide aggregates (PDI). We focus on the intermolecular displacement along the longitudinal translation coordinate, which mimics different amounts of slip-stacking observed in PDI crystals. The analysis, in terms of symmetry-adapted Frenkel excitations (FE) and charge-resonance (CR) states and their interactions, discloses how the interchange of the H/J character for small longitudinal shifts, previously reported for singlet exciton states, also occurs for triplet excitons.

Keywords: molecular aggregates; singlet excitons; triplet excitons; TDDFT; charge-transfer states; charge-resonance states; Frenkel states; localized excitations; diabatic states; adiabatic states

Citation: Dai, Y.; Zubiria-Ulacia, M.; Casanova, D.; Negri, F. Impact of Charge-Resonance Excitations on CT-Mediated J-Type Aggregation in Singlet and Triplet Exciton States of Perylene Di-Imide Aggregates: A TDDFT Investigation. *Computation* **2022**, *10*, 18. https://doi.org/10.3390/computation10020018

Academic Editor: Henry Chermette

Received: 5 January 2022
Accepted: 20 January 2022
Published: 25 January 2022

1. Introduction

The photophysical behavior of organic electronic molecular materials is governed by the nature of their low-lying exciton states [1–17]. Exciton states are superpositions of local excitations (LEs) and charge-transfer (CT) states, namely diabatic states describing electron promotions between occupied molecular orbitals to unoccupied molecular orbitals of the same (neutral) or neighboring (ionic) molecules (or sites), respectively [18–32].

The role of CT states on photo-induced processes is documented by a large number of investigations. For instance, CT states might facilitate intersystem crossing through spin-orbit coupling [33] and are often involved in the first step of singlet fission [34–40] through the process of symmetry-breaking charge separation (SBCS). CT states also have a crucial role in exciton dissociation and charge separation in heterojunctions between electron-donating and electron-accepting materials or in homojunctions between crystalline domains with different orientations [41–43].

While the modelling and analysis of singlet spin exciton states has received considerable attention [20,22,27,44–50], triplet excitons have received comparably less attention and only few investigations have been reported [22,51,52]. A relevant role of triplet states in photoinduced processes is, however, proven and supported by experiments. Enhanced triplet-state generation, following photo-induced charge transfer, was reported by various groups in electron donor–acceptor polymer-blend films in organic photovoltaic devices where nonemissive triplet excitons are responsible for nonradiative-charge recombination [53]. Thus, triplet formation by charge recombination can be a detrimental, unwanted side effect but can also be exploited as a useful triplet-generation method, for instance in photocatalysis and photodynamic therapy [54,55]. Examples are, among others, triplet-state generation in molecular dyads [56] and molecular dimers in which dimerization-induced triplet population has been invoked [57].

A deep insight into the relationship between structural arrangement and triplet exciton-state character is, therefore, essential to identify molecular systems and condensed phase organizations that are most suitable to generate and exploit triplet exciton states.

Perylene-bis(dicarboximide) (PDI) and its derivatives have attracted great interest as chromophores for energy and charge-transport studies, thanks to their propensity to self-organize into ordered assemblies, both in solution and in the solid state via $\pi-\pi$-stacking interactions [11,12,58]. Numerous computational investigations on PDI aggregates have been focused on the prediction of exciton states with different quantum-chemical (QC) approaches, including configuration interaction truncated to single excitations (CIS), time-dependent density functional theory (TDDFT) [22,27,46,47,59,60], and highly accurate levels of theory [61–63].

TDDFT is a cost-effective approach to evaluating exciton states of molecular aggregates and has been shown to provide reliable results when coupled to a suitable choice of the exchange–correlation functional. To analyze the character of excitonic states predicted by QC calculations, a diabatization procedure can be used to determine the superposition of LE and CT diabatic states in each adiabatic exciton state [21,22,31,32,59].

For aggregates characterized by a symmetric arrangement of chromophores, as those investigated here (Figure 1), when the molecules approach each other, intermolecular interactions mix the LE states to form a symmetry-adapted (SA) superposition of (neutral) LE states, that is, Frenkel excitons (FE). Similarly, the ionic states form delocalized charge-resonance (CR) states [18,64,65] of appropriate symmetry. The symmetry point group of a PDI aggregate, where intermolecular displacements along the longitudinal translation coordinate, z, are considered, (Figure 1) is C_{2h}. As a result, the most relevant $\pi\pi^*$ exciton states, along with the FE and CR diabatic states, all belong to A_g and B_u symmetry representations. Symmetry can be exploited to easily identify H- and J- aggregation types, since only B_u exciton states are accessible as dipole-allowed transitions.

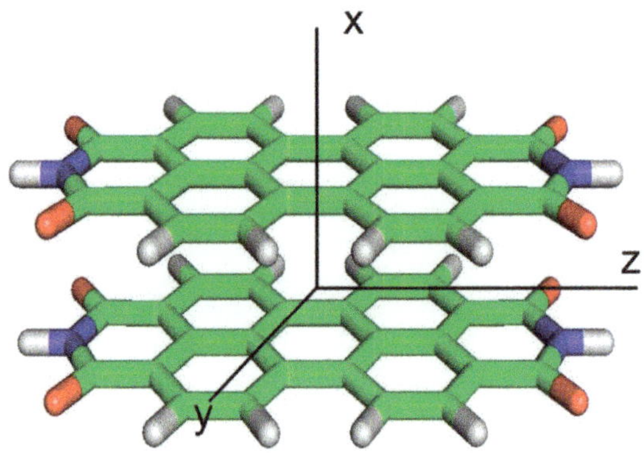

Figure 1. The PDI dimer considered in this work. Singlet and triplet exciton states have been determined at the eclipsed configuration shown here and along the interchromophore longitudinal (z-axis) translation coordinate.

Here we adopt a similar strategy to characterize singlet and triplet exciton states of PDI aggregates. To this end, we analyze exciton states in terms of LE and CT contributions, along the lines described in ref. [60], with three major objectives. First, we validate the protocol of ref. [60] by comparing the computed exciton states character (for both singlet and triplet exciton states) with the results obtained from the character-analysis tool, TheoDORE [66]. Second, we show that for triplet exciton states the modulation of CR/FE interactions along the longitudinal translation coordinate determines a switch in the symmetry (A_g/B_u) of the lowest triplet exciton state, which corresponds to the unconventional H-/J-character alternation previously documented for singlet excitons [8,9,60]. Finally, we demonstrate that the selection of the functional is less critical for triplet excitons compared to singlet excitons and show that the CAM-B3LYP and ωB97X-D results are very close to each other.

2. Computational Methods

The protocol employed to analyze the character of exciton states transforms TDDFT amplitudes from the basis of single excitations between the aggregate's orbitals (the delocalized excitation (DE) basis) to the basis of single excitations between molecular site orbitals, the latter defining the diabatic states. It has been previously described [60] and is briefly summarized here. To analyze the exciton character, we express each relevant exciton state in terms of LEs and CTs. To this end, we select the orbital subspace corresponding to relevant $\pi\pi^*$ exciton states. For PDI aggregates this includes the HOMO and LUMO of each monomer [21,31,59,60,67] and represents the minimal orbital space (MIOS). The next step involves the determination of the linear combination coefficients, $C_{i,j}^{AGGR\_MOB}$, forming the $\boldsymbol{C}_{AGGR\_MOB}$ matrix and describing each aggregate's orbital in the monomer orbital basis (MOB) as [60,68,69]:

$$\boldsymbol{C}_{AGGR\_MOB} = \boldsymbol{C}_{MON\_AOB}^{t} \cdot \boldsymbol{S}_{MON\_AOB} \cdot \boldsymbol{C}_{AGGR\_AOB} \quad (1)$$

where the $\boldsymbol{C}_{MON\_AOB}$ matrix is a block diagonal matrix containing the molecular orbitals' (MOs) coefficients in the atomic orbital basis (AOB) of each monomer, with off-diagonal blocks set to zero, and $\boldsymbol{S}_{MON\_AOB}$ is the overlap matrix of the monomers in the AOB.

In general, monomer orbitals belonging to two different molecules are nonorthogonal to each other. Hence, the aggregate's orbitals, $\boldsymbol{C}_{AGGR\_MOB}^{L}$, expressed in terms of orthogonalized monomer orbitals are obtained as:

$$\boldsymbol{C}_{AGGR\_MOB}^{L} = \boldsymbol{S}_{AGGR\_MOB}^{-\frac{1}{2}} \cdot \boldsymbol{C}_{AGGR\_MOB} \quad (2)$$

where superscript L indicates Löwdin's orthogonalization [70], and the overlap matrix $\boldsymbol{S}_{AGGR\_MOB} = \boldsymbol{C}_{MON\_AOB}^{t} \cdot \boldsymbol{S}_{AGGR\_AOB} \cdot \boldsymbol{C}_{MON\_AOB}$ is obtained from the coefficients of the monomer's orbitals, $\boldsymbol{C}_{MON\_AOB}$, and the overlap of the atomic orbitals in the aggregate configuration, $\boldsymbol{S}_{AGGR\_AOB}$.

From the results of the TDDFT calculations on the aggregate, the subset of the n^2 exciton states originated from the MIOS are then selected out of the full set of computed eigenstates. TDDFT amplitudes are expressed based on the DEs, namely excitations between the aggregate's orbitals, and form the columns of the $\boldsymbol{B}_{DE}^{adia}$ matrix. Thus, the following step is required to expand each DE in terms of diabatic LE and CT states (excitations between monomer orbitals). With the aggregate's orbitals expressed in terms of monomer orbitals via the $\boldsymbol{C}_{AGGR\_MOB}^{L}$ matrix, each excitation from an occupied i to an empty j aggregate's orbital DE($i \to j$) can be expressed as a linear combination of diabatic (LE and CT) excitations ($k \to l$) from an occupied k to an empty l monomer orbital, with the expansion coefficients forming the columns of the unitary matrix $\boldsymbol{U}_{DE \to dia}$ given by

$$U_{k\to l, i\to j}^{DE\to dia} = C_{k,i}^{AGGR\_MOB,\,L} \cdot C_{l,j}^{AGGR\_MOB,L} \quad (3)$$

Exciton states are then readily expressed in the diabatic basis as $\boldsymbol{B}_{dia}^{adia} = \boldsymbol{U}_{DE\to dia} \cdot \boldsymbol{B}_{DE}^{adia}$, and their character is obtained by summing up the contributions from the CT and LE states.

The corresponding n^2 eigenvalues (excitation energies of the selected excitons) form the diagonal H_{adia} matrix, from which the Hamiltonian in the diabatic LE/CT basis, H_{dia}, can be obtained as [21,30,32,71]

$$H_{dia} = B_{dia}^{adia} \cdot H_{adia} \cdot B_{dia}^{adiat} \tag{4}$$

Finally, the H_{dia} is rotated in the SA diabatic basis formed by FE and CR excitations to obtain two matrices, H_{dia}^{sym} (one for B_u states and one for A_g states), whose off-diagonal elements are the interactions between the CR and FE states that ultimately govern the modulation of adiabatic exciton-state energies along the longitudinal translation coordinate. These interaction energies have been shown to correspond to $\pm$ combinations of electron- (D_e) and hole- (D_h) transfer integrals [22,31,59,60].

The PDI-monomer structure was the same one used in previous investigations on PDI aggregates, optimized at the BLYP-D/TZV(P) level of theory [72]. The distance between the planes of different monomers was set to 3.4 Å, which is a distance used in previous investigations on dimers of PDI [59,60]. Exciton states were computed for the eclipsed aggregates and for displacements of 0.5 Å up to 8.0 Å along the longitudinal translation coordinate (z) (Figure 1). For singlet exciton states, trimers and tetramers were also considered beside dimers (Figure S1).

Excitation energies were determined with TDDFT calculations in the Tamm-Dancoff approximation (TDA) [73], using the CAM-B3LYP [74] and ωB97X-D [75] functionals and the 6-31G* basis set. The ωB97X-D functional was previously shown to provide a reliable description of CT character in singlet excitons of PDI dimers [27]. All QC calculations were carried out with the Gaussian16 suite of programs [76]. The CT character of exciton states was determined as described above and using TheoDORE 2.4 [66].

3. Results and Discussion

3.1. Modulation of Singlet Exciton-State Energies and Characters for PDI Aggregates

Previous investigations of excitation energy profiles and CT character in PDI dimers for interchromophore longitudinal shifts have shown that computed exciton states are generally strong mixtures of the two types of diabatic states. The weight of LE and CT characters, however, strongly depends on the energy difference between FE and CR diabatic states, and on their couplings [27,59,60]. This is the reason why the CT character of the lowest singlet exciton state is critically dependent on the chosen functional. In this regard, it has been shown that the adiabatic energy profiles and CT characters computed at the TD-ωB97X-D/6-31G* level match those obtained at higher level of theory [27,59,60]. Here we performed the character analysis based on TDA-computed exciton states rather than the full TDDFT linear response. While the excitation energy profiles of the four exciton states are quite similar (Figure S2), the FE and CT energy profiles of B_u symmetry obtained from TDA-ωB97X-D/6-31G* calculations display a crossing (Figure 2) not seen for TD-ωB97X-D/6-31G* calculations [60], leading ultimately to a CT character of the lowest B_u exciton state (at the eclipsed geometry and for small displacements, see Figure 3) greater than 50%, since the weight of the CT character can be traced back to the energy location of FE and CR states [60] (Figure S3). Such large CT character contrasts with experimental results and was previously reported for other long-range corrected functionals and from CIS calculations [27,59]. Obviously, slightly larger intermonomer separations will reduce the contrasting results between TD and TDA-ωB97X-D/6-31G* calculations, which, however, illustrate the crucial role determined by the energy difference between the CR and FE states.

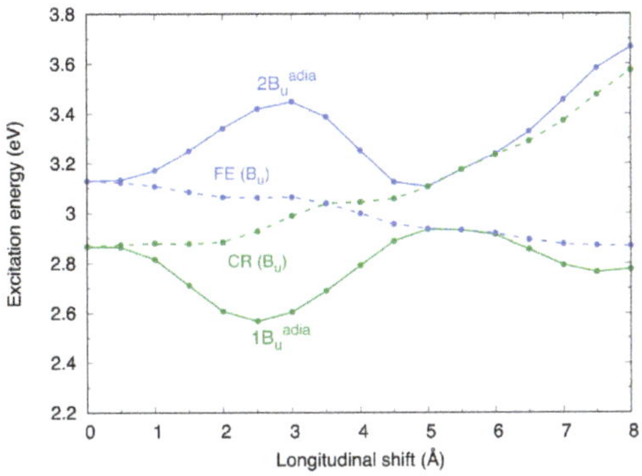

Figure 2. Excitation energy profiles (TDA-ωB97X-D/6-31G*) of B_u adiabatic (adia, solid) singlet exciton states and B_u FE and CR states (dashed) for the PDI dimer along the longitudinal translation coordinate. The computed energy profiles of exciton states are the result of the interactions between CR and FE states of the same symmetry. When the interactions are large, the adiabatic exciton energies differ considerably from the diabatic energies.

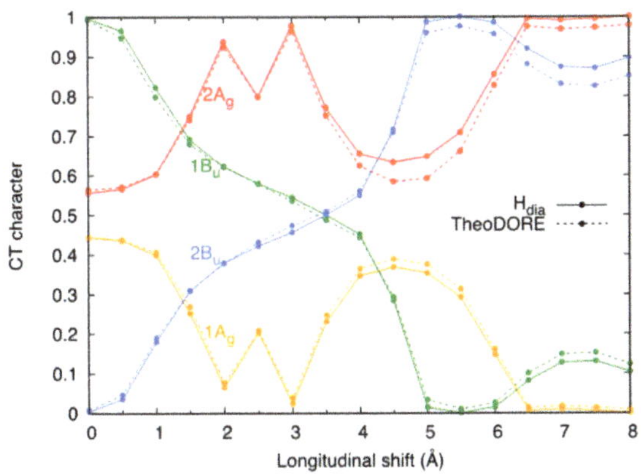

Figure 3. CT character of adiabatic singlet exciton states of PDI dimer along the longitudinal translation coordinate determined with two different approaches (solid) using the procedure described in Section 2 and (dashed) using TheoDORE analysis tool [66].

Finally, we note that the CT character determined following the procedure outlined in Section 2 and the results of TheoDORE [66] are in excellent agreement for all the four low-lying states investigated (Figure 3). A similar correspondence is found for the lowest energy exciton states of larger aggregates (Figure S4).

3.2. Modulation of Triplet Exciton-State Energies and Characters for PDI Dimers

Triplet exciton-state analysis was carried out only on TDA calculations since TD triplet wavefunctions contain relevant contributions from de-excitations. The oscillating trend in adiabatic energy profiles (Figure 4a) are easily rationalized by the oscillating trend of the interactions between the CR and FE states (Figure 4b). As shown previously [22,31,59,60], these interactions are given by the $\pm$ combinations of the D_e and D_h transfer integrals. While couplings between triplet LE states are almost negligible [22], spin triplet CR/FE interactions are of the same magnitude as those computed for singlet exciton states (Figure S5). In analogy with singlet excitons, these interactions determine not only the CT character of the adiabatic triplet exciton states (Figure 4c), but also the symmetry of the lowest energy state (Figure 4a), with an interchange of A_g and B_u along the longitudinal translation coordinate. Such an alternation is well established for singlet exciton states and leads to the appearance of CT-mediated J-type spectroscopic features for small longitudinal displacements of one of the two PDI molecules [8,9,58,60] in a range of displacements where Kasha's theory only predicts H-type aggregation. Therefore, CR states create an effective short-range exciton coupling that can induce unconventional J-type spectroscopic features [8,9,46,58–60,69,77,78].

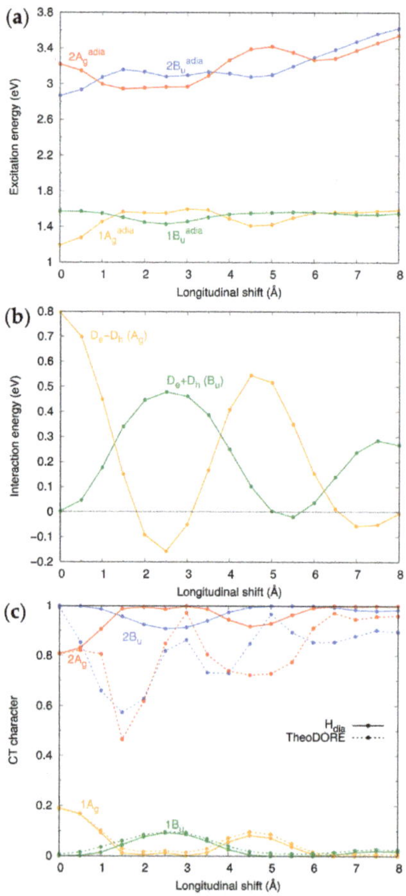

Figure 4. Analysis of triplet exciton states of the PDI dimer (TDA-ωB97X-D/6-31G*). (**a**) Computed adiabatic excitation energy profiles. (**b**) Magnitude and modulation along the longitudinal translation coordinate of the $D_e \pm D_h$ terms for the interaction between CR/FE states of the dimer. (**c**) Comparison between the CT character determined with the procedure described in Section 2 and with TheoDORE [66] for the selected triplet exciton states of PDI dimer.

Notably, here we show that the same mechanism is active in the triplet exciton manifold, albeit with less effective CR/FE mixing. Indeed, as previously reported [22], one of the most relevant differences between singlet and triplet exciton states is the larger energy difference between FE and CR states, which triggers a modest CR/FE mixing. This can be appreciated in Figure 4c, showing that the largest CT contribution for the lowest triplet exciton state does not exceed 20%. At the same time, Figure 5a,b compare the results of CR/FE interactions for singlet and triplet A_g exciton states of the PDI dimer and enlighten the role of the increased energy difference between FE and CR states in the latter. Nevertheless, the CR/FE mixing is sufficient to modulate the triplet exciton energies, ultimately leading to an alternation between B_u and A_g states that parallels what is well documented for singlet exciton states.

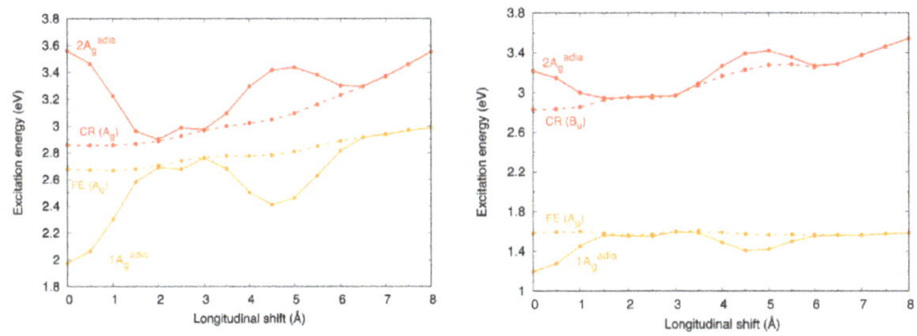

(a) (b)

Figure 5. Energy profiles (TDA-ωB97X-D/6-31G∗) of (solid) A_g adiabatic exciton states and (dashed) A_g FE and CR states for the PDI dimer along the longitudinal translation coordinate: (**a**) singlet exciton states; (**b**) triplet exciton states. The computed exciton energy profiles result from the interactions between CR and FE states of the same symmetry. When the interactions (Figure 4b) are strong, the exciton-state energies deviate from FE and CR energies. The interaction is less effective for triplet states due to the larger energy difference between FE and CR states.

More specifically, the oscillation of the CR/FE interaction (see the $D_e \pm D_h$ profiles shown in Figure 4b), determines an interchange of the lowest exciton states of A_g and B_u symmetry in the dimer for longitudinal shifts in the range of ca. 2–3 Å, where the interaction between A_g symmetry FE and CR states is almost negligible. At the same time, the interaction between the FE and CR B_u states is the maximum and pushes the adiabatic $1B_u$ state energetically below $1A_g$ (see Figure 4a), thereby switching the aggregate character from H- to J-type.

These results suggest that radiative and nonradiative decays from the lowest triplet exciton states of molecular aggregates, intimately related to the nature and symmetry of the lowest energy state, may be modulated by the intermolecular organization, a concept that could be exploited for systems displaying dimerization-induced triplet state populations [57].

We finally compare the CT character analysis obtained from our protocol (solid lines in Figure 4c) with the results of TheoDORE (dashed lines in Figure 4c). The agreement is very good for the two lowest triplet exciton states, while marked differences appear for the remaining two states. The reason is that for several intermolecular configurations the selected $\pi\pi^*$ triplet exciton states (whose wavefunction is dominated by excitations within the selected $\pi\pi^*$ orbital space) are spread over more than one TDDFT-computed exciton state. As a consequence, TheoDORE analysis picks up only the CT contribution corresponding to the fraction of the selected $\pi\pi^*$ exciton state. In contrast, with the procedure described in Section 2, the selected $\pi\pi^*$ exciton states are projected out of the entire space of computed exciton states and renormalized. Interestingly, triplet exciton states (belonging to the selected orbital space) appear to mix more strongly than singlet states with other exciton states of the dimer, as suggested by wavefunction analysis and by the fragmentation of the CT contributions in TheoDORE analysis. Such fragmentation contrasts with the almost perfect match discussed for singlet excitons of the PDI dimer (Figure 2) and suggests that diabatization procedures are likely to be more problematic when analyzing triplet exciton states.

To visualize the nature of exciton states for two representative dimer configurations of the PDI dimer, we have carried out a fragment-based analysis via electron-hole correlation plots using TheoDORE (Figure 6). The two selected fragments correspond to the two molecules forming the dimer. Exciton states are identified by the nonvanishing elements of the 2 × 2 matrix (the Ω-matrix [66]) represented by different levels of grey. Locally excited contributions appear on the main diagonal (going from lower left to upper right), while CT contributions appear off diagonally. In agreement with the character analysis shown in Figure 4c, Figure 6 shows that for the eclipsed configuration, the character of the $1B_u$ state is purely LE while the character of the lowest energy exciton state of $1A_g$ is partially CT, as indicated by the light grey squares in the electron-hole correlation plot. The situation is reversed at the slip-stacked configuration (2.5 Å longitudinal translation), with the $1B_u$ state displaying some CT contribution.

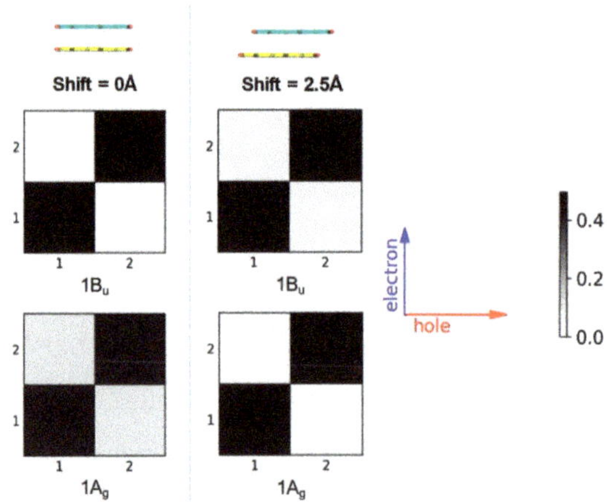

Figure 6. Triplet exciton-state analysis via electron-hole correlation plots [66] for two PDI dimer configurations: eclipsed 0.0 Å and 2.5 Å slip-stacked. The grey scale used is shown on the right panel. From TDA-ωB97X-D/6-31G* calculations.

Influence of Functional on Triplet Exciton Character of the PDI Dimer

While the interplay of CR and FE states is crucial for TDDFT functionals to reproduce higher level results for the calculation of singlet excitons [27,60], the larger energy difference between FE and CR triplet states makes the choice of the functional less critical. This is illustrated by the small differences obtained between TDA-ωB97X-D/6-31G* and TDA-CAM-B3LYP/6-31G* calculations.

The major difference between the two functionals is the lower excitation energies computed at the TDA-CAM-B3LYP level. The difference between the TDA-CAM-B3LYP and TDA-ωB97X-D energies is larger for the two higher energy exciton states, whose CT character is dominant (Figure 7). The lower energy of CT states at the TD-CAM-B3LYP level is not unexpected and was identified as the major source of discrepancy for singlet exciton states computed with the CAM-B3LYP functional compared to ωB97X-D [27]. As a result of the lower energy of CT states with the CAM-B3LYP functional, the CT character of the lowest energy triplet exciton states is slightly larger at the TDA-CAM-B3LYP level, compared to TDA-ωB97X-D, as can be seen in Figure 8. Interestingly, the CR/FE interaction is virtually independent of the chosen functional, with almost identical TD-CAM-B3LYP and TDA-CAM-B3LYP values (Figure 9). While it can be expected that other long-range corrected hybrid variants would provide results similar to those reported here, the use of nonhybrid or hybrid functionals with small contributions of exact (HF) exchange is not recommended because of their tendency to underestimate the energetic position of CT states [46].

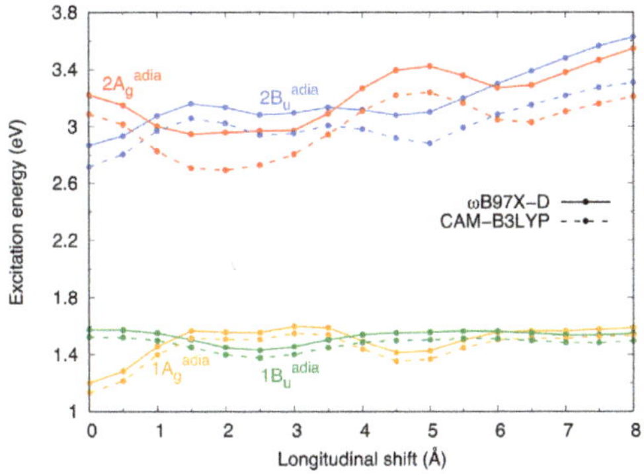

Figure 7. Comparison between excitation energies of triplet exciton states of PDI dimer computed with TDA-CAM-B3LYP/6-31G* (dashed) and TDA-ωB97X-D/6-31G* (solid) along the longitudinal shift.

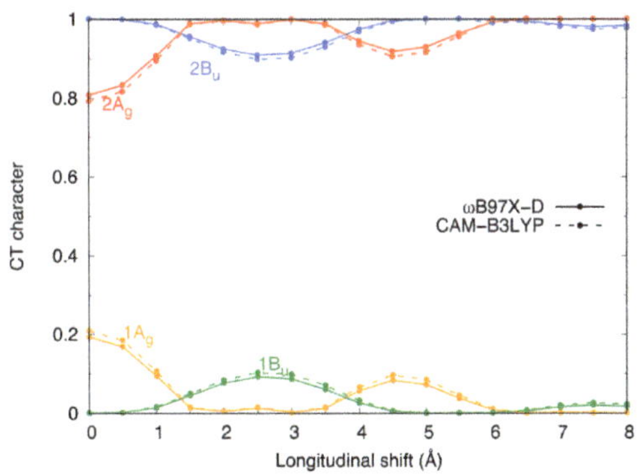

Figure 8. Comparison between triplet exciton CT character computed with TDA-CAM-B3LYP/6-31G* (dashed) and TDA-ωB97X-D/6-31G* (solid) calculations.

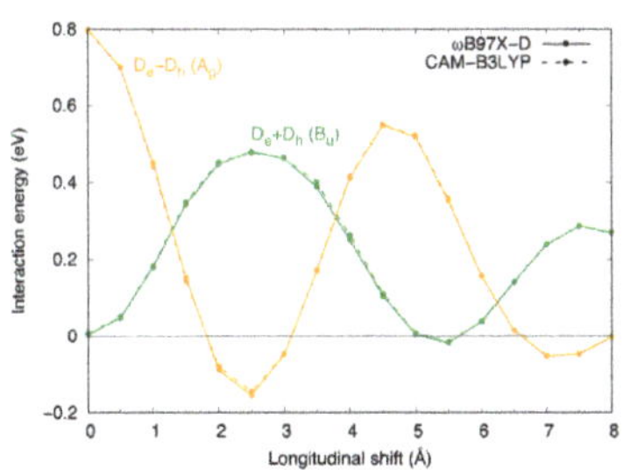

Figure 9. Interaction energies between CR and FE states computed at TDA-CAM-B3LYP/6-31G* (dashed) and TDA-ωB97X-D/6-31G* (solid) levels; (gold) interaction between A_g CR and FE states; (green) interaction between B_u CR and FE states.

4. Conclusions

We have analyzed the modulation of singlet and triplet exciton states of PDI aggregates computed at the TDDFT level, focusing on the intermolecular displacement along the longitudinal translation coordinate, which mimics different amounts of slip stacking observed in PDI crystals.

The CT character of singlet and triplet exciton states has been determined with a diabatization procedure, and it has been shown that the results agree with other character-analysis tools, such as the TheoDORE software.

The study has shown that triplet exciton-state energies can be rationalized in terms of the interactions between CR and FE diabatic states, which are found to be of the same strength as those computed for singlet exciton states. Such CR/FE interactions ultimately lead, not only to a mixed LE/CT character of the low-lying triplet exciton states, but also to a CT-mediated J-aggregation mechanism for small longitudinal displacements, similar to what has been previously documented for singlet exciton states.

Finally, we have compared the results of two long-range corrected hybrid functionals and shown that the magnitude of the CR/FE interactions are almost independent on the selected functional. In contrast with singlet exciton states, the larger energy difference between FE and CR triplet states makes the choice of the functional (among long-range corrected hybrid variants) less critical to define the CT contributions in low-lying triplet exciton states.

Overall, these results demonstrate that the same interactions responsible for the CT-mediated J-aggregation in singlet exciton states of PDI aggregates are operative in triplet exciton states and show that the role of CR states must be carefully considered as a factor influencing the processes following triplet exciton generation. These findings might be exploited to design chromophores and interchromophore organizations to optimize dimer-induced triplet state population.

Supplementary Materials: The following supporting information can be downloaded at: https://www.mdpi.com/article/10.3390/computation10020018/s1, Figure S1:The PDI trimer and tetramer considered in this work; Figure S2:Comparison between adiabatic energy profiles of the four singlet exciton states of PDI dimer along the longitudinal shift calculated at TDA-ωB97X-D/6-31G* (dashed) and TD-ωB97X-D/6-31G* (solid) levels; Figure S3:Comparison between FE and CR singlet states of Bu symmetry calculated at TDA-ωB97X-D (dashed) and TD-ωB97X-D (solid) levels. 6-31G* basis set is used; Figure S4:CT character of adiabatic singlet exciton states of PDI trimer (a,b) and tetramer (c,d) along the longitudinal translation coordinate determined with two different approaches; Figure S5:Comparison between the modulation of the $D_e \pm D_h$ terms of singlet (dashed) and triplet (solid) exciton states of the dimer along the longitudinal translation coordinate.

Author Contributions: Conceptualization, F.N.; investigation and formal analysis, F.N., D.C., Y.D., and M.Z.-U.; writing—original draft preparation, F.N.; writing—review and editing, F.N., D.C., and Y.D. All authors have read and agreed to the published version of the manuscript.

Funding: This research was funded by Ministerio de Economía y Competitividad of Spain, grant number PID2019-109555GB-I00 and RED2018-102815-T and by Eusko Jaurlaritza, grant number PIBA19-0004.

Institutional Review Board Statement: Not applicable.

Informed Consent Statement: Not applicable.

Data Availability Statement: The data presented in this study are available in the Supplementary Material.

Acknowledgments: The authors acknowledge financial support from the Ministerio de Economía y Competitividad of Spain, from the Eusko Jaurlaritza, and from "Valutazione della Ricerca di Ateneo" (VRA)—University of Bologna, Italy. Y.D. acknowledges Ministero dell'Università e della Ricerca (MUR), Italy, for her PhD fellowship.

Conflicts of Interest: The authors declare no conflict of interest.

References

1. Ostroverkhova, O. Organic Optoelectronic Materials: Mechanisms and Applications. *Chem. Rev.* **2016**, *116*, 13279–13412. [CrossRef] [PubMed]
2. Xu, Y.; Xu, P.; Hu, D.; Ma, Y. Recent progress in hot exciton materials for organic light-emitting diodes. *Chem. Soc. Rev.* **2021**, *50*, 1030–1069. [CrossRef] [PubMed]
3. Samanta, S.; Chaudhuri, D. Suppressing Excimers in H-Aggregates of Perylene Bisimide Folda-Dimer: Role of Dimer Conformation and Competing Assembly Pathways. *J. Phys. Chem. Lett.* **2017**, *8*, 3427–3432. [CrossRef]

4. Hong, Y.; Kim, J.; Kim, W.; Kaufmann, C.; Kim, H.; Würthner, F.; Kim, D. Efficient Multiexciton State Generation in Charge-Transfer-Coupled Perylene Bisimide Dimers via Structural Control. *J. Am. Chem. Soc.* **2020**, *142*, 7845–7857. [CrossRef] [PubMed]
5. Haldar, R.; Mazel, A.; Krstić, M.; Zhang, Q.; Jakoby, M.; Howard, I.A.; Richards, B.S.; Jung, N.; Jacquemin, D.; Diring, S.; et al. A de novo strategy for predictive crystal engineering to tune excitonic coupling. *Nat. Commun.* **2019**, *10*, 2048. [CrossRef] [PubMed]
6. Haldar, R.; Mazel, A.; Joseph, R.; Adams, M.; Howard, I.A.; Richards, B.S.; Tsotsalas, M.; Redel, E.; Diring, S.; Odobel, F.; et al. Excitonically Coupled States in Crystalline Coordination Networks. *Chem.-Eur. J.* **2017**, *23*, 14316–14322. [CrossRef] [PubMed]
7. Spano, F.C. The Spectral Signatures of Frenkel Polarons in H- and J-Aggregates. *Acc. Chem. Res.* **2010**, *43*, 429–439. [CrossRef]
8. Hestand, N.J.; Spano, F.C. Molecular Aggregate Photophysics beyond the Kasha Model: Novel Design Principles for Organic Materials. *Acc. Chem. Res.* **2017**, *50*, 341–350. [CrossRef]
9. Hestand, N.J.; Spano, F.C. Expanded Theory of H- and J-Molecular Aggregates: The Effects of Vibronic Coupling and Intermolecular Charge Transfer. *Chem. Rev.* **2018**, *118*, 7069–7163. [CrossRef]
10. Yu, P.; Zhen, Y.; Dong, H.; Hu, W. Crystal Engineering of Organic Optoelectronic Materials. *Chem* **2019**, *5*, 2814–2853. [CrossRef]
11. Würthner, F.; Saha-Möller, C.R.; Fimmel, B.; Ogi, S.; Leowanawat, P.; Schmidt, D. Perylene Bisimide Dye Assemblies as Archetype Functional Supramolecular Materials. *Chem. Rev.* **2016**, *116*, 962–1052. [CrossRef] [PubMed]
12. Hecht, M.; Würthner, F. Supramolecularly Engineered J-Aggregates Based on Perylene Bisimide Dyes. *Acc. Chem. Res.* **2021**, *54*, 642–653. [CrossRef] [PubMed]
13. Hoeben, F.J.M.; Jonkheijm, P.; Meijer, E.W.; Schenning, A.P.H.J. About supramolecular assemblies of pi-conjugated systems. *Chem. Rev.* **2005**, *105*, 1491–1546. [CrossRef] [PubMed]
14. Su, F.; Chen, G.; Korevaar, P.A.; Pan, F.; Liu, H.; Guo, Z.; Schenning, A.P.H.J.; Zhang, H.; Lin, J.; Jiang, Y. Discrete π-Stacks from Self-Assembled Perylenediimide Analogues. *Angew. Chem. Int. Ed. Engl.* **2019**, *58*, 15273–15277. [CrossRef] [PubMed]
15. Yang, L.; Langer, P.; Davies, E.S.; Baldoni, M.; Wickham, K.; Besley, N.A.; Besley, E.; Champness, N.R. Synthesis and characterisation of rylene diimide dimers using molecular handcuffs. *Chem. Sci.* **2019**, *10*, 3723–3732. [CrossRef] [PubMed]
16. Kaufmann, C.; Bialas, D.; Stolte, M.; Würthner, F. Discrete π-Stacks of Perylene Bisimide Dyes within Folda-Dimers: Insight into Long- and Short-Range Exciton Coupling. *J. Am. Chem. Soc.* **2018**, *140*, 9986–9995. [CrossRef]
17. Kaufmann, C.; Kim, W.; Nowak-Król, A.; Hong, Y.; Kim, D.; Würthner, F. Ultrafast Exciton Delocalization, Localization, and Excimer Formation Dynamics in a Highly Defined Perylene Bisimide Quadruple π-Stack. *J. Am. Chem. Soc.* **2018**, *140*, 4253–4258. [CrossRef]
18. Bardeen, C.J. The Structure and Dynamics of Molecular Excitons. *Annu. Rev. Phys. Chem.* **2014**, *65*, 127–148. [CrossRef]
19. Mewes, S.A.; Dreuw, A. Density-based descriptors and exciton analyses for visualizing and understanding the electronic structure of excited states. *Phys. Chem. Chem. Phys.* **2019**, *21*, 2843–2856. [CrossRef]
20. Darghouth, A.A.M.H.M.; Correa, G.C.; Juillard, S.; Casida, M.E.; Humeniuk, A.; Mitrić, R. Davydov-type excitonic effects on the absorption spectra of parallel-stacked and herringbone aggregates of pentacene: Time-dependent density-functional theory and time-dependent density-functional tight binding. *J. Chem. Phys.* **2018**, *149*, 134111. [CrossRef]
21. Casanova, D. Theoretical investigations of the perylene electronic structure: Monomer, dimers, and excimers. *Int. J. Quantum Chem.* **2015**, *115*, 442–452. [CrossRef]
22. Zubiria-Ulacia, M.; Matxain, J.M.; Casanova, D. The role of CT excitations in PDI aggregates. *Phys. Chem. Chem. Phys.* **2020**, *22*, 15908–15918. [CrossRef] [PubMed]
23. Mao, Y.; Montoya-Castillo, A.; Markland, T.E. Accurate and efficient DFT-based diabatization for hole and electron transfer using absolutely localized molecular orbitals. *J. Chem. Phys.* **2019**, *151*, 164114. [CrossRef] [PubMed]
24. Mao, Y.; Montoya-Castillo, A.; Markland, T.E. Excited state diabatization on the cheap using DFT: Photoinduced electron and hole transfer. *J. Chem. Phys.* **2020**, *153*, 244111. [CrossRef] [PubMed]
25. Mewes, S.A.; Plasser, F.; Krylov, A.; Dreuw, A. Benchmarking Excited-State Calculations Using Exciton Properties. *J. Chem. Theory Comput.* **2018**, *14*, 710–725. [CrossRef] [PubMed]
26. Jurinovich, S.; Cupellini, L.; Guido, C.A.; Mennucci, B. EXAT: EXcitonic analysis tool. *J. Comput. Chem.* **2018**, *39*, 279–286. [CrossRef]
27. Walter, C.; Krämer, V.; Engels, B. On the applicability of time-dependent density functional theory (TDDFT) and semiempirical methods to the computation of excited-state potential energy surfaces of perylene-based dye-aggregates. *Int. J. Quantum Chem.* **2017**, *117*, e25337. [CrossRef]
28. Accomasso, D.; Persico, M.; Granucci, G. Diabatization by Localization in the Framework of Configuration Interaction Based on Floating Occupation Molecular Orbitals (FOMO−CI). *ChemPhotoChem* **2019**, *3*, 933–944. [CrossRef]
29. Tamura, H. Diabatization for Time-Dependent Density Functional Theory: Exciton Transfers and Related Conical Intersections. *J. Phys. Chem. A* **2016**, *120*, 9341–9347. [CrossRef]
30. Liu, W.; Lunkenheimer, B.; Settels, V.; Engels, B.; Fink, R.F.; Köhn, A. A general ansatz for constructing quasi-diabatic states in electronically excited aggregated systems. *J. Chem. Phys.* **2015**, *143*, 084106. [CrossRef]
31. Carreras, A.; Uranga-Barandiaran, O.; Castet, F.; Casanova, D. Photophysics of Molecular Aggregates from Excited State Diabatization. *J. Chem. Theory Comput.* **2019**, *15*, 2320–2330. [CrossRef] [PubMed]
32. Shirai, S.; Iwata, S.; Tani, T.; Inagaki, S. Ab Initio Studies of Aromatic Excimers Using Multiconfiguration Quasi-Degenerate Perturbation Theory. *J. Phys. Chem. A* **2011**, *115*, 7687–7699. [CrossRef] [PubMed]

33. Lv, M.; Lu, X.; Jiang, Y.; Sandoval-Salinas, M.E.; Casanova, D.; Sun, H.; Sun, Z.; Xu, J.; Yang, Y.; Chen, J. Near-Unity Triplet Generation Promoted via Spiro-Conjugation. *Angew. Chemie Int. Ed.* **2021**, *61*, 2–10. [CrossRef]
34. Margulies, E.A.; Logsdon, J.L.; Miller, C.E.; Ma, L.; Simonoff, E.; Young, R.M.; Schatz, G.C.; Wasielewski, M.R. Direct Observation of a Charge-Transfer State Preceding High-Yield Singlet Fission in Terrylenediimide Thin Films. *J. Am. Chem. Soc.* **2017**, *139*, 663–671. [CrossRef]
35. Margulies, E.A.; Miller, C.E.; Wu, Y.; Ma, L.; Schatz, G.C.; Young, R.M.; Wasielewski, M.R. Enabling singlet fission by controlling intramolecular charge transfer in π-stacked covalent terrylenediimide dimers. *Nat. Chem.* **2016**, *8*, 1120–1125. [CrossRef]
36. Young, R.M.; Wasielewski, M.R. Mixed Electronic States in Molecular Dimers: Connecting Singlet Fission, Excimer Formation, and Symmetry-Breaking Charge Transfer. *Acc. Chem. Res.* **2020**, *53*, 1957–1968. [CrossRef] [PubMed]
37. Beljonne, D.; Yamagata, H.; Brédas, J.L.; Spano, F.C.; Olivier, Y. Charge-Transfer Excitations Steer the Davydov Splitting and Mediate Singlet Exciton Fission in Pentacene. *Phys. Rev. Lett.* **2013**, *110*, 226402. [CrossRef]
38. Berkelbach, T.C.; Hybertsen, M.S.; Reichman, D.R. Microscopic theory of singlet exciton fission. II. Application to pentacene dimers and the role of superexchange. *J. Chem. Phys.* **2013**, *138*, 114103. [CrossRef]
39. Zeng, T.; Hoffmann, R.; Ananth, N. The Low-Lying Electronic States of Pentacene and Their Roles in Singlet Fission. *J. Am. Chem. Soc.* **2014**, *136*, 5755–5764. [CrossRef]
40. Casanova, D. Theoretical Modeling of Singlet Fission. *Chem. Rev.* **2018**, *118*, 7164–7207. [CrossRef]
41. Han, G.; Yi, Y. Local Excitation/Charge-Transfer Hybridization Simultaneously Promotes Charge Generation and Reduces Nonradiative Voltage Loss in Nonfullerene Organic Solar Cells. *J. Phys. Chem. Lett.* **2019**, *10*, 2911–2918. [CrossRef] [PubMed]
42. Coropceanu, V.; Chen, X.-K.; Wang, T.; Zheng, Z.; Brédas, J.-L. Charge-transfer electronic states in organic solar cells. *Nat. Rev. Mater.* **2019**, *4*, 689–707. [CrossRef]
43. Dong, Y.; Nikolis, V.C.; Talnack, F.; Chin, Y.-C.; Benduhn, J.; Londi, G.; Kublitski, J.; Zheng, X.; Mannsfeld, S.C.B.; Spoltore, D.; et al. Orientation dependent molecular electrostatics drives efficient charge generation in homojunction organic solar cells. *Nat. Commun.* **2020**, *11*, 4617. [CrossRef] [PubMed]
44. Shi, B.; Nachtigallová, D.; Aquino, A.J.A.; Machado, F.B.C.; Lischka, H. Excited states and excitonic interactions in prototypic polycyclic aromatic hydrocarbon dimers as models for graphitic interactions in carbon dots. *Phys. Chem. Chem. Phys.* **2019**, *21*, 9077–9088. [CrossRef] [PubMed]
45. Fliegl, H.; You, Z.; Hsu, C.; Sundholm, D. The Excitation Spectra of Naphthalene Dimers: Frenkel and Charge-transfer Excitons. *J. Chin. Chem. Soc.* **2016**, *63*, 20–32. [CrossRef]
46. Liu, W.; Settels, V.; Harbach, P.H.P.; Dreuw, A.; Fink, R.F.; Engels, B. Assessment of TD-DFT- and TD-HF-based approaches for the prediction of exciton coupling parameters, potential energy curves, and electronic characters of electronically excited aggregates. *J. Comput. Chem.* **2011**, *32*, 1971–1981. [CrossRef]
47. Settels, V.; Liu, W.; Pflaum, J.; Fink, R.F.; Engels, B. Comparison of the electronic structure of different perylene-based dye-aggregates. *J. Comput. Chem.* **2012**, *33*, 1544–1553. [CrossRef]
48. Mukazhanova, A.; Malone, W.; Negrin-Yuvero, H.; Fernandez-Alberti, S.; Tretiak, S.; Sharifzadeh, S. Photoexcitation dynamics in perylene diimide dimers. *J. Chem. Phys.* **2020**, *153*, 244117. [CrossRef]
49. Bellinger, D.; Pflaum, J.; Brüning, C.; Engel, V.; Engels, B. The electronic character of PTCDA thin films in comparison to other perylene-based organic semi-conductors: Ab initio-, TD-DFT and semi-empirical computations of the opto-electronic properties of large aggregates. *Phys. Chem. Chem. Phys.* **2017**, *19*, 2434–2448. [CrossRef]
50. Engels, B.; Engel, V. The dimer-approach to characterize opto-electronic properties of and exciton trapping and diffusion in organic semiconductor aggregates and crystals. *Phys. Chem. Chem. Phys.* **2017**, *19*, 12604–12619. [CrossRef]
51. Hartzler, D.A.; Slipchenko, L.V.; Savikhin, S. Triplet-Triplet Coupling in Chromophore Dimers: Theory and Experiment. *J. Phys. Chem. A* **2018**, *122*, 6713–6723. [CrossRef] [PubMed]
52. You, Z.-Q.; Hsu, C.-P.; Fleming, G.R. Triplet-triplet energy-transfer coupling: Theory and calculation. *J. Chem. Phys.* **2006**, *124*, 044506. [CrossRef] [PubMed]
53. Gillett, A.J.; Privitera, A.; Dilmurat, R.; Karki, A.; Qian, D.; Pershin, A.; Londi, G.; Myers, W.K.; Lee, J.; Yuan, J.; et al. The role of charge recombination to triplet excitons in organic solar cells. *Nature* **2021**, *597*, 666–671. [CrossRef] [PubMed]
54. Bassan, E.; Gualandi, A.; Cozzi, P.G.; Ceroni, P. Design of BODIPY dyes as triplet photosensitizers: Electronic properties tailored for solar energy conversion, photoredox catalysis and photodynamic therapy. *Chem. Sci.* **2021**, *12*, 6607–6628. [CrossRef]
55. Gibbons, D.J.; Farawar, A.; Mazzella, P.; Leroy-Lhez, S.; Williams, R.M. Making triplets from photo-generated charges: Observations, mechanisms and theory. *Photochem. Photobiol. Sci.* **2020**, *19*, 136–158. [CrossRef]
56. Cakmak, Y.; Kolemen, S.; Duman, S.; Dede, Y.; Dolen, Y.; Kilic, B.; Kostereli, Z.; Yildirim, L.T.; Dogan, A.L.; Guc, D.; et al. Designing Excited States: Theory-Guided Access to Efficient Photosensitizers for Photodynamic Action. *Angew. Chem. Int. Ed.* **2011**, *50*, 11937–11941. [CrossRef]
57. Galán, L.A.; Andrés Castán, J.M.; Dalinot, C.; Marqués, P.S.; Galiana, J.; Blanchard, P.; Andraud, C.; Dumont, E.; Maury, O.; Cabanetos, C.; et al. Exploring the Concept of Dimerization-Induced Intersystem Crossing: At the Origins of Spin–Orbit Coupling Selection Rules. *J. Phys. Chem. B* **2021**, *125*, 8572–8580. [CrossRef]
58. Oleson, A.; Zhu, T.; Dunn, I.S.; Bialas, D.; Bai, Y.; Zhang, W.; Dai, M.; Reichman, D.R.; Tempelaar, R.; Huang, L.; et al. Perylene Diimide-Based Hj- and hJ-Aggregates: The Prospect of Exciton Band Shape Engineering in Organic Materials. *J. Phys. Chem. C* **2019**, *123*, 20567–20578. [CrossRef]

59. Liu, W.; Canola, S.; Köhn, A.; Engels, B.; Negri, F.; Fink, R.F. A model hamiltonian tuned toward high level ab initio calculations to describe the character of excitonic states in perylenebisimide aggregates. *J. Comput. Chem.* **2018**, *39*, 1979–1989. [CrossRef]
60. Canola, S.; Bagnara, G.; Dai, Y.; Ricci, G.; Calzolari, A.; Negri, F. Addressing the Frenkel and charge transfer character of exciton states with a model Hamiltonian based on dimer calculations: Application to large aggregates of perylene bisimide. *J. Chem. Phys.* **2021**, *154*, 124101. [CrossRef]
61. Hellweg, A.; Grün, S.A.; Hättig, C. Benchmarking the performance of spin-component scaled CC2 in ground and electronically excited states. *Phys. Chem. Chem. Phys.* **2008**, *10*, 4119. [CrossRef] [PubMed]
62. Winter, N.O.C.; Graf, N.K.; Leutwyler, S.; Hättig, C. Benchmarks for 0–0 transitions of aromatic organic molecules: DFT/B3LYP, ADC(2), CC2, SOS-CC2 and SCS-CC2 compared to high-resolution gas-phase data. *Phys. Chem. Chem. Phys.* **2013**, *15*, 6623–6630. [CrossRef] [PubMed]
63. Dreuw, A.; Wormit, M. The algebraic diagrammatic construction scheme for the polarization propagator for the calculation of excited states. *Wiley Interdiscip. Rev. Comput. Mol. Sci.* **2015**, *5*, 82–95. [CrossRef]
64. Plasser, F.; Lischka, H. Analysis of Excitonic and Charge Transfer Interactions from Quantum Chemical Calculations. *J. Chem. Theory Comput.* **2012**, *8*, 2777–2789. [CrossRef] [PubMed]
65. Casanova, D.; Krylov, A.I. Quantifying local exciton, charge resonance, and multiexciton character in correlated wave functions of multichromophoric systems. *J. Chem. Phys.* **2016**, *144*, 014102. [CrossRef] [PubMed]
66. Plasser, F. TheoDORE: A toolbox for a detailed and automated analysis of electronic excited state computations. *J. Chem. Phys.* **2020**, *152*, 084108. [CrossRef] [PubMed]
67. Petelenz, P. Mixing of Frenkel excitons with charge transfer states in the neighbourhood of charged defect. *Chem. Phys. Lett.* **1977**, *47*, 603–605. [CrossRef]
68. Norton, J.E.; Brédas, J.-L. Theoretical characterization of titanyl phthalocyanine as a p-type organic semiconductor: Short intermolecular π-π interactions yield large electronic couplings and hole transport bandwidths. *J. Chem. Phys.* **2008**, *128*, 034701. [CrossRef]
69. Kim, D. A Theoretical Analysis of the Excited State of Oligoacene Aggregates: Local Excitation vs. Charge-Transfer Transition. *Bull. Korean Chem. Soc.* **2015**, *36*, 2284–2289. [CrossRef]
70. Löwdin, P. On the Non-Orthogonality Problem Connected with the Use of Atomic Wave Functions in the Theory of Molecules and Crystals. *J. Chem. Phys.* **1950**, *18*, 365–375. [CrossRef]
71. Nottoli, M.; Jurinovich, S.; Cupellini, L.; Gardiner, A.T.; Cogdell, R.; Mennucci, B. The role of charge-transfer states in the spectral tuning of antenna complexes of purple bacteria. *Photosynth. Res.* **2018**, *137*, 215–226. [CrossRef]
72. Fink, R.F.; Seibt, J.; Engel, V.; Renz, M.; Kaupp, M.; Lochbrunner, S.; Zhao, H.-M.; Pfister, J.; Würthner, F.; Engels, B. Exciton trapping in pi-conjugated materials: A quantum-chemistry-based protocol applied to perylene bisimide dye aggregates. *J. Am. Chem. Soc.* **2008**, *130*, 12858–12859. [CrossRef] [PubMed]
73. Hirata, S.; Head-Gordon, M. Time-dependent density functional theory within the Tamm–Dancoff approximation. *Chem. Phys. Lett.* **1999**, *314*, 291–299. [CrossRef]
74. Yanai, T.; Tew, D.P.; Handy, N.C. A new hybrid exchange–correlation functional using the Coulomb-attenuating method (CAM-B3LYP). *Chem. Phys. Lett.* **2004**, *393*, 51–57. [CrossRef]
75. Chai, J.-D.; Head-Gordon, M. Long-range corrected hybrid density functionals with damped atom–atom dispersion corrections. *Phys. Chem. Chem. Phys.* **2008**, *10*, 6615. [CrossRef]
76. Frisch, M.J.; Trucks, G.W.; Schlegel, H.B.; Scuseria, G.E.; Robb, M.A.; Cheeseman, J.R.; Scalmani, G.; Barone, V.; Petersson, G.A.; Nakatsuji, H.; et al. *Gaussian 16*; Gaussian, Inc.: Wallingford, CT, USA, 2016.
77. Hestand, N.J.; Kazantsev, R.V.; Weingarten, A.S.; Palmer, L.C.; Stupp, S.I.; Spano, F.C. Extended-Charge-Transfer Excitons in Crystalline Supramolecular Photocatalytic Scaffolds. *J. Am. Chem. Soc.* **2016**, *138*, 11762–11774. [CrossRef]
78. Stehr, V.; Engels, B.; Deibel, C.; Fink, R.F. Anisotropy of singlet exciton diffusion in organic semiconductor crystals from ab initio approaches. *J. Chem. Phys.* **2014**, *140*, 024503. [CrossRef]

computation

MDPI

Article

Kinetic Energy Density Functionals Based on a Generalized Screened Coulomb Potential: Linear Response and Future Perspectives

Eduardo Fabiano [1,2], Fulvio Sarcinella [2,3], Lucian A. Constantin [4] and Fabio Della Sala [1,2,*]

1 Institute for Microelectronics and Microsystems (CNR-IMM), Via Monteroni, Campus Unisalento, 73100 Lecce, Italy; eduardo.fabiano@cnr.it
2 Center for Biomolecular Nanotechnologies@UNILE, Istituto Italiano di Tecnologia, Via Barsanti 14, 73010 Arnesano, Italy; fulvio.sarcinella@iit.it
3 Department of Mathematics and Physics "E. De Giorgi", University of Salento, Via Arnesano, 73100 Lecce, Italy
4 Istituto di Nanoscienze, Consiglio Nazionale delle Ricerche (CNR-NANO), 41125 Modena, Italy; lucian.constantin68@gmail.com
* Correspondence: fabio.dellasala@iit.it

Abstract: We consider kinetic energy functionals that depend, beside the usual semilocal quantities (density, gradient, Laplacian of the density), on a generalized Yukawa potential, that is the screened Coulomb potential of the density raised to some power. These functionals, named Yukawa generalized gradient approximations (yGGA), are potentially efficient real-space semilocal methods that include significant non-local effects and can describe different important exact properties of the kinetic energy. In this work, we focus in particular on the linear response behavior for the homogeneous electron gas (HEG). We show that such functionals are able to reproduce the exact Lindhard function behavior with a very good accuracy, outperforming all other semilocal kinetic functionals. These theoretical advances allow us to perform a detailed analysis of a special class of yGGAs, namely the linear yGGA functionals. Thus, we show how the present approach can generalize the yGGA functionals improving the HEG linear behavior and leading to an extended formula for the kinetic functional. Moreover, testing on several jellium cluster model systems allows highlighting advantages and limitations of the linear yGGA functionals and future perspectives for the development of yGGA kinetic functionals.

Keywords: density functional theory; kinetic functional; Yukawa potential

Citation: Fabiano, E.; Sarcinella, F.; Constantin, L.A.; Della Sala, F. Kinetic Energy Density Functionals Based on a Generalized Screened Coulomb Potential: Linear Response and Future Perspectives. *Computation* **2022**, *10*, 30. https://doi.org/10.3390/computation10020030

Academic Editor: Henry Chermette

Received: 22 January 2022
Accepted: 10 February 2022
Published: 15 February 2022

1. Introduction

The non-interacting kinetic energy (KE) functional is one of the main quantities of interest in density functional theory [1,2]. Its exact formal definition is readily obtained, following the Levy constrained search formalism [3], as

$$T_s[n] = \min_{\Psi \to n} \langle \Phi | -\frac{1}{2}\nabla^2 | \Phi \rangle \,, \tag{1}$$

where n is the electron density and Φ is a Kohn-Sham (KS) [4] Slater determinant yielding the density n. This formula allows us to study different important properties of the KE functional and provides an explicit expression for T_s in terms of KS orbitals. However, it does not allow us to obtain an explicit expression in terms of the electron density. Therefore, the quest for the KE density functional is still open, also considering the importance of this quantity in many contexts including orbital-free density functional theory (OF-DFT) [5–8], subsystem DFT [9–15], and quantum hydrodynamic theory [16–19]. In addition, semilocal KE functionals have been used in meta-GGA exchange-correlation functionals to remove their orbital dependence [20–24].

Numerous investigations have been dedicated in the last decades to the study of KE functionals [25–72]. Nevertheless, accurate approximations of $T_s[n]$ are hard to obtain because this quantity usually gives a dominant contribution to the ground-state energy [3] and because of the highly non-local nature of the KE functional [5,51,73–78]. For this reason, more recently, machine-learning methods have also been used to develop KE functionals [79–87].

A KE functional approximation can be written

$$T_s[n] = \int \tau[n](\mathbf{r})d\mathbf{r}\,, \tag{2}$$

where $\tau(\mathbf{r})$ is the KE density and, actually, there are two main strategies to approximate $\tau(\mathbf{r})$. The simplest one considers the KE density to be a semilocal functional of the density, that is

$$\tau[n](\mathbf{r}) = \tau^{semilocal}(n(\mathbf{r}), \nabla n(\mathbf{r}), \nabla^2 n(\mathbf{r}), \ldots)\,. \tag{3}$$

This approach, which traces back to the nearsightedness principle [88], is computationally efficient because of the local nature of $\tau(\mathbf{r})$. However, since it is not explicitly including non-local effects, that are quite relevant for KE functionals, semilocal functionals face several limitations [46,89].

To overcome this problem, the other popular approach used to describe the KE density makes explicit use of a non-local ansatz [47–63]

$$\tau^{non-local}(\mathbf{r}) = \int n^{\alpha}(\mathbf{r})K(\mathbf{r},\mathbf{r}')n^{\beta}(\mathbf{r}')d\mathbf{r}'\,, \tag{4}$$

where $K(\mathbf{r},\mathbf{r}')$ is a proper non-local kernel. The presence of the non-local kernel strongly increases the computational cost of the method and raises several practical difficulties. On the other hand, non-local KE functionals are much more accurate [47–63].

In particular, the kernel can be designed in order to reproduce the correct linear response behavior of the homogeneous electron gas (HEG), which has been shown to be a very important property for the KE functional [47,60]. In fact, the KE functional and the linear response function χ have a close relation, given by the equation [5]

$$\mathcal{F}\left(\frac{\delta^2 T_s[n]}{\delta n(\mathbf{r})\delta n(\mathbf{r}')}\right) = -\frac{1}{\chi(\mathbf{k})}\,, \tag{5}$$

where $\mathcal{F}$ denotes the Fourier transform and $\mathbf{k}$ is the momentum vector. For the HEG, the linear response function can be computed analytically [5] and is related to the Lindhard function [90]. Hence,

$$\chi^{HEG}(\mathbf{k}) = -\frac{(3\pi^2 n)^{1/3}}{\pi^2}F^{-1}_{Lind}(\eta)\,, \tag{6}$$

$$F^{-1}_{Lind}(\eta) = \frac{1}{2} + \frac{1-\eta^2}{4\eta}\ln\left|\frac{1+\eta}{1-\eta}\right|\,, \tag{7}$$

with $\eta = |\mathbf{k}|/[2(3\pi^2 n)^{1/3}]$ being a dimensionless momentum. The Lindhard function cannot be accurately mimicked by any semilocal functional, because these all have a polynomial Fourier transform [71]. On the other hand, it is the main property for most of the non-local KE functionals.

Recently, a new class of KE functionals [71] has been proposed to join the advantages of the semilocal methods and the good features of the non-local functionals. These functionals, named Yukawa generalized gradient approximation (yGGA), use as input ingredients, beside the density and its gradients, a Yukawa potential

$$u_{\alpha}(\mathbf{r}) = \int \frac{n(\mathbf{r}')e^{-\alpha k_F(\mathbf{r})|\mathbf{r}-\mathbf{r}'|}}{|\mathbf{r}-\mathbf{r}'|}d\mathbf{r}'\,, \tag{8}$$

i.e., a screened Coulomb potential with $\alpha k_F(\mathbf{r}) = \alpha(3\pi^2)^{1/3}n(\mathbf{r})^{1/3}$ as the screening length. In this way, it is possible to include efficiently non-local effects and improve the description of the HEG response function without resorting to the reciprocal space [71] (as it is instead accustomed in non-local functionals).

In this work, we will take a step further in this direction and we will consider a modification of the basic input ingredient of the functional, allowing the Yukawa potential to be computed for a power of the density, i.e.,

$$u_{\alpha\beta}(\mathbf{r}) = \int \frac{n^{\beta}(\mathbf{r}')e^{-\alpha k_F(\mathbf{r})|\mathbf{r}-\mathbf{r}'|}}{|\mathbf{r}-\mathbf{r}'|} d\mathbf{r}' \,. \tag{9}$$

The computational cost of Equation (9) is, at least for the spherical systems considered in this work, the same as the one of the conventional Yukawa potential in Equation (8); thus, it is interesting to investigate if and how the β parameter will impact the accuracy of the linear response and of the resulting KE functional.

Therefore, we will consider yGGA functionals of the general form

$$\tau[n](\mathbf{r}) = \tau(n(\mathbf{r}), \nabla n(\mathbf{r}), \nabla^2 n(\mathbf{r}), u_{\alpha\beta}(\mathbf{r})) \,. \tag{10}$$

For these functionals, we will provide a full analytical derivation of the linear response function and we will consider the exact constraints required to reproduce the Lindhard behavior. Finally, we will analyze the role of the β parameter for the description of jellium spheres and we will give insights for the further development of yGGA functionals.

2. Theory

We consider a KE density of the form

$$\tau = C_F n^{5/3} F_s(p, q, y_{\alpha\beta}) \,, \tag{11}$$

where F_s is the KE enhancement factor and

$$C_F = \frac{3}{10}k_0^2 \quad \text{with} \quad k_0 = (3\pi^2)^{1/3} \,, \tag{12}$$

$$p(\mathbf{r}) = \frac{|\nabla n(\mathbf{r})|^2}{4k_0^2 n(\mathbf{r})^{8/3}} \,, \tag{13}$$

$$q(\mathbf{r}) = \frac{\nabla^2 n(\mathbf{r})}{4k_0^2 n(\mathbf{r})^{5/3}} \,, \tag{14}$$

$$y_{\alpha\beta}(\mathbf{r}) = \frac{3\pi\alpha^2}{4k_0 n(\mathbf{r})^{\beta-2/3}} \int \frac{n^{\beta}(\mathbf{r}')e^{-\alpha k_0 n(\mathbf{r})^{1/3}|\mathbf{r}-\mathbf{r}'|}}{|\mathbf{r}-\mathbf{r}'|} d\mathbf{r}' \,. \tag{15}$$

The quantity $y_{\alpha\beta}$ is the main ingredient of the yGGA functional, whereas p (the reduced gradient) and q (the reduced Laplacian) are the conventional ingredients of meta-GGA functionals. The ingredient $y_{\alpha\beta}$ is proportional to the potential $u_{\alpha\beta}$, where the normalization constant has been choosen so that $y_{\alpha\beta}$ is adimensional and invariant under uniform scaling, as shown in Ref. [71]. Moreover, in the case of a large number of electrons, where the Thomas–Fermi (TF) limit is exact, we have also $y_{\alpha\beta} \to 1$: thus, the enhancement factor F_s must be 1 when $p = 0, q = 0$, and $y_{\alpha\beta} = 1$.

While the case of $\beta = 1$ has been deeply investigated in Ref. [71], when $\beta \neq 1$, the properties of $y_{\alpha\beta}$ change. In particular, in the tail of finite systems we have:

$$y_{\alpha\beta} \to \frac{3\pi\alpha^2}{4k_0}\frac{Q_\beta}{n^{\beta-2/3}r} \,, \tag{16}$$

$$Q_\beta = \int n^{\beta}(\mathbf{r}')d\mathbf{r}' \,, \tag{17}$$

(for $\beta = 1$, Q_β is just the number of electrons). Thus, $y_{\alpha\beta}$ diverges for $\beta > 2/3$ (as the density vanishes exponentially), whereas for $\beta \leq 2/3$, it vanishes.

2.1. Kinetic Energy Potential

Following the same derivation as in Ref. [71], the KE potential of a general modified yGGA functional is

$$\frac{\delta T_s^{\rm yGGA}}{\delta n(\mathbf{r})} = v_k^{MGGA}(\mathbf{r}) + v_k^{\rm yGGA,1}(\mathbf{r}) + v_k^{\rm yGGA,2}(\mathbf{r}) \,, \tag{18}$$

with

$$v_k^{MGGA}(\mathbf{r}) = \frac{\partial \tau}{\partial n}(\mathbf{r}) - \nabla \cdot \frac{\partial \tau}{\partial \nabla n} + \nabla^2 \frac{\partial \tau}{\partial \nabla^2 n} \,, \tag{19}$$

$$v_k^{\rm yGGA,1}(\mathbf{r}) = \beta n(\mathbf{r})^{\beta-1} \int \frac{\partial \tau}{\partial u_{\alpha\beta}(\mathbf{r}')} \frac{e^{-\alpha k_F(\mathbf{r}')|\mathbf{r}-\mathbf{r}'|}}{|\mathbf{r}-\mathbf{r}'|} \mathrm{d}^3\mathbf{r}' \,, \tag{20}$$

$$v_k^{\rm yGGA,2}(\mathbf{r}) = -\frac{\alpha k_F(\mathbf{r})}{3n(\mathbf{r})} \frac{\partial \tau}{\partial u_{\alpha\beta}(\mathbf{r})} \int n^\beta(\mathbf{r}') e^{-\alpha k_F(\mathbf{r})|\mathbf{r}-\mathbf{r}'|} \mathrm{d}^3\mathbf{r}'. \tag{21}$$

The potential can also be expressed as a function of the enhancement factor using:

$$\frac{d\tau}{du_{\alpha\beta}} = \frac{9\pi\alpha^2}{40} k_F(\mathbf{r}) n(\mathbf{r})^{2-\beta} \frac{dF_s}{dy_{\alpha\beta}} \,. \tag{22}$$

Note that the potential in Equation (20) is a non-local function of the expression in Equation (22). This means, for example, that a divergence of $\frac{d\tau}{du_{\alpha\beta}}$ in the tail of a finite system will have an impact everywhere in the space. Thus, the enhancement factor F_s must be properly defined in all points.

In the limit of a large number of electrons ($k_F \to \infty$), we have

$$\begin{aligned} v_k^{\rm yGGA,1}(\mathbf{r}) &= \beta n(\mathbf{r})^{\beta-1} \frac{9\pi\alpha^2}{40} k_F(\mathbf{r}) n(\mathbf{r})^{2-\beta} \frac{dF_s}{dy_{\alpha\beta}} \frac{4\pi}{\alpha^2 k_F(\mathbf{r})^2} \\ &= \beta \frac{3}{10} k_F(\mathbf{r})^2 \frac{dF_s}{dy_{\alpha\beta}(\mathbf{r})} \,, \end{aligned} \tag{23}$$

$$\begin{aligned} v_k^{\rm yGGA,2}(\mathbf{r}) &= -\frac{\alpha k_F(\mathbf{r})}{3n(\mathbf{r})} \frac{9\pi\alpha^2}{40} k_F(\mathbf{r}) n(\mathbf{r})^{2-\beta} \frac{dF_s}{dy_{\alpha\beta}} n^\beta(\mathbf{r}) \frac{8\pi}{\alpha^3 k_F(\mathbf{r})^3} \\ &= -\frac{2}{10} k_F(\mathbf{r})^2 \frac{dF_s}{dy_{\alpha\beta}(\mathbf{r})} \,, \end{aligned} \tag{24}$$

which are stable expressions for all values of β. In this case, we also have that the MGGA term of Equation (19) reduces to the first LDA term only. Hence, for the total KE potential, we find

$$\frac{\partial T_s^{\rm yGGA}}{\partial n}(\mathbf{r}) = \frac{5}{10} k_F(\mathbf{r})^2 F_s(y_{\alpha\beta}(\mathbf{r})) - (\frac{3}{10}\beta - \frac{2}{10}) k_F(\mathbf{r})^2 y_{\alpha\beta}(\mathbf{r}) \frac{dF_s}{dy_{\alpha\beta}(\mathbf{r})} \,. \tag{25}$$

With the condition $F_s \to 1$, we have that the total KE potential of Equation (25) reduces to $\partial\tau/\partial n = (5/10) k_F(\mathbf{r}) = v_{TF}(\mathbf{r})$, where v_{TF} is the Thomas–Fermi potential. Thus, the condition that in the TF limit $F_s \to y_{\alpha\beta} \to 1$ yields also a correct potential.

2.2. Linear Response of a yGGA Functional

Following Refs. [71,91], we compute the linear response considering the perturbed density $n = n_0 + n_k e^{-i\mathbf{k}\cdot\mathbf{r}}$. Note that the derivation in Ref. [71] was limited to a specific class of yGGA functionals (the linear yGGA functional, see Section 3 in the following) and with $\beta = 1$. Hereafter, a completely general derivation is presented.

Expanding the resulting perturbed KE density in power series, we obtain the linear response $\chi = 1/F$ as twice the coefficient of the inverse of the second-order term. For simplicity, we can evaluate the whole expression in $\mathbf{r} = 0$, since the HEG is homogeneous and isotropic. Then, we have $n = n_0 + n_k$, $|\nabla n|^2 = n_k^2 k^2$ and $\nabla^2 n = -n_k k^2$. Therefore, we need to consider

$$\tau = C_F (n_0 + n_k)^{5/3} F_s(p, q, y_{\alpha\beta}) \,, \tag{26}$$

with

$$p = \frac{n_k^2 k^2}{4k_0^2 (n_0 + n_k)^{8/3}} \,, \tag{27}$$

$$q = -\frac{n_k k^2}{4k_0^2 (n_0 + n_k)^{5/3}} \,, \tag{28}$$

$$\begin{aligned} y_{\alpha\beta} &\approx 1 - \frac{\beta k^2}{A^2 + k^2} \frac{n_k}{n_0} + \\ &+ \left[\frac{\beta(\beta-1)}{2} \frac{A^2}{A^2 + 4k^2} + \frac{2\beta A^2 k^2 - 3\beta^2 A^2 (A^2 + k^2)}{3(A^2 + k^2)^2} + \frac{\beta(\beta+1)}{2} \right] \left(\frac{n_k}{n_0} \right)^2 , \end{aligned} \tag{29}$$

where we have set $A = \alpha k_0 n_0^{1/3}$. For the derivation of the expression for $y_{\alpha\beta}$, see Appendix A.

To proceed, we can expand in powers of n_k both factors in Equation (26). Thus, using the notation $\partial_{n_k} = \partial/\partial n_k|_{n_k=0}$, we can write

$$\begin{aligned} \tau &\approx C_F \left(n_0^{5/3} + \frac{5}{3} n_0^{2/3} n_k + \frac{5}{9} n_0^{-1/3} n_k^2 + \cdots \right) \times \\ &\times \left(F_s(0,0,1) + \partial_{n_k} F_s n_k + \frac{1}{2} \partial^2_{n_k} F_s n_k^2 + \cdots \right) = \\ &= C_F n_0^{5/3} F_s(0,0,1) + \left[\frac{5 C_F n_0^{2/3}}{3} F_s(0,0,1) + C_F n_0^{5/3} \partial_{n_k} F_s \right] n_k + \\ &+ \left[\frac{5 C_F n_0^{-1/3}}{9} F_s(0,0,1) + \frac{5 C_F n_0^{2/3}}{3} \partial_{n_k} F_s + \frac{C_F n_0^{5/3}}{2} \partial^2_{n_k} F_s \right] n_k^2 + \mathcal{O}(n_k^3) \,. \end{aligned} \tag{30}$$

The linear response $\chi = 1/F$ is given by twice the coefficient of the inverse of the second-order term. Therefore,

$$\begin{aligned} F &= 2 \left[\frac{5 C_F n_0^{-1/3}}{9} F_s(0,0,1) + \frac{5 C_F n_0^{2/3}}{3} \partial_{n_k} F_s + \frac{C_F n_0^{5/3}}{2} \partial^2_{n_k} F_s \right] = \\ &= k_0^2 n_0^{2/3} \left[\frac{F_s(0,0,1)}{3 n_0} + \partial_{n_k} F_s + \frac{3 n_0}{10} \partial^2_{n_k} F_s \right] . \end{aligned} \tag{31}$$

The corresponding Thomas–Fermi-renormalized linear response is $\tilde{F} = k_F F/\pi^2 = k_0 n^{1/3} F/\pi^2$. Then,

$$\tilde{F} = 3 n_0 \left[\frac{F_s(0,0,1)}{3 n_0} + \partial_{n_k} F_s + \frac{3 n_0}{10} \partial^2_{n_k} F_s \right] . \tag{32}$$

For simplicity of notation in the following, we neglect the subscript $\alpha\beta$ in the $y_{\alpha\beta}$ ingredient.

To obtain a more explicit expression for Equation (32), we use the chain rule

$$\partial_{n_k} F_s = D_p \partial_{n_k} p + D_q \partial_{n_k} q + D_y \partial_{n_k} y \,, \tag{33}$$

where we have employed the notation ($i = p, q, y$)

$$D_i \equiv \left. \frac{\partial F_s}{\partial i} \right|_{(p,q,y)=(0,0,1)} . \tag{34}$$

Hence, substituting the values for $\partial_{n_k} p$, $\partial_{n_k} q$, and $\partial_{n_k} y$, we find

$$\partial_{n_k} F_s = -D_q \frac{k^2}{4k_0^2 n_0^{5/3}} - D_y \frac{1}{n_0} \frac{\beta k^2}{A^2 + k^2} . \tag{35}$$

For the second derivative, we have

$$\begin{aligned} \partial^2_{n_k} F_s &= \partial_{n_k} \left(\frac{\partial p}{\partial n_k} \frac{\partial F_s}{\partial p} + \frac{\partial q}{\partial n_k} \frac{\partial F_s}{\partial q} + \frac{\partial y}{\partial n_k} \frac{\partial F_s}{\partial y} \right) = \\ &= \partial^2_{n_k} p D_p + \partial^2_{n_k} q D_q + \partial^2_{n_k} y D_y + \left(\partial_{n_k} p\right)^2 D_{pp} + \left(\partial_{n_k} q\right)^2 D_{qq} + \left(\partial_{n_k} y\right)^2 D_{yy} + \\ &\quad + 2\partial_{n_k} p \partial_{n_k} q D_{pq} + 2\partial_{n_k} p \partial_{n_k} y D_{py} + 2\partial_{n_k} q \partial_{n_k} y D_{qy} . \end{aligned} \tag{36}$$

Since $\partial_{n_k} p = 0$, this immediately simplifies to

$$\partial^2_{n_k} F_s = \partial^2_{n_k} p D_p + \partial^2_{n_k} q D_q + \partial^2_{n_k} y D_y + \left(\partial_{n_k} q\right)^2 D_{qq} + \left(\partial_{n_k} y\right)^2 D_{yy} + 2\partial_{n_k} q \partial_{n_k} y D_{qy} . \tag{37}$$

Substituting the values for the various derivatives, we find

$$\begin{aligned} \partial^2_{n_k} F_s &= \frac{k^2}{2k_0^2 n_0^{8/3}} D_p + \frac{5k^2}{6k_0^2 n_0^{8/3}} D_q + \\ &\quad + \frac{2}{n_0^2} \left[\frac{\beta(\beta-1)}{2} \frac{A^2}{A^2 + 4k^2} + \frac{2\beta A^2 k^2 - 3\beta^2 A^2 (A^2 + k^2)}{3(A^2 + k^2)^2} + \frac{\beta(\beta+1)}{2} \right] D_y + \\ &\quad + \frac{k^4}{16 k_0^4 n_0^{10/3}} D_{qq} + \frac{1}{n_0^2} \frac{\beta^2 k^4}{(A^2 + k^2)^2} D_{yy} + \frac{\beta k^4}{2k_0^2 n_0^{8/3} (A^2 + k^2)} D_{qy} . \end{aligned} \tag{38}$$

Finally, using Equations (35) and (38) into Equation (32), we obtain the formula

$$\begin{aligned} \bar{F} &= F_s(0,0,1) + \frac{9}{20} \frac{k^2}{k_0^2 n_0^{2/3}} D_p + \frac{9}{160} \frac{k^4}{n_0^{4/3} k_0^4} D_{qq} + \\ &\quad + \frac{3}{5} \left[\frac{3\beta(\beta-1)}{2} \frac{A^2}{A^2 + 4k^2} - \frac{3\beta A^2 (\beta A^2 + k^2) + bk^2 (3\beta A^2 + 5k^2)}{(A^2 + k^2)^2} + \frac{3\beta(\beta+1)}{2} \right] D_y + \\ &\quad + \frac{9}{10} \frac{\beta^2 k^4}{(A^2 + k^2)^2} D_{yy} + \frac{9}{20} \frac{\beta k^4}{k_0^2 n_0^{2/3} (A^2 + k^2)} D_{qy} , \end{aligned} \tag{39}$$

which, with the substitution $\eta = \alpha k / (2A)$ and after some algebra, assumes the final form

$$\begin{aligned} \bar{F} &= F_s(0,0,1) + \frac{9}{5} \eta^2 D_p + \frac{9}{10} \eta^4 D_{qq} + \frac{36}{5} \frac{\beta \eta^4}{\alpha^2 + 4\eta^2} D_{qy} \\ &+ \frac{9}{10} \beta(\beta - 1) \left(1 + \frac{\alpha^2}{\alpha^2 + 16\eta^2} - \frac{2\alpha^2}{\alpha^2 + 4\eta^2} \right) D_y + \\ &+ \frac{24}{5} \beta \frac{\eta^4}{(\alpha^2 + 4\eta^2)^2} (3\beta D_{yy} - 4D_y) . \end{aligned} \tag{40}$$

We note that this expression is well defined and continuous for any value of α and β. However, the case $\beta = 1$ is a special one. In fact, in this case, $\bar{F}$ loses its dependence on D_y, being dependent only on the linear combination $3\beta D_{yy} - 4D_y$ as well as on $F_s(0,0,1)$, D_p, D_{qq}, D_{qy} (here, we only consider the dependence on the degrees of freedom related to the modeling of the enhancement factor; α and β, which are related to the definition of y, are considered as additional parameters). Therefore, when $\beta = 1$, there is a reduced parametric flexibility to optimize the linear response function through the modeling of the enhancement factor. As we will see, this has consequences for the possibility to impose the correct asymptotic behavior to $\bar{F}$.

Asymptotic Behavior

The asymptotic expansions of the exact response function are [5]

$$\bar{F}^{Lind} = 1 + \frac{\eta^2}{3} + \frac{8}{45}\eta^4 + \mathcal{O}(\eta^6) \qquad \eta \to 0\,, \tag{41}$$

$$\bar{F}^{Lind} = 3\eta^2 - \frac{3}{5} + \mathcal{O}(\eta^{-2}) \qquad \eta \to \infty\,. \tag{42}$$

From Equation (40), we get for the small-η limit

$$\begin{aligned} \bar{F} &= F_s(0,0,1) + \frac{9}{5}\left(D_p - 4\frac{\beta(\beta-1)}{\alpha^2}D_y\right)\eta^2 + \\ &+ \frac{3}{5}\left(\frac{3}{2}D_{qq} + \frac{12\beta}{\alpha^2}D_{qy} + \frac{16}{\alpha^4}\beta(21\beta-23)D_y + \frac{24}{\alpha^4}\beta^2 D_{yy}\right)\eta^4\,, \end{aligned} \tag{43}$$

and in the large-η limit

$$\begin{aligned} \bar{F} &= \frac{9}{10}D_{qq}\eta^4 + \frac{9}{5}(D_p + \beta D_{qy})\eta^2 + \\ &+ \frac{3}{5}\left(\frac{5}{3}F_s(0,0,1) + \frac{\beta(3\beta-7)}{2}D_y + \frac{3\beta^2}{2}D_{yy} - \frac{3}{4}\alpha^2\beta D_{qy}\right). \end{aligned}$$

Therefore, we have the following asymptotic conditions

$$F_s(0,0,1) = 1 \qquad \eta \to 0 \quad \mathcal{O}(\eta^0)\,, \tag{44}$$

$$D_p - 4\frac{\beta(\beta-1)}{\alpha^2}D_y = \frac{5}{27} \qquad \eta \to 0 \quad \mathcal{O}(\eta^2)\,, \tag{45}$$

$$\begin{aligned} &\frac{3}{2}D_{qq} + \frac{12\beta}{\alpha^2}D_{qy} + \\ &+\frac{16}{\alpha^4}\beta(21\beta-23)D_y + \frac{24}{\alpha^4}\beta^2 D_{yy} = \frac{8}{27} \qquad \eta \to 0 \quad \mathcal{O}(\eta^4)\,, \end{aligned} \tag{46}$$

$$\frac{9}{10}D_{qq} = 0 \qquad \eta \to \infty \quad \mathcal{O}(\eta^4)\,, \tag{47}$$

$$\frac{9}{5}(D_p + \beta D_{qy}) = 3 \qquad \eta \to \infty \quad \mathcal{O}(\eta^2)\,, \tag{48}$$

$$\begin{aligned} &\frac{5}{3}F_s(0,0,1) + \frac{\beta(3\beta-7)}{2}D_y + \\ &+\frac{3\beta^2}{2}D_{yy} - \frac{3}{4}\alpha^2\beta D_{qy} = -1 \qquad \eta \to \infty \quad \mathcal{O}(\eta^0)\,. \end{aligned} \tag{49}$$

As we saw, in the most general case (i.e., for any value of β), the linear response function has *only five* degrees of freedom ($F_s(0,0,1), D_{qq}, D_p, D_{qy}, 3\beta D_{yy} - 4D_y$) to satisfy these conditions. Thus, initially, we chose to consider Equations (44)–(48), neglecting for the moment the η^0 order in the large-η limit. With this choice, we obtain

$$
\begin{aligned}
F_s(0,0,1) &= 1\,, && (50)\\
D_{qq} &= 0\,, && (51)\\
D_p &= \frac{5}{27}+\frac{4}{\alpha^2}\beta(\beta-1)D_y\,, && (52)\\
D_{qy} &= \frac{40}{27\beta}-\frac{4}{\alpha^2}(\beta-1)D_y\,, && (53)\\
3\beta D_{yy}-4D_y &= \frac{\alpha^2(\alpha^2-60)}{27\beta}-36(\beta-1)D_y\,. && (54)
\end{aligned}
$$

These conditions fix the asymptotic behavior of the response function up to the η^4 order in the short-range limit and to the η^2 order in the long-range one. Using these values into Equation (40), we find

$$
\begin{aligned}
\bar{F} &= \frac{34560\eta^8+Q_6\eta^6+Q_4\eta^4+15\alpha^4\eta^2(\alpha^2+72)+45\alpha^6}{45(\alpha^2+4\eta^2)^2(\alpha^2+16\eta^2)}\,, && (55)\\
Q_6 &= 16\Big(8\alpha^4+255\alpha^2-5832\beta(\beta-1)D_y+720\Big)\,, && (56)\\
Q_4 &= 8\alpha^2\Big(\alpha^4+45\alpha^2+810\Big)\,. && (57)
\end{aligned}
$$

For large η values, Equation (55) behaves as

$$
\begin{aligned}
\bar{F} &= 3\eta^2+\Delta\,, && (58)\\
\Delta &= (1/90)\alpha^4-(81/10)D_y\beta(\beta-1)-(4/3)\alpha^2+1\,. && (59)
\end{aligned}
$$

For $\beta=1$, the exact condition $\Delta=-3/5$ can only be obtained for specific values of α, as the dependence from D_y vanishes. For $\beta\neq 1$, instead, we can solve for D_y obtaining

$$
D_y=\frac{\alpha^4-120\alpha^2+144}{729\beta(\beta-1)}\,. \qquad (60)
$$

In this general case, we have also

$$
D_{yy}=\frac{13\alpha^4-3180\alpha^2+5760+(-9\alpha^4+2700\alpha^2-5184)\beta}{2187\beta^2(\beta-1)}\,. \qquad (61)
$$

Equations (60) and (61), together with Equations (50)–(53), fix the asymptotic behavior of the response function up to the η^4 order in the small-η limit and down to the η^0 order in the large-η one.

We note that the expressions for D_y and D_{yy} diverge at $\beta=1$: This reflects the fact that, as discussed above, for $\beta=1$, it is not always possible to fulfill all the asymptotic conditions. Nevertheless, all equations are well defined for any other value of β as well in the limit $\beta\to 1$. In fact, the divergence in the denominator in this limit, the contribution D_y, becomes negligibly small in Equation (40) as well as in Equations (52) and (53); moreover, the linear combination $3\beta D_{yy}-4D_y$ is always well defined, since it does not diverge for any $\beta>0$.

Substitution of the asymptotic conditions into Equation (40) finally yields the general yGGA response

$$
\bar{F}^{gen}=\frac{34560\eta^8+432\eta^6(45\alpha^2-16)+8\alpha^2\eta^4(\alpha^4+45\alpha^2+810)+15\alpha^4\eta^2(\alpha^2+72)+45\alpha^6}{45(\alpha^2+4\eta^2)^2(\alpha^2+16\eta^2)}\,. \qquad (62)
$$

Note that, remarkably, this formula displays no dependence on β. Thus, β can be used to optimize the functional beyond the linear response regime. The value of α can instead be optimized by fitting to match as close as possible the exact Lindhard function. To this purpose, we can minimize the quantity

$$\sigma = \int w(\eta) \left| \frac{1}{\bar{F}^{lind}(\eta)} - \frac{1}{\bar{F}(\eta)} \right| d\eta \,, \tag{63}$$

where $\bar{F}^{lind}$ is the normalized Lindhard linear response function and

$$w(\eta) = e^{-\mu(\eta-1)^2} \,, \tag{64}$$

with $\mu = 2$, is a weighting function. The use of this weighting function allows to focus the fit in the region close to $\eta = 1$ instead of the asymptotic ones that are already included by construction. We remark that the results are only weakly dependent on the choice of μ.

Figure 1 reports the values of the errors with respect to the exact Lindhard function for different values of α. We see that the best reproduction of the Lindhard function is achieved for $\alpha = 3.31$, with $\sigma = 0.0513$. However, a quite broad range of values allows us to attain a low error; in particular, for $\alpha \in [2.26, 4.66]$, the response is always better than the one obtained by the yuk2 functional [71].

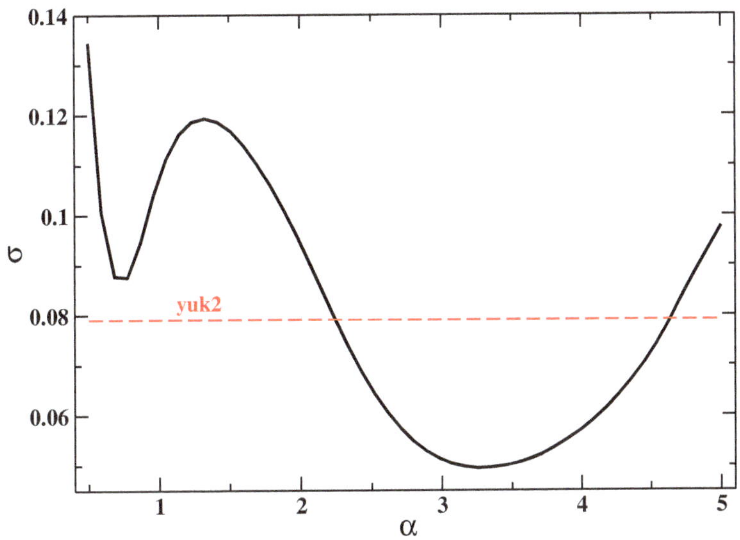

Figure 1. Values of the error function σ (Equation (63)) for the function $\bar{F}^{gen}$ for different values of the α parameter. For comparison, the σ value corresponding to the yuk2 functional, ref. [71] is also reported.

The linear response functions for various cases are reported in Figure 2. Inspection of the plot shows that they are all quite similar, as already suggested by the considerations above. For $\alpha = 3.31$, we obtain at $\eta = 1$, where the Lindhard function shows a derivative singularity, that $1/\bar{F} = 0.47$, which is very close to the exact value $1/\bar{F} = 1/2$ [5]. Actually, it is also possible also to satisfy $1/\bar{F} = 1/2$ exactly using $\alpha = 3.64$, even if the global accuracy is somehow smaller in this case ($\sigma = 0.0527$).

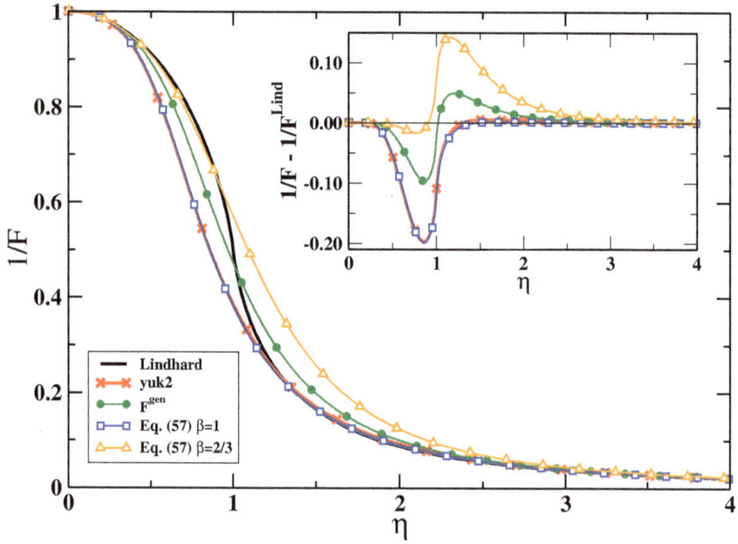

Figure 2. Linear response functions as computed by Equations (55) and (62) [$\bar{F}^{gen}$] for $\alpha = 3.31$. For Equation (55), two different choices of β have been considered; for each one, the value of D_y has been chosen such to minimize the error σ in Equation (63). The exact Lindhard and the yuk2 responses are also reported for comparison. The inset shows instead the difference between the Lindhard function and the various response functions reported in the plot.

3. Linear yGGA Functionals

We consider the simplest case of yGGA functional by taking functionals that have an enhancement factor with the general form

$$F_s^{\text{linyGGA}}(p,q,y) = 1 - G_0 + \frac{5}{3}p + y\,(G_0 + G(p,q))\,. \tag{65}$$

i.e., a linear dependence on y. Note that these are closely related with the yGGAs defined in Ref. [71], which are recovered if $G_0 = 1$ and $\beta = 1$. However, in this work, we lift this restriction.

From Equation (65), we have that $F_s(0,0,1) = 1$ is satisfied by construction if $G(0,0) = 0$ and $D_{qq} = G_{qq}$, $D_p = 5/3 + G_p$, $D_{qy} = G_q$, $D_{yy} = 0$, $D_y = G_0$, where we have used the short-hand notation $G_i = \partial G(0,0,1)/\partial i$, and $G_{ij} = \partial G(0,0,1)/\partial i \partial j$. Using Equations (50)–(53), we then obtain

$$G_{qq} = 0 \tag{66}$$

$$G_p = -\frac{40}{27} + \frac{4}{\alpha^2}\beta(\beta-1)G_0 = -\beta G_q\,. \tag{67}$$

The simplest functional satisfying these conditions is the one with

$$G(p,q) = \left(\frac{40}{27\beta} - \frac{4}{\alpha^2}(\beta-1)G_0\right)(q - \beta p) \equiv x \tag{68}$$

$$G_0 = \frac{\alpha^2(\alpha^2-60)}{108\beta(9\beta-10)}\,. \tag{69}$$

Equation (68) has been derived from Equation (54) using the fact that, for functionals defined by Equation (65), $D_{yy} = 0$. The Equations (65), (68), and (69) define the yuk2β

functional, which reproduces the Lindhard functional at small η up to fourth order and for large η behaves as

$$\bar{F}^{yuk2\beta} \rightarrow 3\eta^2 + \Delta^{yuk2\beta} \tag{70}$$

$$\Delta^{yuk2\beta} = 1 + \frac{(-2700\beta + 3180)\alpha^2}{360(9\beta - 10)} + \frac{(9\beta - 13)\alpha^4}{360(9\beta - 10)} . \tag{71}$$

Although the exact value $\Delta^{yuk2\beta} = -3/5$ can be obtained for specific values of α and β, this condition does not have a big impact on the overall accuracy of the linear response. We consider the general error indicator of the linear response of yuk2β with respect to the Lindhard function (Equation (63)). Results for different values of the α and β are reported as a colormap in Figure 3.

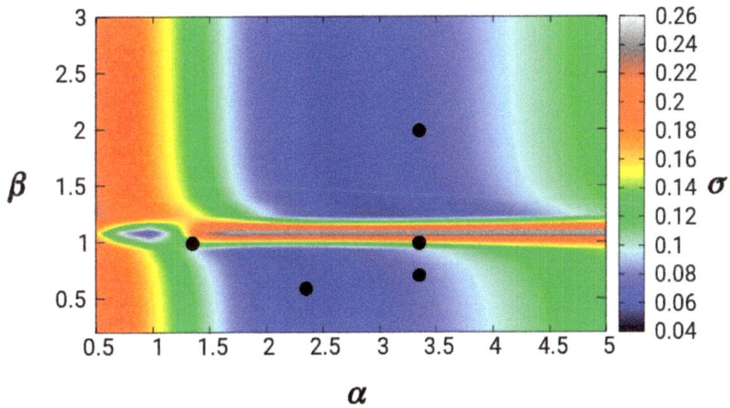

Figure 3. Error with respect to the Lindhard function (Equation (63)) for the yuk2β functional at various values of the parameters α and β. The black dots denote the positions of the pairs of parameters listed in Table 1.

Table 1. Considered values of α and β, with the corresponding values of G_0, see Equation (69), G_p, see Equation (67), the error defined in Equation (63), and the value of $1/F$ at $\eta = 1$.

α	β	G_0	G_p	σ	$1/\bar{F}(\eta = 1)$
1.36	1	1	1.48	0.0791	0.385
3.31	1	4.97	1.48	0.1049	0.617
3.31	2	−0.31	0.85	0.0676	0.545
3.31	2/3	1.86	2.45	0.0758	0.567
2.34	5/9	1	2.98	0.0588	0.469

We find that, as seen in the previous section, there is a quite broad range of α values that yield small errors (blue areas in the plot). Moreover, the errors are almost independent on β, except for values $\beta \approx 1.1$, where a discontinuity in the linear response behavior occurs.

Moreover, from the results of the previous section, we can fix $\alpha = 3.31$, such that $G_0 \approx 4.484/[\beta(10 - 9\beta)]$, which is shown in Figure 4.

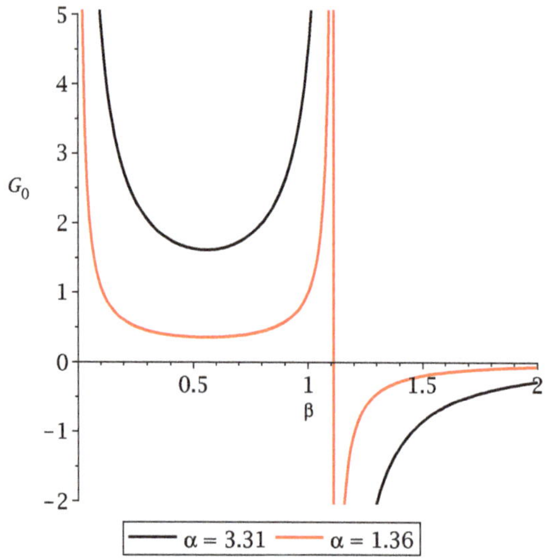

Figure 4. Value of G_0 as a function of β for different values of α.

Note that G_0 is positive up to $\beta = 10/9$, where it has a pole, and it has a minimum at $\beta = 5/9$. We remark that instead setting $G_0 = 1$ and $\beta = 1$, then Equation (68) readily yields $\alpha = 1.36$. Thus, the yuk2β functional immediately reduces to the yuk2 functional of Ref. [71].

The yuk2β functional is just a simple ansatz recovering an accurate linear response behavior. However, because it uses a linear dependence on p and q, it may display severe drawbacks in real applications. In particular, the positivity of the Pauli KE density must be ensured [38,92], which is not the case for Equation (65). In fact, the quantity

$$1 + w = 1 - G_0 + y(G_0 + x) \tag{72}$$

is not always positive. Thus, we define the yuk3β functional as

$$F_s^{\text{yuk3}\beta} = \frac{5}{3}p + T(w)\,, \tag{73}$$

with $T(w)$ being a positive function such that for $w \approx 0$, we have $T(w) = 1 + w + O(x^3)$, such as the one considered in Ref. [71]. Thus, lacking any quadratic term, the yuk3β has the same linear response of yuk2β. In this way, although the functional $F_s^{\text{yuk3}\beta}$ is not truly a linear function of y, it practically behaves as a linear function of y because $D_{yy} = 0$. True non-linear yGGA functionals, with $D_{yy} \neq 0$, are much more complicated, and they will be considered elsewhere.

4. Computational Details

Densities and enhancement factors were computed in post-SCF fashion using the orbitals of self-consistent Kohn–Sham calculations. All Kohn–Sham calculations have been performed using an in-house developed code solving numerically the spherical symmetry Kohn–Sham problem with the local density approximation being used for the exchange-correlation functional.

Neutral jellium spheres with $N = 40, 92, 254$ electrons and Wigner–Seitz radius $r_s = 2$, 3, 4, 5, and 6 were considered. They are characterized by the positive background density

$$n_+(\vec{r}) = \begin{cases} 3/(4\pi r_s^3), & r < R \\ 0, & r \geq R\,. \end{cases} \tag{74}$$

The post-processing of the Kohn–Sham orbitals to obtain the quantities $G(p,q)$, y, and w has been carried out by an additional in-house software which evaluates the required derivatives and integrals in real space using the same grid as the one used in the Kohn–Sham calculations.

5. Results

We considered different pairs of α and β values for the y ingredient, as reported in Table 1 (see also Figure 3). The first pair ($\alpha = 1.36$, $\beta = 1$) is the one of the original yuk2 functional of Ref. [71]. The second, third, and fourth pairs consider the α value suggested by Figure 1 and several values of β. The last pair considers the minimizing β value of Figure 4 and a corresponding α such that $G_0 = 1$. All these pairs give a quite accurate description of the HEG linear response: the highest accuracy is obtained for $(\alpha, \beta) = (3.31, 2)$ and $(2.34, 5/9)$, as reported in the last column of Table 1.

In Figure 5, we report, in panels (a), (b), (c) and (d) respectively, the electronic density, the function x defined in Equation (68), as well as the values of y and w for a jellium sphere with $N = 254$ electrons and $r_s = 4$.

Looking at Figure 5, we see in panel (b) that for $\beta \geq 1$, the values of x are negative in the tail, as also discussed in Appendix C. Moreover, panel (c) shows that the ingredient y is close to 1 inside the jellium spheres for all (α,β) pairs but in the tail, it diverges for $\beta \geq 1$. On the other hand, for $\beta = 2/3$, it reaches a constant (but it will slowly vanish in very far regions), while for $\beta \geq 2/3$, it soon decreases to zero, as discussed in Section 2. Finally, panel (d) shows the ingredient w. Inside the sphere, the values of w are in the range $|w| < 0.3$ for all the (α, β) combinations. The behavior in the tail follows the one of x but for $\beta = 1$ and $\alpha = 3.31$, there is a peak before the negative divergence.

These results indicate that for the total kinetic energy, which is mainly influenced by the functional behavior inside the jellium sphere, all the cases considered here can be expected to yield similar results, which are in line with the performance of the yuk3 functional [71]. On the other hand, for the potential, we have a contribution $\partial F_s/\partial y \propto \partial F_s/\partial w \cdot \partial w/\partial y$. However, this latter term for functionals with $\beta < 1$ is largely divergent ($w(\mathbf{r})$ is diverging but $y(\mathbf{r})$ is not); similarly, it diverges in the ($\alpha = 3.31$, $\beta = 1$) case, where w strongly oscillates around $r = 30$ bohr. Thus, we can expect, for these values of the parameters, major problems on the KE potential.

To understand better the role of β on the KE functional, we can perform a further analysis of the "exact" Pauli KE enhancement factor

$$F_p^{\text{exact}} \equiv \tau_p^{KS}(\mathbf{r})/\tau^{TF}(\mathbf{r})\,, \tag{75}$$

where τ_p^{KS} is the Pauli KE density corresponding to the Kohn–Sham positive-defined KE density and τ^{TF} is the Thomas–Fermi density. Hence, we have computed F_p^{exact} for several jellium clusters with different numbers of electrons ($N = 40, 92, 254$) and r_s = 2, 3, 4, 5, and 6 on all grid points, and we have plotted it versus the corresponding values of w. In fact, Equation (73) implies that that the Pauli enhancement factor can be written as a universal single-valued function of w. However, this is an ansatz, which must be verified; if it is correct, then we must obtain a unique value of F_p for each w, also for different systems.

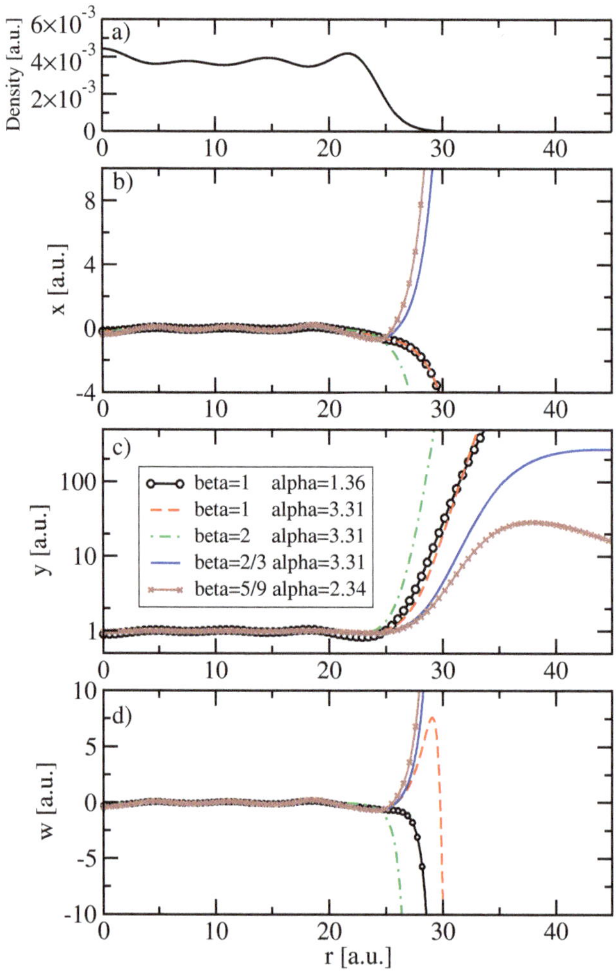

Figure 5. Electronic density (**a**) and KE ingredients (**b**–**d**) for a jellium sphere with $N = 254$ electrons and $r_s = 4$. The ingredients x, y, and w are reported for different choices of the β and α parameters.

Figure 6 shows the plot of F_p^{exact} vs. w for the various pairs of parameters considered in this work. It is evident that the cases with $\beta < 1$ and the case with ($\alpha = 3.31$, $\beta = 1$) do not yield single-valued functions but rather display multiple branches of F_p^{exact} for all values of w. This confirms that using these parameters, it is not possible to obtain an accurate description of the kinetic functional. On the contrary, the yuk2 parameterization ($\alpha = 1.35$, $\beta = 1$) and the one with ($\alpha = 3.31$, $\beta = 2$) show a nice single line for $w \geq 0$. For negative w values, they display a multi-valued behavior; however, for these cases, the $w < 0$ region corresponds to the density tail region. Therefore, the effect, at least on the computation of the energy, is minor. Thus, we can infer that both functionals will be accurate for the kinetic energies (this is indeed the case for yuk2 [71]) but may yield some oscillations in the asymptotic part of the potential because of the multi-valued behavior at $w < 0$. This latter feature is possibly significantly reduced for ($\alpha = 3.31$, $\beta = 2$) that displays the lowest grade of multi-valuedness at negative w values.

In a future work, we will develop an approach that will allow us to obtain an analytical $F(w)$ function in order to build up an accurate KE functional and KE potential. However, the results reported in Figure 6 also indicate that the ansatz in Equation (73) may be not accurate enough and further studies are required, in particular, to investigate the effect of the non-linear term (y^2), which is already important for the HEG linear-response and can be even more important for the description of finite systems.

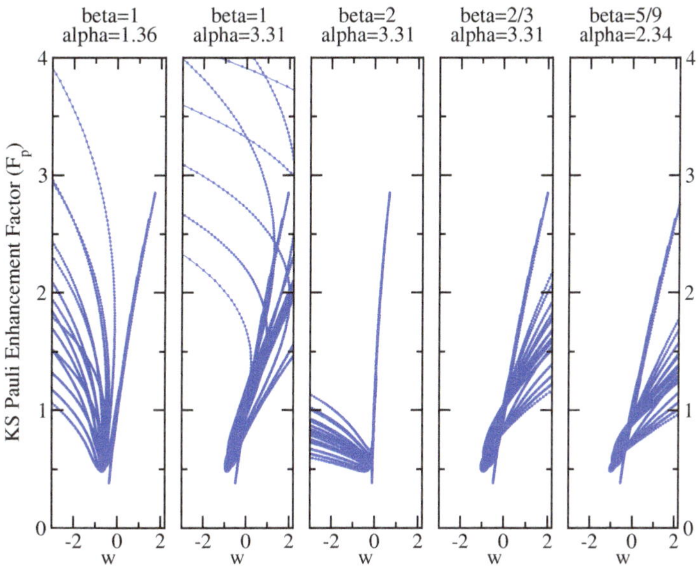

Figure 6. Exact Pauli Kohn–Sham enhancement factor (Equation (75)) vs. the w ingredient as computed on all grid points for several jellium spheres (see text for details). Each panel corresponds to a different choice of the β and α parameters.

6. Conclusions

In this paper, we have investigated non-interacting kinetic energy functionals depending on a generalized Yukawa potential, i.e., a screened Coulomb potential of the electronic density raised to the power of β. The use of this input ingredient allows us to introduce in an efficient way non-local features into the functional.

In particular, we have derived the exact homogeneous electron gas linear response behavior of generalized yGGA functionals and, by comparing to the Lindhard function, we have derived exact asymptotic constraints for the functional.

In particular, it turned out that $\beta = 1$ is a very special case, and the Lindhard asymptotic constraints can only be satisfied for a specific value of the screening parameter (α). Moreover, the final linear response of yGGA functionals satisfying the low and high-wavevector Lindhard properties does not depend on β, which can be then used as an additional degree of freedom to model systems beyond the linear-response regime.

We have used the developed theory to extend the work reported in Ref. [71] and investigate in detail the simplest class of yGGA functionals, namely the linear yGGAs, i.e., those yGGA functionals depending only linearly on y. We have found that although this class of functionals can satisfy rather accurately the linear response constraints, it is not flexible enough to perform very accurately for both the kinetic energy and potential computation. This is mainly due to the fact that imposing the linear response behavior implies, in these simple functionals, that the Pauli enhancement factor is a function of a well-defined combination of the density ingredients (the w ingredient of Equation (72)), but this is not sufficient to describe the non-local nature of the Pauli kinetic energy. Although for some wise choices of the parameters, this effect can be minimized; this is an intrinsic limitation of the linear yGGA functional class.

Therefore, future work will focus on the development of more sophisticated functionals forms using an explicit non-linear dependence on the Yukawa potential. The theoretical framework established in this paper will possibly allow us to develop more efficient and broadly applicable kinetic energy functionals.

Author Contributions: Conceptualization and methodology, E.F. and F.D.S.; software, F.S.; investigation, E.F., F.S. and F.D.S.; writing, E.F., F.S., L.A.C. and F.D.S. All authors have read and agreed to the published version of the manuscript.

Funding: This research received no external funding.

Data Availability Statement: Data sharing is not applicable to this article.

Conflicts of Interest: The authors declare no conflict of interest.

Appendix A. Weak Perturbation of y

To compute the perturbation of y defined in Equation (15), we first consider the generalized Yukawa potential

$$u_\alpha(\mathbf{r}) = \int \frac{n^\beta(r')e^{-A(r)|\mathbf{r}-\mathbf{r}'|}}{|\mathbf{r}-\mathbf{r}'|} d\mathbf{r}' \,, \tag{A1}$$

where we have set $A = \alpha k_0 n_0^{1/3}$. Upon the density perturbation, this becomes

$$u_\alpha = n_0^\beta \int \frac{\left(1+e^{-i\mathbf{k}\cdot\mathbf{r}'}\left(\frac{n_k}{n_0}\right)\right)^\beta e^{-A\left(1+\frac{n_k}{n_0}\right)^{1/3} r'}}{r'} d\mathbf{r}' \,. \tag{A2}$$

Expanding in series and keeping only terms up to second order, we find

$$u_\alpha \approx n_0^\beta \left[I_1 + \left(\beta I_2 - \frac{1}{3}AI_3\right)\frac{n_k}{n_0} + \left(\frac{\beta(\beta-1)}{2}I_4 + \frac{A}{18}(AI_5 + 2I_3) - \frac{A\beta}{3}I_6\right)\left(\frac{n_k}{n_0}\right)^2 \right] ,$$

with

$$\begin{aligned}
I_1 &\equiv \int \frac{e^{-Ar'}}{r'} d\mathbf{r}' && \text{(A3)}\\
I_2 &\equiv \int \frac{e^{-i\mathbf{k}\cdot\mathbf{r}'-Ar'}}{r'} d\mathbf{r}' && \text{(A4)}\\
I_3 &\equiv \int e^{-Ar'} d\mathbf{r}' && \text{(A5)}\\
I_4 &\equiv \int \frac{e^{-2i\mathbf{k}\cdot\mathbf{r}'-Ar'}}{r'} d\mathbf{r}' && \text{(A6)}\\
I_5 &\equiv \int r' e^{-Ar'} d\mathbf{r}' && \text{(A7)}\\
I_6 &\equiv \int e^{-i\mathbf{k}\cdot\mathbf{r}'-Ar'} d\mathbf{r}'. && \text{(A8)}
\end{aligned}$$

Substituting the values of the integrals (see Appendix B), we find

$$\begin{aligned}
u_\alpha \approx\ & \frac{4\pi n_0^\beta}{A^2} + 4\pi n_0^\beta \left(\frac{\beta}{A^2+k^2} - \frac{2}{3A^2}\right)\frac{n_k}{n_0} + \\
& + 4\pi n_0^\beta \left(\frac{\beta(\beta-1)}{2}\frac{1}{A^2+4k^2} + \frac{5}{9A^2} - \frac{2A^2\beta}{3(A^2+k^2)^2}\right)\left(\frac{n_k}{n_0}\right)^2.
\end{aligned} \tag{A9}$$

Now, we can consider the perturbation of y, that is

$$\begin{aligned} y &= \frac{3\pi\alpha^2}{4k_0(n_0+n_k)^{\beta-2/3}} u_\alpha[n_0+n_k](\mathbf{r}) = \\ &= \frac{3\pi\alpha^2 n_0^{2/3-\beta}}{4k_0}\left(1+\frac{n_k}{n_0}\right)^{2/3-\beta} u_\alpha[n_0+n_k](\mathbf{r})\,. \end{aligned} \tag{A10}$$

Using the result of Equation (A9) and expanding to second-order, we find

$$\begin{aligned} y &\approx 1 - \frac{\beta k^2}{A^2+k^2}\frac{n_k}{n_0} + \\ &+ \left[\frac{\beta(\beta-1)}{2}\frac{A^2}{A^2+4k^2} + \frac{2\beta A^2k^2 - 3\beta^2 A^2(A^2+k^2)}{3(A^2+k^2)^2} + \frac{\beta(\beta+1)}{2}\right]\left(\frac{n_k}{n_0}\right)^2. \end{aligned} \tag{A11}$$

Appendix B. Integrals for the Generalized Yukawa Expansion

For integral I_1, I_3 and I_5, we use that

$$E[A,n] = \int_0^\infty (r')^2 r^n e^{-Ar'} dr' = \frac{\Gamma(3+n)}{A^{3+n}}. \tag{A12}$$

Thus, $I_1 = 4\pi E[A,-1] = \frac{4\pi}{A^2}$, $I_3 = 4\pi E[A,0] = \frac{8\pi}{A^3}$, $I_5 = 4\pi E[A,1] = \frac{24\pi}{A^4}$.

To compute the I_2, I_4, and I_6 integrals, we can choose the axis such that $\mathbf{k}$ is aligned with the z axis; then, we have that $\mathbf{k}\cdot\mathbf{r}' = kr'\cos\theta$. Hence, we can write

$$\begin{aligned} F[k,n] &= \int_0^{2\pi} d\phi \int_0^{\pi} \sin\theta d\theta \int_0^\infty dr' r'^2 (r')^n e^{-Ar'} e^{-ikr'\cos\theta} = \\ &= 2\pi \int_0^\infty dr' r'^2 (r')^2 e^{-Ar'} \int_0^\pi d\theta \sin\theta e^{-ikr'\cos\theta} = \\ &= \frac{2\pi}{ik}\int_0^\infty (r')^n r' e^{-Ar'}\left(e^{ikr'} - e^{-ikr'}\right) dr' = \frac{2\pi}{ik}(E[A-ik,n] - E[A+ik,n]) \\ &= \frac{4\pi}{k}\Im[E[A-ik,n-1]]. \end{aligned} \tag{A13}$$

Thus, $I_2 = F[k,-1] = \frac{4\pi}{A^2+k^2}$ and $I_6 = F[k,0] = \frac{8\pi A}{(A^2+k^2)^2}$.

The I_4 integral is similar to the integral I_2 with the substitution $\mathbf{k} \to 2\mathbf{k}$. Then,

$$I_4 = \frac{4\pi}{A^2+4k^2}\,.$$

Appendix C. Asymptotics

For an exponential spherical density $\rho(r) = A\exp(-2Zr)$, where $Z = \sqrt{-2\epsilon_H}$, we have

$$p \to Z^2 \frac{1}{k_0^2}\frac{1}{\rho^{2/3}} \tag{A14}$$

$$q \to \frac{Z(Zr-1)}{r}\frac{1}{k_0^2}\frac{1}{\rho^{2/3}} \tag{A15}$$

so that

$$q - p \to -\frac{Z}{r}\frac{1}{k_0^2}\frac{1}{\rho^{2/3}} = -\frac{p}{rZ}. \tag{A16}$$

Thus, $q - p$ is negative in the tail; i.e., q is very large in the tail but smaller than p. If we consider $q - \beta p$ with $\beta > 1$, then $q - \beta p$ will be more negative. If we consider $q - \beta p$ with $\beta < 1$, then $q - \beta p$ will be positive.

References

1. Hohenberg, P.; Kohn, W. Inhomogeneous Electron Gas. *Phys. Rev.* **1964**, *136*, B864–B871. [CrossRef]
2. Dreizler, R.M.; Gross, E.K.U. *Density Functional Theory*; Springer: Berlin/Heidelberg, Germany, 1990.
3. Levy, M. Universal variational functionals of electron densities, first-order density matrices, and natural spin-orbitals and solution of the v-representability problem. *Proc. Nat. Acad. Sci. USA* **1979**, *76*, 6062–6065. [CrossRef] [PubMed]
4. Kohn, W.; Sham, L.J. Self-Consistent Equations Including Exchange and Correlation Effects. *Phys. Rev.* **1965**, *140*, A1133–A1138. [CrossRef]
5. Wang, Y.A.; Carter, E.A. Orbital-free kinetic-energy density functional theory. In *Theoretical Methods in Condensed Phase Chemistry*; Springer: Berlin/Heidelberg, Germany, 2002; pp. 117–184.
6. Wesolowski, T.A.; Wang, Y.A. (Eds.) *Recent Progress in Orbital-Free Density Functional Theory*; World Scientific: Singapore, 2013. [CrossRef]
7. Gavini, V.; Bhattacharya, K.; Ortiz, M. Quasi-continuum orbital-free density-functional theory: A route to multi-million atom non-periodic DFT calculation. *J. Mech. Phys. Sol.* **2007**, *55*, 697–718. [CrossRef]
8. Witt, W.C.; Beatriz, G.; Dieterich, J.M.; Carter, E.A. Orbital-free density functional theory for materials research. *J. Mater. Res.* **2018**, *33*, 777–795. [CrossRef]
9. Götz, A.W.; Beyhan, S.M.; Visscher, L. Performance of Kinetic Energy Functionals for Interaction Energies in a Subsystem Formulation of Density Functional Theory. *J. Chem. Theory Comput.* **2009**, *5*, 3161–3174. [CrossRef]
10. Jacob, C.R.; Neugebauer, J. Subsystem density-functional theory. *Wiley Interdiscip. Rev. Comput. Mol. Sci.* **2014**, *4*, 325–362. [CrossRef]
11. Wesolowski, T.A.; Shedge, S.; Zhou, X. Frozen-Density Embedding Strategy for Multilevel Simulations of Electronic Structure. *Chem. Rev.* **2015**, *115*, 5891–5928. [CrossRef]
12. Neugebauer, J. Chromophore-specific theoretical spectroscopy: From subsystem density functional theory to mode-specific vibrational spectroscopy. *Phys. Rep.* **2010**, *489*, 1–87. [CrossRef]
13. Krishtal, A.; Sinha, D.; Genova, A.; Pavanello, M. Subsystem density-functional theory as an effective tool for modeling ground and excited states, their dynamics and many-body interactions. *J. Phys. Condens. Matter* **2015**, *27*, 183202. [CrossRef]
14. Laricchia, S.; Fabiano, E.; Della Sala, F. Frozen density embedding with hybrid functionals. *J. Chem. Phys.* **2010**, *133*, 164111. [CrossRef] [PubMed]
15. Śmiga, S.; Fabiano, E.; Laricchia, S.; Constantin, L.A.; Della Sala, F. Subsystem density functional theory with meta-generalized gradient approximation exchange-correlation functionals. *J. Chem. Phys.* **2015**, *142*, 154121. [CrossRef]
16. Toscano, G.; Straubel, J.; Kwiatkowski, A.; Rockstuhl, C.; Evers, F.; Xu, H.; Asger Mortensen, N.; Wubs, M. Resonance Shifts and Spill-out Effects in Self-Consistent Hydrodynamic Nanoplasmonics. *Nat. Commun.* **2015**, *6*, 7132. [CrossRef] [PubMed]
17. Ciracì, C.; Della Sala, F. Quantum hydrodynamic theory for plasmonics: Impact of the electron density tail. *Phys. Rev. B* **2016**, *93*, 205405. [CrossRef]
18. Moldabekov, Z.A.; Bonitz, M.; Ramazanov, T.S. Theoretical Foundations of Quantum Hydrodynamics for Plasmas. *Phys. Plasmas* **2018**, *25*, 031903. [CrossRef]
19. Baghramyan, H.M.; Della Sala, F.; Ciracì, C. Laplacian-Level Quantum Hydrodynamic Theory for Plasmonics. *Phys. Rev. X* **2021**, *11*, 011049. [CrossRef]
20. Mejia-Rodriguez, D.; Trickey, S.B. Deorbitalization strategies for meta-generalized-gradient-approximation exchange-correlation functionals. *Phys. Rev. A* **2017**, *96*, 052512. [CrossRef]
21. Mejia-Rodriguez, D.; Trickey, S.B. Deorbitalized meta-GGA exchange-correlation functionals in solids. *Phys. Rev. B* **2018**, *98*, 115161. [CrossRef]
22. Mejía-Rodríguez, D.; Trickey, S.B. Meta-GGA performance in solids at almost GGA cost. *Phys. Rev. B* **2020**, *102*, 121109. [CrossRef]
23. Tran, F.; Kovács, P.; Kalantari, L.; Madsen, G.K.H.; Blaha, P. Orbital-free approximations to the kinetic-energy density in exchange-correlation MGGA functionals: Tests on solids. *J. Chem. Phys.* **2018**, *149*, 144105. [CrossRef]
24. Tran, F.; Baudesson, G.; Carrete, J.; Madsen, G.K.H.; Blaha, P.; Schwarz, K.; Singh, D.J. Shortcomings of meta-GGA functionals when describing magnetism. *Phys. Rev. B* **2020**, *102*, 024407. [CrossRef]
25. Acharya, P.K.; Bartolotti, L.J.; Sears, S.B.; Parr, R.G. An atomic kinetic energy functional with full Weizsacker correction. *Proc. Nat. Acad. Sci. USA* **1980**, *77*, 6978–6982. [CrossRef] [PubMed]
26. Thakkar, A.J. Comparison of kinetic-energy density functionals. *Phys. Rev. A* **1992**, *46*, 6920. [CrossRef]
27. Lembarki, A.; Chermette, H. Obtaining a gradient–corrected kinetic–energy functional from the Perdew–Wang exchange functional. *Phys. Rev. A* **1994**, *50*, 5328. [CrossRef]
28. Tran, F.; Wesołowski, T.A. Link between the kinetic-and exchange-energy functionals in the generalized gradient approximation. *Int. J. Quant. Chem.* **2002**, *89*, 441–446. [CrossRef]
29. Constantin, L.A.; Fabiano, E.; Laricchia, S.; Della Sala, F. Semiclassical neutral atom as a reference system in density functional theory. *Phys. Rev. Lett.* **2011**, *106*, 186406. [CrossRef]

30. Laricchia, S.; Fabiano, E.; Constantin, L.A.; Della Sala, F. Generalized Gradient Approximations of the Noninteracting Kinetic Energy from the Semiclassical Atom Theory: Rationalization of the Accuracy of the Frozen Density Embedding Theory for Nonbonded Interactions. *J. Chem. Theory Comput.* **2011**, *7*, 2439–2451. [CrossRef]
31. Karasiev, V.V.; Chakraborty, D.; Shukruto, O.A.; Trickey, S.B. Nonempirical generalized gradient approximation free-energy functional for orbital-free simulations. *Phys. Rev. B* **2013**, *88*, 161108(R). [CrossRef]
32. Borgoo, A.; Tozer, D.J. Density scaling of noninteracting kinetic energy functionals. *J. Chem. Theory Comput.* **2013**, *9*, 2250–2255. [CrossRef]
33. Trickey, S.B.; Karasiev, V.V.; Chakraborty, D. Comment on "Single-point kinetic energy density functionals: A pointwise kinetic energy density analysis and numerical convergence investigation". *Phys. Rev. B* **2015**, *92*, 117101. [CrossRef]
34. Xia, J.; Carter, E.A. Reply to "Comment on 'Single-point kinetic energy density functionals: A pointwise kinetic energy density analysis and numerical convergence investigation'". *Phys. Rev. B* **2015**, *92*, 117102. [CrossRef]
35. Constantin, L.A.; Fabiano, E.; Śmiga, S.; Della Sala, F. Jellium-with-gap model applied to semilocal kinetic functionals. *Phys. Rev. B* **2017**, *95*, 115153. [CrossRef]
36. Luo, K.; Karasiev, V.V.; Trickey, S. A simple generalized gradient approximation for the noninteracting kinetic energy density functional. *Phys. Rev. B* **2018**, *98*, 041111(R). [CrossRef]
37. Lehtomäki, J.; Lopez-Acevedo, O. Semilocal kinetic energy functionals with parameters from neutral atoms. *Phys. Rev. B* **2019**, *100*, 165111. [CrossRef]
38. Perdew, J.P.; Constantin, L.A. Laplacian-level density functionals for the kinetic energy density and exchange-correlation energy. *Phys. Rev. B* **2007**, *75*, 155109. [CrossRef]
39. Karasiev, V.V.; Jones, R.S.; Trickey, S.B.; Harris, F.E. Properties of constraint-based single-point approximate kinetic energy functionals. *Phys. Rev. B* **2009**, *80*, 245120. [CrossRef]
40. Laricchia, S.; Constantin, L.A.; Fabiano, E.; Della Sala, F. Laplacian-Level kinetic energy approximations based on the fourth-order gradient expansion: Global assessment and application to the subsystem formulation of density functional theory. *J. Chem. Theory Comput.* **2013**, *10*, 164–179. [CrossRef]
41. Cancio, A.C.; Stewart, D.; Kuna, A. Visualization and analysis of the Kohn-Sham kinetic energy density and its orbital-free description in molecules. *J. Chem. Phys.* **2016**, *144*, 084107. [CrossRef]
42. Cancio, A.C.; Redd, J.J. Visualisation and orbital-free parametrisation of the large-Z scaling of the kinetic energy density of atoms. *Mol. Phys.* **2017**, *115*, 618–635. [CrossRef]
43. Constantin, L.A.; Fabiano, E.; Della Sala, F. Semilocal Pauli–Gaussian Kinetic Functionals for Orbital-Free Density Functional Theory Calculations of Solids. *J. Phys. Chem. Lett.* **2018**, *9*, 4385–4390. [CrossRef]
44. Constantin, L.A.; Fabiano, E.; Della Sala, F. Performance of Semilocal Kinetic Energy Functionals for Orbital-Free Density Functional Theory. *J. Chem. Theory Comput.* **2019**, *15*, 3044–3055. [CrossRef] [PubMed]
45. Śmiga, S.; Constantin, L.A.; Della Sala, F.; Fabiano, E. The Role of the Reduced Laplacian Renormalization in the Kinetic Energy Functional Development. *Computation* **2019**, *7*, 65. [CrossRef]
46. Xia, J.; Carter, E.A. Single-point kinetic energy density functionals: A pointwise kinetic energy density analysis and numerical convergence investigation. *Phys. Rev. B* **2015**, *91*, 045124. [CrossRef]
47. Wang, L.W.; Teter, M.P. Kinetic-energy functional of the electron density. *Phys. Rev. B* **1992**, *45*, 13196. [CrossRef]
48. Perrot, F. Hydrogen-hydrogen interaction in an electron gas. *J. Phys. Condens. Matter* **1994**, *6*, 431. [CrossRef]
49. Smargiassi, E.; Madden, P.A. Orbital-free kinetic-energy functionals for first-principles molecular dynamics. *Phys. Rev. B* **1994**, *49*, 5220. [CrossRef]
50. García-González, P.; Alvarellos, J.E.; Chacón, E. Nonlocal kinetic-energy-density functionals. *Phys. Rev. B* **1996**, *53*, 9509. [CrossRef]
51. García-González, P.; Alvarellos, J.E.; Chacón, E. Kinetic-energy density functional: Atoms and shell structure. *Phys. Rev. A* **1996**, *54*, 1897. [CrossRef]
52. García-González, P.; Alvarellos, J.E.; Chacón, E. Nonlocal symmetrized kinetic-energy density functional: Application to simple surfaces. *Phys. Rev. B* **1998**, *57*, 4857. [CrossRef]
53. Wang, Y.A.; Govind, N.; Carter, E.A. Orbital-free kinetic-energy functionals for the nearly free electron gas. *Phys. Rev. B* **1998**, *58*, 13465. [CrossRef]
54. Wang, Y.A.; Govind, N.; Carter, E.A. Orbital-free kinetic-energy density functionals with a density-dependent kernel. *Phys. Rev. B* **1999**, *60*, 16350. [CrossRef]
55. Zhou, B.; Ligneres, V.L.; Carter, E.A. Improving the orbital-free density functional theory description of covalent materials. *J. Chem. Phys.* **2005**, *122*, 044103. [CrossRef]
56. Garcia-Aldea, D.; Alvarellos, J. Fully nonlocal kinetic energy density functionals: A proposal and a general assessment for atomic systems. *J. Chem. Phys.* **2008**, *129*, 074103. [CrossRef]
57. Garcia-Aldea, D.; Alvarellos, J.E. Approach to kinetic energy density functionals: Nonlocal terms with the structure of the von Weizsäcker functional. *Phys. Rev. A* **2008**, *77*, 022502. [CrossRef]
58. Huang, C.; Carter, E.A. Nonlocal orbital-free kinetic energy density functional for semiconductors. *Phys. Rev. B* **2010**, *81*, 045206. [CrossRef]

59. Shin, I.; Carter, E.A. Enhanced von Weizsäcker Wang-Govind-Carter kinetic energy density functional for semiconductors. *J. Chem. Phys.* **2014**, *140*, 18A531. [CrossRef]
60. Constantin, L.A.; Fabiano, E.; Della Sala, F. Nonlocal kinetic energy functional from the jellium-with-gap model: Applications to orbital-free density functional theory. *Phys. Rev. B* **2018**, *97*, 205137. [CrossRef]
61. Mi, W.; Genova, A.; Pavanello, M. Nonlocal kinetic energy functionals by functional integration. *J. Chem. Phys.* **2018**, *148*, 184107. [CrossRef]
62. Mi, W.; Pavanello, M. Orbital-free density functional theory correctly models quantum dots when asymptotics, nonlocality, and nonhomogeneity are accounted for. *Phys. Rev. B* **2019**, *100*, 041105. [CrossRef]
63. Xu, Q.; Lv, J.; Wang, Y.; Ma, Y. Nonlocal kinetic energy density functionals for isolated systems obtained via local density approximation kernels. *Phys. Rev. B* **2020**, *101*, 045110. [CrossRef]
64. Finzel, K. Shell-structure-based functionals for the kinetic energy. *Theor. Chem. Acc.* **2015**, *134*, 106. [CrossRef]
65. Finzel, K. Local conditions for the Pauli potential in order to yield self-consistent electron densities exhibiting proper atomic shell structure. *J. Chem. Phys.* **2016**, *144*, 034108. [CrossRef] [PubMed]
66. Ludeña, E.V.; Salazar, E.X.; Cornejo, M.H.; Arroyo, D.E.; Karasiev, V.V. The Liu-Parr power series expansion of the Pauli kinetic energy functional with the incorporation of shell-inducing traits: Atoms. *Int. J. Quantum Chem.* **2018**, *118*, e25601. [CrossRef]
67. Salazar, E.X.; Guarderas, P.F.; Ludena, E.V.; Cornejo, M.H.; Karasiev, V.V. Study of some simple approximations to the non-interacting kinetic energy functional. *Int. J. Quantum Chem.* **2016**, *116*, 1313–1321. [CrossRef]
68. Yao, K.; Parkhill, J. Kinetic Energy of Hydrocarbons as a Function of Electron Density and Convolutional Neural Networks. *J. Chem. Theory Comput.* **2016**, *12*, 1139–1147. [CrossRef]
69. Alonso, J.A.; Girifalco, L.A. Nonlocal approximation to the exchange potential and kinetic energy of an inhomogeneous electron gas. *Phys. Rev. B* **1978**, *17*, 3735. [CrossRef]
70. Chacón, E.; Alvarellos, J.E.; Tarazona, P. Nonlocal kinetic energy functional for nonhomogeneous electron systems. *Phys. Rev. B* **1985**, *32*, 7868. [CrossRef]
71. Sarcinella, F.; Fabiano, E.; Constantin, L.A.; Della Sala, F. Nonlocal kinetic energy functionals in real space using a Yukawa-potential kernel: Properties, linear response, and model functionals. *Phys. Rev. B* **2021**, *103*, 155127. [CrossRef]
72. Kumar, S.; Borda, E.L.; Sadigh, B.; Zhu, S.; Hamel, S.; Gallagher, B.; Bulatov, V.; Klepeis, J.; Samanta, A. Accurate parameterization of the kinetic energy functional. *J. Chem. Phys.* **2022**, *156*, 024110. [CrossRef]
73. Karasiev, V.V.; Chakraborty, D.; Trickey, S.B. Progress on New Approaches to Old Ideas: Orbital-Free Density Functionals. In *Many-Electron Approaches in Physics, Chemistry and Mathematics*; Bach, V., Delle Site, L., Eds.; Springer: Cham, Switzerland, 2014; pp. 113–134. [CrossRef]
74. March, N.; Santamaria, R. Non-local relation between kinetic and exchange energy densities in Hartree–Fock theory. *Int. J. Quantum Chem.* **1991**, *39*, 585–592. [CrossRef]
75. Della Sala, F.; Fabiano, E.; Constantin, L.A. Kohn-Sham kinetic energy density in the nuclear and asymptotic regions: Deviations from the von Weizsäcker behavior and applications to density functionals. *Phys. Rev. B* **2015**, *91*, 035126. [CrossRef]
76. Howard, I.A.; March, N.H.; Van Doren, V.E. *r*- and *p*-space electron densities and related kinetic and exchange energies in terms of s states alone for the leading term in the $1/Z$ expansion for nonrelativistic closed-shell atomic ions. *Phys. Rev. A* **2001**, *63*, 062501. [CrossRef]
77. Constantin, L.A.; Fabiano, E.; Della Sala, F. Kinetic and Exchange Energy Densities near the Nucleus. *Computation* **2016**, *4*, 19. [CrossRef]
78. Śmiga, S.; Siecińska, S.; Fabiano, E. Methods to generate reference total and Pauli kinetic potentials. *Phys. Rev. B* **2020**, *101*, 165144. [CrossRef]
79. Snyder, J.C.; Rupp, M.; Hansen, K.; Müller, K.R.; Burke, K. Finding density functionals with machine learning. *Phys. Rev. Lett.* **2012**, *108*, 253002. [CrossRef]
80. Li, L.; Snyder, J.C.; Pelaschier, I.M.; Huang, J.; Niranjan, U.N.; Duncan, P.; Rupp, M.; Müller, K.R.; Burke, K. Understanding machine-learned density functionals. *Int. J. Quant. Chem.* **2016**, *116*, 819–833. [CrossRef]
81. Alharbi, F.H.; Kais, S. Kinetic energy density for orbital-free density functional calculations by axiomatic approach. *Int. J. Quant. Chem.* **2017**, *117*, e25373. [CrossRef]
82. Seino, J.; Kageyama, R.; Fujinami, M.; Ikabata, Y.; Nakai, H. Semi-local machine-learned kinetic energy density functional with third-order gradients of electron density. *J. Chem. Phys.* **2018**, *148*, 241705. [CrossRef]
83. Golub, P.; Manzhos, S. Kinetic energy densities based on the fourth order gradient expansion: Performance in different classes of materials and improvement via machine learning. *Phys. Chem. Chem. Phys.* **2019**, *21*, 378–395. [CrossRef]
84. Manzhos, S.; Golub, P. Data-driven kinetic energy density fitting for orbital-free DFT: Linear vs Gaussian process regression. *J. Chem. Phys.* **2020**, *153*, 074104. [CrossRef]
85. Meyer, R.; Weichselbaum, M.; Hauser, A.W. Machine Learning Approaches toward Orbital-free Density Functional Theory: Simultaneous Training on the Kinetic Energy Density Functional and Its Functional Derivative. *J. Chem. Theory Comput.* **2020**, *16*, 5685–5694. [CrossRef] [PubMed]
86. Imoto, F.; Imada, M.; Oshiyama, A. Order-*N* orbital-free density-functional calculations with machine learning of functional derivatives for semiconductors and metals. *Phys. Rev. Res.* **2021**, *3*, 033198. [CrossRef]

87. Ryczko, K.; Wetzel, S.J.; Melko, R.G.; Tamblyn, I. Toward Orbital-Free Density Functional Theory with Small Data Sets and Deep Learning. *J. Chem. Theory Comput.* **2022**, *18*, 1122–1128. [CrossRef] [PubMed]
88. Prodan, E.; Kohn, W. Nearsightedness of electronic matter. *Proc. Nat. Acad. Sci. USA* **2005**, *102*, 11635–11638. [CrossRef]
89. Constantin, L.A.; Fabiano, E.; Della Sala, F. Modified Fourth-Order Kinetic Energy Gradient Expansion with Hartree Potential-Dependent Coefficients. *J. Chem. Theory Comput.* **2017**, *13*, 4228–4239. [CrossRef]
90. Lindhard, J. On the properties of a gas of charged particles. *Dan. Vid. Selsk Mat.-Fys. Medd.* **1954**, *28*, 8.
91. Tao, J.; Perdew, J.P.; Almeida, L.M.; Fiolhais, C.; Kümmel, S. Nonempirical density functionals investigated for jellium: Spin-polarized surfaces, spherical clusters, and bulk linear response. *Phys. Rev. B* **2008**, *77*, 245107. [CrossRef]
92. Levy, M.; Ou-Yang, H. Exact properties of the Pauli potential for the square root of the electron density and the kinetic energy functional. *Phys. Rev. A* **1988**, *38*, 625–629. [CrossRef]

computation

MDPI

Article

Periodic DFTB for Supported Clusters: Implementation and Application on Benzene Dimers Deposited on Graphene

Mathias Rapacioli [1,*] and Nathalie Tarrat [2,*]

1 Laboratoire de Chimie et Physique Quantiques LCPQ/FERMI, UMR5626, Université de Toulouse (UPS) and CNRS, 118 Route de Narbonne, 31062 Toulouse, France
2 CEMES, CNRS, Université de Toulouse, 29 Rue Jeanne Marvig, 31055 Toulouse, France
* Correspondence: mathias.rapacioli@irsamc.ups-tlse.fr (M.R.); nathalie.tarrat@cemes.fr (N.T.)

Abstract: The interest for properties of clusters deposited on surfaces has grown in recent years. In this framework, the Density Functional based Tight Binding (DFTB) method appears as a promising tool due to its ability to treat extended systems at the quantum level with a low computational cost. We report the implementation of periodic boundary conditions for DFTB within the deMonNano code with **k**-points formalism and corrections for intermolecular interactions. The quality of DFTB results is evaluated by comparison with dispersion-corrected DFT calculations. Optimized lattice properties for a graphene sheet and graphite bulk are in agreement with reference data. The deposition of both benzene monomer and dimers on graphene are investigated and the observed trends are similar at the DFT and DFTB levels. Moreover, interaction energies are of similar orders of magnitude for these two levels of calculation. This study has evidenced the high stability of a structure made of two benzene molecules deposited close to each other on the graphene sheet. This work demonstrates the ability of the new implementation to investigate surface-deposited molecular clusters properties.

Keywords: periodic DFTB; deMonNano; graphene; graphite; benzene dimers; deposited benzene; supported clusters; weighted mulliken charges

Citation: Rapacioli, M.; Tarrat, N. Periodic DFTB for Supported Clusters: Implementation and Application on Benzene Dimers Deposited on Graphene. *Computation* **2022**, *10*, 39. https://doi.org/10.3390/computation10030039

Academic Editor: Henry Chermette

Received: 10 February 2022
Accepted: 8 March 2022
Published: 14 March 2022

1. Introduction

The modeling of functional extended surfaces has grown in past decades to investigate, for fundamental and engineering purposes, a large number of phenomena or applications such as, e.g., deposition [1], growth and migration [2], 2D assembly [3], catalysis [4], electrocatalysis [5], photocatalysis [6], molecular electronics [7], depollution [8], sensing [9], etc. Many of these studies have focused on deposited clusters, i.e., finite aggregations of basis elements (atoms or molecules) adsorbed on surfaces. Indeed, the physico-chemical properties of a cluster are distinct from the ones of both the single entities and the infinite cluster (bulk) and strongly depend on the size and structure of the cluster [10]. Understanding and controlling the structure of deposited clusters could thus allow the precise tuning of their properties.

The theoretical study of clusters deposited on extended surfaces is very challenging due to the size of the space of structural and electronic configurations to be explored and the high level of computational methods that has to be implemented. Indeed, the a priori unknown nature of the interactions between the cluster building blocks and between the cluster and the surface (with potential charge transfers at the cluster–surface interface) prevents the use of empirical force fields in favor of methods in which the electronic structure is explicitly considered. The very high computational cost of ab initio and Density Functional Theory (DFT) methods prohibit their use for the study of such systems, particularly when dealing with global optimisation or finite-temperature molecular dynamics, and one strategy consists in implementing, in a periodic formulation, approximate quantum mechanical methods. Among them, DFTB is an approximated DFT scheme with a much lower computational cost enabled by the use of parameterized integrals in a minimal valence basis set [11–14]. There have been several implementations of DFTB within periodic conditions in various codes [15–21], allowing the computation of structural, mechanical and electronic properties. In the present work, we report a new implementation of the DFTB scheme for periodic systems within the deMonNano code [22] combined with corrections to describe long range intermolecular interactions.

After assessing the performance of this implementation on graphene and graphite, we apply this method to characterize model systems consisting in benzene and benzene dimers deposited on a graphene surface. Such a system is relevant in an astrophysical context as it can be seen as a simple model of Polycyclic Aromatic Hydrocarbons (PAH) clusters adsorbed on large carbonaceous grains or on very large PAHs. Indeed, despite the fact that PAHs are expected to be ubiquitous in the interstellar medium [23] and their clusters have been proposed to play a significant role in interstellar physics and chemistry [24], the structural and energetic property changes induced by their deposition on a surface remain, to a large extent, unknown. The second motivation for selecting this benchmark system is that a reasonable description of the benzene dimers potential energy surface is challenging even with ab initio schemes [25], making it a system of choice to address the quality of our approach. This is due to the fine equilibrium between Pauli repulsion, dispersion and coulomb interaction, which drives the competition between parallel and T-shaped structures. In the past, we have shown that the combination of DFTB with empirical dispersion and atomic charges corrections allowed for a proper description of such systems [26].

In this paper, the periodic formulation of DFTB that has been implemented in deMon-Nano is presented in Section 2, with a special focus on the originality of the present scheme with respect to other periodic implementations, i.e., its combination with the WMull charge-correction approach. Computational details are provided in Section 3, and the benchmarks on graphene and graphite are presented in Section 4. The applications to benzene monomer and dimers deposited on graphene are discussed in Section 5. Finally, a conclusion is provided in Section 6.

2. Methods

2.1. DFTB

The Density Functional-based Tight-Binding method (DFTB) can be derived from DFT from several approximations [11,13,14,16,27].

The first one relies on the expression of molecular orbitals (MOs) $\phi_i(\mathbf{r})$ as linear combinations of atomic orbital (LCAO)-type basis sets using minimal valence bases χ_μ:

$$\phi_i(\mathbf{r}) = \sum_\mu c_{i\mu}\chi_\mu(\mathbf{r}) \tag{1}$$

where $c_{i\mu}$ is the coefficient of the molecular orbital i on the atomic orbital χ_μ.

A Taylor expansion of the DFT energy is performed as a function of the electronic density, and the real density ρ of the system minimizing the Kohn–Sham energy is searched as a perturbation with respect to reference density ρ_0 ($\rho = \rho_0 + \delta\rho$).

$$\begin{aligned} E[\rho(\mathbf{r})] = E[\rho_0(\mathbf{r})] + &\int \frac{\delta E[\rho(\mathbf{r})]}{\delta\rho(\mathbf{r})}\bigg|_{\rho_0} \delta\rho(\mathbf{r}) + \frac{1}{2}\int\int \frac{\delta^2 E[\rho(\mathbf{r})]}{\delta\rho(\mathbf{r})\delta\rho(\mathbf{r}')}\bigg|_{\rho_0} \delta\rho(\mathbf{r})\delta\rho(\mathbf{r}') + \\ &\cdots + \frac{1}{p!}\int\int\cdots\int \frac{\delta^p E[\rho(\mathbf{r})]}{\delta\rho(\mathbf{r})\delta\rho(\mathbf{r}')\ldots\delta\rho(\mathbf{r}^{(\mathbf{p})})}\bigg|_{\rho_0} \delta\rho(\mathbf{r})\delta\rho(\mathbf{r}')\ldots.\delta\rho(\mathbf{r}^{(\mathbf{p})}) \end{aligned} \tag{2}$$

In the original version, also known as the non self-consistent DFTB (sometimes referred to as zeroth-order DFTB or simply DFTB [11,12]), only the zeroth and first-order terms of the Taylor expansion are retained. In the DFTB2 scheme [27], also known as self-consistent charge (SCC) DFTB and in the DFTB3 scheme [28], the second-order and third-order terms are also taken into account, respectively.

At the DFTB0 level, the electronic energy reads as follows:

$$E^{DFTB0} = \sum_{\alpha<\beta} E_{rep}(\mathbf{r}_{\alpha\beta}) + \sum_{i\mu\nu} n_i c_{i\mu} c_{i\nu} H^0_{\mu\nu} \tag{3}$$

with $\mathbf{r}_{\alpha\beta} = \mathbf{r}_\beta - \mathbf{r}_\alpha$, where $\mathbf{r}_\alpha$ and $\mathbf{r}_\beta$ are the positions of atoms α and β, respectively, $E_{rep}(\mathbf{r}_{\alpha\beta})$ is a repulsive contribution between atoms α and β, n_i is the occupation of the orbital i and $H^0_{\mu\nu}$ is the matrix element associated with the Kohn–Sham operator at the reference density expressed in the atomic basis. These matrix elements, as well as those of the atomic basis overlap matrix S, can be parameterized as only one-body or two-body terms. This is allowed by the definition of the reference density as a superposition of atomic densities $\rho_0 = \rho_0^\alpha + \rho_0^\beta + \rho_0^\gamma + \ldots$, and the reduction of integrals to one-center or two-center terms:

- $H^0_{\mu,\nu\in\alpha}(\rho_0) \approx H^0_{\mu\nu}(\rho_0^\alpha) \approx \delta_{\mu\nu}\epsilon_{\mu\alpha}$; the atomic orbital energies of the isolated atom α;
- $H^0_{\mu\in\alpha,\nu\in\beta}(\rho_0) \approx H^0_{\mu\nu}(\rho_0^\alpha + \rho_0^\beta)$, which only depends on the distance between the two corresponding atomic centers: $H^0_{\mu\in\alpha,\nu\in\beta}(\mathbf{r}_\alpha - \mathbf{r}_\beta)$.

Focusing from now on the SCC-DFTB level [27], the energy expression is :

$$E^{SCC-DFTB} = E^{DFTB0} + \frac{1}{2}\sum_{\alpha\beta} \gamma_{\alpha\beta} q_\alpha q_\beta \tag{4}$$

The last term corresponds to the second-order contribution and depends on the electronic density fluctuation $\delta\rho$ represented by atomic charges q_α. $\gamma_{\alpha\beta}$ is a matrix for which diagonal terms are equal to the atomic Hubbard parameters and off-diagonal terms contain the $1/R$ coulomb interaction between atomic charges and an exchange-correlation energy E_{xc} contribution:

$$\gamma_{\alpha\beta} = \int\int \left(\frac{1}{|\mathbf{r} - \mathbf{r}'|} + \frac{\delta^2 E_{xc}}{\delta\rho(\mathbf{r})\delta\rho(\mathbf{r}')}\bigg|_{\rho_0} \right) F_0^\alpha(\mathbf{r} - \mathbf{r}_\alpha) F_0^\beta(\mathbf{r}' - \mathbf{r}_\beta) d\mathbf{r} d\mathbf{r}' \tag{5}$$

where F_0^α is the normalized spatial extension for the excess/default of electrons around atom α with respect to the neutral atom, and it is assumed here to have no angular dependence.

Since the second-order term contains atomic charges, this introduces a term depending on the charges of $H^1(q)$ into the TB operator:

$$(H^0 + H^1(q))C_i = \epsilon_i S C_i \tag{6}$$

with the following:

$$H^1_{\mu\nu} = \frac{1}{2} S_{\mu\nu} \sum_\xi q_\xi (\gamma_{\alpha\xi} + \gamma_{\beta\xi})$$

where μ and ν belong to atoms α and β, respectively, and C_i is the column vector containing the coefficients for the ith MO. As the charges depend on MO coefficients $c_{i\mu}$, the new secular equation must be solved self-consistently with respect to atomic charges, at the origin of the method's name self-consistent charge (SCC-)DFTB.

In the standard SCC-DFTB version [27], the atomic charges are computed from the density matrix P and the atomic basis overlap S matrix within the Mulliken approximation:

$$q_\alpha = \sum_{\mu\in\alpha} \sum_\beta \sum_{\nu\in\beta} P_{\mu\nu} S_{\mu\nu} \tag{7}$$

making use of the density matrix:

$$P_{\mu\nu} = \sum_{\mu\nu} n_i c_{i\mu} c_{i\nu}$$

In previous works, we have shown that atomic charges can be improved by taking into account the bond polarisation, adapting the Charge Model class IV scheme for DFTB [26,29]. This approach, which requires the calculation of Mayer's bond order, is computationally expensive and hardly transferable to a periodic implementation. To circumvent this bottleneck, we have recently introduced a simpler scheme, named in the following WMull for Weighted Mulliken charges [30], to correct atomic charges with the following expression.

$$q_\alpha = \sum_{\mu\in\alpha}\sum_{\beta}\sum_{\nu\in\beta} P_{\mu\nu}S_{\mu\nu}(1+t_{\alpha\beta}) \tag{8}$$

$t_{\alpha\beta} = -t_{\beta\alpha}$ is an empirical parameter accounting for a non-symmetric repartition of the electrons between different atomic types, where the Mulliken symmetric repartition is recovered for $t_{\alpha\beta} = 0$. The second order contribution to the Kohn–Sham operator matrix is modified as follows.

$$H^1_{\mu\nu} = \frac{1}{2}S_{\mu\nu}\sum_{\xi} q_\xi(\gamma_{\alpha\xi}(1+t_{\alpha\beta}) + \gamma_{\beta\xi}(1-t_{\alpha\beta})) \tag{9}$$

We have shown that this simple scheme provides similar results to those obtained with the Charge Model approach to model clusters of PAHs [31] and water [32].

2.2. DFTB for Periodic Systems

The former implementation of periodic DFTB within deMonNano was restricted to Γ-point approximation only [22]. In the present implementation, after defining a set of **k**-points in the reciprocal space, the electronic problem is searched self consistently. A step of the self-consistent scheme consists in solving separately the secular equations for each **k**-point to obtain molecular orbitals $\phi_i^{\mathbf{k}}$. Molecular orbitals obtained for all **k**-point are then used to build the total electronic density, the latter being used as an input for the next self-consistent step.

For a given **k**-point, the molecular orbitals $\phi_i^{\mathbf{k}}$ must fulfill Bloch's theorem:

$$\hat{T}_{\mathbf{R}}\phi_i^{\mathbf{k}} = e^{i\mathbf{k}\mathbf{R}}\phi_i^{\mathbf{k}}(\mathbf{r}) \tag{10}$$

where $\hat{T}_{\mathbf{R}}$ is the operator associated to a translation of **R** and where **R** is a vector connecting two unit cells. This is achieved by expanding MOs on a basis of Bloch functions built from real space atomic orbitals:

$$\phi_i^{\mathbf{k}}(\mathbf{r}) = \sum_{\mathbf{k}} c_{i\mu}^{\mathbf{k}}\chi_\mu^{\mathbf{k}}(\mathbf{r})$$

$$\chi_\mu^{\mathbf{k}}(\mathbf{r}) = \frac{1}{\sqrt{N}}\sum_N e^{i\mathbf{k}\mathbf{R}_N}\chi_\mu(\mathbf{r}-\mathbf{R}_N) \tag{11}$$

where the infinite sum relies on all N possible translations from the main unit cell to any of the other ones through translation vectors $\mathbf{R}_N$. The overlap and Kohn–Sham operator matrices expressed in this basis can be written from their real-space equivalent by making use of the following transformation:

$$H^{\mathbf{k}}_{\mu\nu} = \sum_N e^{i\mathbf{k}\mathbf{R}_N}H_{\mu\nu}(\mathbf{r}_\alpha-\mathbf{r}_\beta-\mathbf{R}_N) = \sum_N e^{i\mathbf{k}\mathbf{R}_N}(H^0_{\mu\nu}(\mathbf{r}_\alpha-\mathbf{r}_\beta-\mathbf{R}_N) + H^1_{\mu\nu}(\mathbf{r}_\alpha-\mathbf{r}_\beta-\mathbf{R}_N))$$

and

$$S^{\mathbf{k}}_{\mu\nu} = \sum_N e^{i\mathbf{k}\mathbf{R}_N}S_{\mu\nu}(\mathbf{r}_\alpha-\mathbf{r}_\beta-\mathbf{R}_N) \tag{12}$$

where μ and ν belong to atoms α and β, respectively.

In the previous expressions, the matrix elements of H^0 and S are easily obtained from the DFTB Slater–Koster tables [33] and rapidly vanish for large values of $\mathbf{R}_N$. The first order contribution to $H_{\mu\nu}$ is also short range with respect to $(\mathbf{r}_\alpha - \mathbf{r}_\beta - \mathbf{R}_N)$ but contains an infinite long range coulomb sum.

$$H^{1,\mathbf{k}}_{\mu\nu}(\mathbf{r}_\alpha - \mathbf{r}_\beta - \mathbf{R}_N) = \frac{1}{2}S_{\mu\nu}(\mathbf{r}_\alpha - \mathbf{r}_\beta - \mathbf{R}_N)\sum_{\xi}\sum_{N} q_\xi(\gamma_{\alpha\xi}(\mathbf{r}_\alpha - \mathbf{r}_\xi - \mathbf{R}_N) + \gamma_{\beta\xi}(\mathbf{r}_\beta - \mathbf{r}_\xi - \mathbf{R}_N)) \quad (13)$$

In practice, this infinite sum is replaced by an Ewald summation. The secular equation is solved for each **k**-point.

$$H^{\mathbf{k}}C^{\mathbf{k}}_i = \epsilon^{\mathbf{k}}_i S^{\mathbf{k}}C^{\mathbf{k}}_i \quad (14)$$

The eigenvalues $\epsilon^{\mathbf{k}}_i$ resulting from all the **k**-point secular equations are then sorted in ascending order to drive the determination of the orbital occupation numbers $n^{\mathbf{k}}_i$ following either a canonical occupation or a Fermi distribution.

The density matrix can, therefore, be computed for each **k**-point.

$$P_{\mu\nu}(\mathbf{k}) = \sum_i n^{\mathbf{k}}_i c^{\mathbf{k}*}_{i\mu} c^{\mathbf{k}}_{i\nu} \quad (15)$$

We follow the approach of reference [15], which consists in building the real space density matrix and computing atomic charges in the real space. The real space density matrix is obtained by summing over **k**-points:

$$P_{\mu\nu}(\mathbf{R}_N) = \sum_{\mathbf{k}} P_{\mu\nu}(\mathbf{k})e^{-i\mathbf{k}\mathbf{R}_N} \quad (16)$$

and the Mulliken charges are then computed as follows:

$$q_\alpha = \sum_{R_N} P_{\mu\nu}(\mathbf{R}_N)S_{\mu\nu}(\mathbf{r}_\alpha - \mathbf{r}_\beta - \mathbf{R}_N) \quad (17)$$

and they are used as inputs for the next SCC cycle.

The simple WMull correction to Mulliken charges can be generalized to the periodic equations replacing Equation (17) by the following:

$$q_\alpha = \sum_{R_N} P_{\mu\nu}(\mathbf{R}_N)S_{\mu\nu}(\mathbf{r}_\alpha - \mathbf{r}_\beta - \mathbf{R}_N)(1 + t_{\alpha\beta}) \quad (18)$$

and Equation (13) by the following.

$$\begin{aligned} H^{1,\mathbf{k}}_{\mu\nu}(\mathbf{r}_\alpha - \mathbf{r}_\beta - \mathbf{R}_N) = \frac{1}{2}S_{\mu\nu}(\mathbf{r}_\alpha - \mathbf{r}_\beta - \mathbf{R}_N)\sum_{\xi}\sum_{N} \; & q_\xi \;\; (\gamma_{\alpha\xi}(\mathbf{r}_\alpha - \mathbf{r}_\xi - \mathbf{R}_N)(1 + t_{\alpha\beta}) \\ & + \;\; \gamma_{\beta\xi}(\mathbf{r}_\beta - \mathbf{r}_\xi - \mathbf{R}_N)(1 - t_{\alpha\beta})) \end{aligned} \quad (19)$$

3. Computational Details

3.1. DFTB Calculations

Different DFTB parameters are available in the literature (cf. website www.dftb.org) depending on the choices made during the parameterization procedure such as the DFT functional, the basis sets type (Gaussian and Lorentzian) used to generate the atomic orbitals, the confinement imposed on these orbitals, the reference data used to compute the repulsive contribution E_{rep} and, for the second and third order DFTB, the values of the atomic Hubbard parameters and their derivatives. In this work, we are working with the BIO DFTB set of parameters [27] provided within the deMonNano code.

Dispersion interaction corrections can be introduced in the DFTB Hamiltonian using an empirical atom-atom formula. Two types of corrections are available in the deMonNano code and will be tested in the next section. The first one (hereafter labelled D1) is a Lennard–

Jones type potential with short range corrections introduced by Zhechkov et al. [34]. The second one (hereafter labelled D2 [26]) is given by the following expression:

$$E_{\text{disp}} = -\sum_{N}\sum_{\alpha,\beta} f_{\text{damp}}(|\mathbf{r}_\alpha - \mathbf{r}_\beta - \mathbf{R}_N|)\frac{C^6_{\alpha\beta}}{|\mathbf{r}_\alpha - \mathbf{r}_\beta - \mathbf{R}_N|^6} \tag{20}$$

where f_{damp} is a damping function screening the short range contribution and $C^6_{\alpha\beta}$ is an empirical parameter (see [26] for details). In both cases, only the van der Waals contributions larger than 10^{-5} Hartree are taken into account in order to limit the number N of boxes involved in the sum.

When calculations are performed with the WMull scheme, a value of $t_{CH} = 0.245$ has been determined to provide the atomic charges for the benzene molecule in agreement with reference calculations (see tables and discussion in reference [26]).

Regarding convergency criterions, we have used a tolerance of 10^{-8} for the atomic charges during the SCC process and 5×10^{-6} Hartree/Bohr for the largest gradient for local optimizations.

3.2. DFT Calculations

Dispersion-corrected DFT calculations were performed under periodic boundary conditions using the Vienna ab initio simulation package (VASP [35–37]) together with PAW pseudopotentials [38,39] and the PBE-D3 semiempirical dispersion-corrected functional in its zero-damping formalism [40]. This functional has been chosen as it has been reported as a relevant choice for studies involving graphene [41]. A conjugate-gradient algorithm was used to relax the ions and the convergence criterion was set up so that the maximum atomic force was less than 0.01 eV Å^{-1}, and all atoms were allowed to relax unconstrained. To avoid interactions between the benzene monomers/dimers and their periodic images, a cubic box measuring 50 Å on a side was used for isolated systems. For supported ones, the calculations were performed on a 29.92 Å $\times$ 34.55 Å graphene surface (these values having been calculated on the basis of the graphene equilibrium lattice parameter reported in Section 4) containing 392 carbon atoms placed in a 50 Å high simulation box to avoid any interaction between the adsorbed molecule and the underside of the graphene sheet of the upper periodic box. Since the size of the supercell was large enough, the Brillouin zone sampling in reciprocal space restricted to the Γ-point was sufficient to ensure good convergence of the total energy, except for the calculations aimed at determining the equilibrium parameters of the graphite bulk, which required a $1 \times 1 \times 5$ **k**-points grid. A plane-wave kinetic energy cutoff of 450 eV was employed. For dealing with the partial occupancies around the Fermi level, Methfessel–Paxton smearing was used with $\sigma = 0.2$ eV [42].

4. Benchmark Calculations: Graphene and Graphite

The goal of this section is twofold. First, **k**-points grid size tests are performed on two systems of interest, namely the 392-atom graphene sheet that will be further used in Section 5 and 2×392-atom graphite built from periodic repetition of two graphene layers. Second, we aim at evaluating the quality of the DFTB potential to reproduce the main features of graphene and graphite. In order to model graphene, we have first optimized the DFTB lattice parameter, working with a periodic box containing 392 atoms (~30 Å $\times$ 35 Å $\times$ 50 Å). For such a large simulation box, the Γ-point approximation remains valid as the energy varies by less than 3.4×10^{-4} eV/atom (0.0008%) when increasing from one to three **k**-points in the x and y directions and by less than 1.3×10^{-4} eV/atom (0.0003%) when changing from three to five **k**-points. The equilibrium C-C bond lengths determined with one or three **k**-points in x and y directions are the same at the precision of 10^{-3} Å . Values of 1.430 ± 0.001 Å and 1.426 ± 0.001 Å were obtained with DFTB-D1 and DFTB-D2 methods, respectively (see Table 1). These values are slightly larger than the value of 1.421 Å previously reported by Zhechkov et al. using the Γ-point approximation and a smaller unit cell [34]. The DFTB-D2 values produces the best agreement with the C-C bond length obtained at the dispersion-corrected DFT level (1.425 ± 0.001 Å with DFT-D3 and 1.4226 Å with PW91-OBS [43]) as well as with the experimental values (1.42 Å).

The graphite bulk has been modeled by including two layers of the previously defined graphene sheet in the periodic box. In order to determine the appropriate number of **k**-points in the z direction (perpendicular to the graphene planes), we have performed DFTB single point energy calculations for an interlayer distance of 3.5 Å, chosen because it corresponds to the DFT-D3 one (3.488 Å, see Table 1), with one k-point in the x and y directions. The total energy varies by 2×10^{-4} eV/atom (0.0004%) when the number of **k**-points increased from one to three in the z direction, and then it remains constant for calculations performed with five, seven and nine **k**-points in the z-direction. We have determined the equilibrium parameters of the graphite bulk with three **k**-points in the z directions (Table 1). Using either one or three **k**-points in the x and y directions led to the sames results at the target precision of 0.001 Å. DFTB-D1 and DFTB-D2 C-C bond lengths are reduced by 0.001 Å with respect to their values in the graphene sheet, which is a trend also observed at the DFT-D3 level. The DFTB-D1 interlayer equilibrium distance (3.383 ± 0.001 Å) is in agreement with both the value of reference [34] with a four layers model in the Γ-point approximation (3.38 Å) and the experimental data (3.356 Å). The interlayer distance is reduced to 3.131 ± 0.001 Å at the DFTB-D2 level. With respect to theoretical references (DFT-D3, RPA and QMC) and experimental values, we can conclude that the DFTB-D1 method produces better quality results for graphite interlayer distances, while the DFTB-D2 method prevails for the C-C bond length.

Table 1. Graphene and Graphite structural data (in Å). * In these computational studies, d_{C-C} distance was fixed to the one determined experimentally. ** Differences between the two DFTB-D1 calculations are detailed in the text.

Methods	$d_{C-C}^{Graphene}$	$d_{C-C}^{Graphite}$	$d_{interlayer}^{Graphite}$
DFTB-D1 **	1.430 ± 0.001	1.429 ± 0.001	3.383 ± 0.001
DFTB-D2	1.426 ± 0.001	1.425 ± 0.001	3.131 ± 0.001
DFTB-D1 **	1.421 [34]	1.421 [34]	3.38 [34]
DFT-D3	1.425 ± 0.001	1.424 ± 0.001	3.488 ± 0.001
PW91-OBS	1.4226 [43]		
RPA		1.42 * [44]	3.34 * [44]
QMC		1.42 * [45]	3.426 * [45]
Expt	1.42 [46]	1.422 [47,48]	3.356 [47,48]

5. Graphene Supported Benzene and Benzene Dimers

5.1. Benzene Supported on Graphene

We performed local structural optimization for systems consisting of an isolated benzene molecule deposited on top of a graphene monolayer. On the basis of the results obtained in Section 4, the calculations have been performed in the Γ-point approximation, with the initial structures corresponding to a benzene molecule deposited in the proper orientation on the optimized graphene layer. Four different configurations have been probed and labeled a_1, a_2, a_3 and a_4, which can be visualized in Figure 1. The three first ones correspond to the structures labeled a_1, a_2 and a_3 in reference [49]: hollow, bridge and top in reference [50] and AA, SP and AB in reference [51]. The last structure a_4 was named top-rot in reference [50] and was also studied in reference [52].

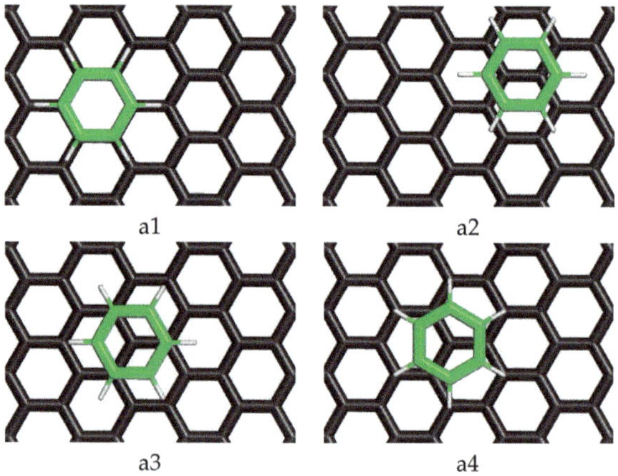

Figure 1. Four configurations for an isolated benzene molecule deposited on graphene (**a1**–**a4**). The atoms are depicted with sticks and the color labels are dark grey for C atoms of the graphene layer, green for C atoms of the benzene molecule and white for H atoms.

The binding energies of benzene on graphene, reported on Table 2, were computed with the following formula:

$$E_{bind}^{graphene@bz} = E_{tot}^{graphene@bz} - E_{tot}^{graphene} - E_{tot}^{bz} \tag{21}$$

where all energies correspond to relaxed structures. DFTB-D1 and DFTB-D2 results agree on the main trends which are also present at the DFT-D3 level: three almost degenerated structures, namely a_2, a_3 and a_4, and the a_1 structure being less stable by about 0.012(DFTB-D2)/0.015(DFTB-D1)/0.025(DFT-D3) eV (see Table 2). The absolute binding energies provided by the DFTB-D2 scheme are in very good agreement with DFT results (apart from LDA) and experimental measurements. The DFTB-D1 scheme provides poorer results, with an overestimation of the binding energies of about 35% (∼0.2 eV).

Table 2. Binding energies of benzene on graphene in eV. The experimental binding energy of a benzene molecule on a graphite surface is −0.50 ± 0.08 eV [53].

Method	a1	a2	a3	a4
DFTB-D1	−0.639	−0.652	−0.654	−0.651
DFTB-D2	−0.439	−0.448	−0.447	−0.451
DFT-D3	−0.428	−0.450	−0.453	−0.450
LDA [49]	−0.16	−0.23	−0.24	
ωB97X-D [51]			−0.47	
optB86b-vdw [52]				−0.5
vdW-DF1 [50]			−0.49	
vdW-DF2 [50]			−0.43	
Expt. Saturated Adsorption Enthalpy [52]				−0.5

Regarding z-separation between the benzene monomer and the graphene sheet (see Table 3), DFTB-D1 and DFTB-D2 benzene-graphene distances are shorter than DFT ones by ∼0.35 Å. However, it should be noted that z-separations calculated with dispersion-corrected DFT functionals also significantly differ from each other by up to 0.25 Å and experimental data are missing in the literature to our knowledge. Concerning the a_4 structure, z-separation can be compared with the MP2 one reported in reference [54], which confirms the slight underestimation at the DFTB-D1 and DFTB-D2 levels.

The binding energy and the z-separation of a graphene sheet with a deposited benzene monomer can be compared with the ones of larger aromatic molecules. DFTB-D2 and DFT-D3 computed benzene binding energies are in the range 0.43–0.45 eV, which is consistent with the almost three-times larger binding energies of molecules containing three aromatic rings such as anthraquinone (1.30 eV [55]) and phenanthraquinone (1.31 eV [56]). For these systems, the reported z-separations are 3.19 Å [55] for anthraquinone and 3.14 Å [56]) for phenanthraquinone, and these values last between the DFTB-D2 and DFT-D3 values obtained for the graphene-supported benzene molecule. One can also mention the slightly larger distance reported for indole (tryptophan model, 3.50 Å [54]).

Table 3. Z-separation of benzene on graphene (in Å).

Method	a1	a2	a3	a4
DFTB-D1	3.15	3.13	3.13	3.14
DFTB-D2	3.08	3.07	3.08	3.05
DFT-D3	3.47	3.38	3.36	3.42
ωB97X-D [51]	3.36	3.30	3.35	
vdW-DF1 [50]			3.6	
vdW-DF2 [50]			3.5	
MP2 [54]				3.33

5.2. Benzene Dimers in Vacuum

Reproducing the benzene dimer potential energy surface is a challenging task for DFT schemes, due to the fine competition between the various contributions to the total energy. This is even more true for approximated schemes such as the DFTB method. Briefly, three characteristic structural families can be identified, namely sandwich (S), parallel-displaced (PD) and T-shaped (T), each one presenting several minima. In this work, the sandwich-eclipsed (SE) structure has been selected to represent the sandwich family. The PD family is represented by the isomer shown in Figure 2 as it was previously reported to be the most stable of this family at the DFTB level. Two additional structures were considered to account for the T-shaped family, namely T and Csoa isomers (corresponding to T4 and Csoa in reference [26]), which only differ by a slight displacement of the top benzene from a symetric position toward a position over a carbon atom. These two structures were previously reported to be degenerated as their energies differed by less than 10^{-3} eV at the DFTB level, and the present DFT-D3 calculations show a difference of 8×10^{-3} eV in favor of the Csoa isomer.

The binding energy of benzene dimer were computed with the following formula:

$$E_{bind}^{bz_2} = E_{tot}^{bz_2} - 2 \times E_{tot}^{bz} \tag{22}$$

where all energies correspond to relaxed structures. These energies are gathered in Table 4 for the different structures optimized with both DFTB and DFT-D3 methods, together with the ones reported in the literature for ab initio reference calculations (CCSD(T) and SAPT). In the case of the T-shaped family, only one of the two studied isomers could be located on DFTB potential energy surfaces, namely Csoa with the DFTB-D1 method and T with the DFTB-D2 one. DFT-D3 and ab initio reference calculations agree on the fact that T-shaped and PD structures are close in energy and by far more stable than the SE structure. This ordering is not reproduced at the DFTB-D1 nor at the DFTB-D2 levels, i.e., for the two dispersion corrections investigated in the absence of atomic charge corrections, because the SE structure is found to be almost degenerated with the PD structure and the T-shaped isomer is found to be the less stable one in both cases. Introducing the WMull charge correction detailed in Section 2 with the D1 dispersion (DFTB-D1-WMull) renders the PD structure the most stable, but the T-shaped structure remains the least stable. Finally, the DFTB-D2-WMull method provides a correct picture, with T-shaped and PD isomers being close in energy and more stable than the SE isomer. In addition, the binding energies are of the same order as those of the reference calculations.

Table 4. Binding energies for benzene dimer bz_2 in eV.

Method	T-Shaped		PD	SE
DFTB-D1	Csoa	−0.126	−0.194	−0.192
DFTB-D1-WMull	Csoa	−0.140	−0.162	−0.148
DFTB-D2	T	−0.099	−0.135	−0.132
DFTB-D2-WMull	T	−0.113	−0.104	−0.086
DFT-D3	Csoa	−0.146	−0.152	−0.106
	T	−0.138		
CCSD(T) [25]	Csoa	−0.12	−0.12	−0.07
SAPT [25]	Csoa	−0.12	−0.12	−0.08

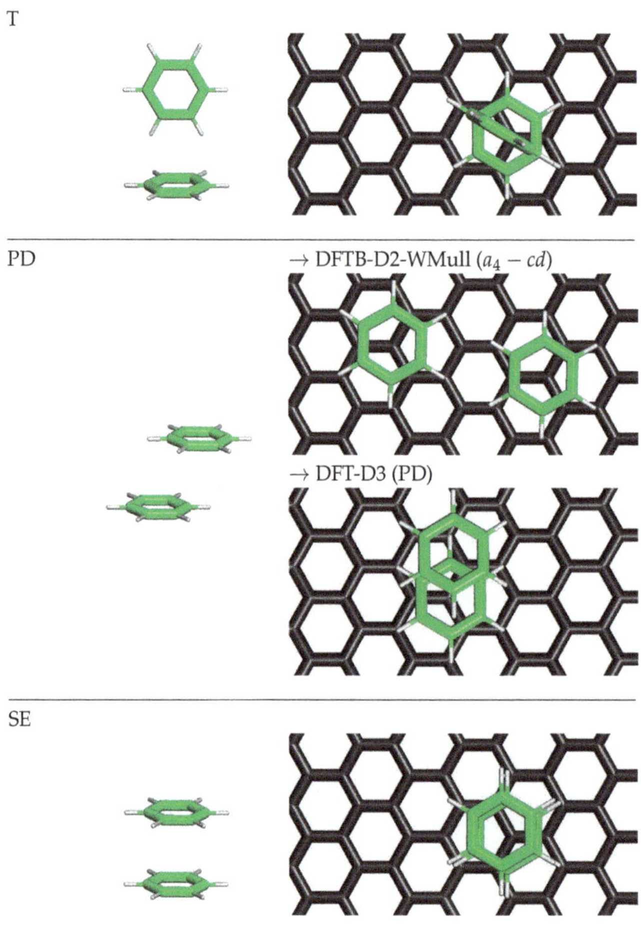

Figure 2. Benzene dimers bz_2 in vacuum and deposited on graphene.

5.3. Benzene Dimers Supported on Graphene

It appeared from the previous sections that the DFTB-D2-WMull method is the best DFTB choice for modeling both an isolated benzene molecule deposited on a graphene layer and a benzene dimer in vacuum. This level of theory has been chosen to investigate the deposition of a benzene dimer on a graphene sheet. Among the possible adsorption modes of a benzene molecule on a graphene monolayer, we selected the a_4 one as it was found to be the most stable one at this level of theory (see Section 5.1) and added a second benzene unit to form T, PD or SE configurations. The optimized structures, obtained in the Γ-point approximation, are shown in Figure 2.

It can be observed that T and SE structures were preserved during optimization. On the opposite side, the deposited PD dimer was stable at the DFT-D3 level only and resulted in a dissociated configuration at the DFTB-D2-WMull level in which the two benzene units are close to each other, both exhibiting a a_4 adsorption configuration on the graphene sheet. This configuration is hereafter named $a_4 - cd$ (close deposition on a_4 adsorption sites). In this latter configuration, the hydrogen atom of each benzene molecule is pointing in between two hydrogen atoms of the other benzene unit. Such a configuration limits the coulomb repulsion between the positively charged hydrogen atoms while preserving some attractive dispersion interactions.

Z-separations between the graphene sheet and the benzene units parallel to it are reported in Table 5. At both the DFT and DFTB levels, the distance between the graphene sheet and the closest benzene unit is very slightly reduced (less than 0.05 Å with DFTB-D2-WMull and 0.03 Å with DFT) with respect to the z-separation values obtained for the deposited benzene monomer (see Table 3). In the SE structure, the presence of the graphene surface results in a slight reduction in intermolecular distance from 3.55 Å to 3.52 Å at the DFTB level, whereas it slightly increases from 3.88 Å to 3.94 Å at the DFT level.

Table 5. Binding energy of benzene dimers bz_2 on graphene in eV and z-separation between the graphene sheet and the benzene units parallel to it in Å. * No value is reported at the DFTB-D2-WMull level for PD as the optimization resulted in the $a_4 - cd$ structure.

$E_{tot}^{ref} =$	DFTB-D2-WMull				DFT-D3			
	T	PD *	$a_4 - cd$	SE	T	PD	$a_4 - cd$	SE
$E_{tot}^{graphene} + 2 \times E_{tot}^{bz}$	−0.586	↳	−0.929	−0.573	−0.605	−0.624	−0.938	−0.583
$E_{tot}^{graphene} + E_{tot}^{bz_2}$	−0.473	↳	-	−0.487	−0.467	−0.473	-	−0.478
$E_{tot}^{graphene@bz} + E_{tot}^{bz}$	−0.135	↳	−0.478	−0.122	−0.156	−0.175	−0.489	−0.134
$2 \times E_{tot}^{graphene@bz} - E_{tot}^{graphene}$	0.316	↳	−0.027	0.329	0.294	0.275	−0.039	0.316
Z-separation	3.03	↳	3.00	3.02/6.53	3.38	3.39/6.91	3.40/3.40	3.39/7.33

The binding energies associated with the optimized structures are computed with the following formula:

$$E_{bind}^{graphene@bz_2} = E_{tot}^{graphene@bz_2} - E_{tot}^{ref} \quad (23)$$

and reported in Table 5, making use of various choices for E_{tot}^{ref}, which is the reference potential energy. In the first line (graphene + 2bz), E_{tot}^{ref} is the energy of an optimized graphene monolayer plus that of two isolated benzene molecules. It appears that, at DFT and DFTB levels, the most stable configuration relies on the dissociation of the benzene dimer to form the $a_4 - cd$ structure. The energetic difference between the T-shaped structure and the less stable SE dimer is twice smaller (0.013 eV vs. 0.027 eV at the DFTB level and 0.022 eV vs. 0.052 eV at the DFT level) when the dimer is deposited with respect to the gas phase condition. This is probably due to the interaction between the graphene surface and the benzene molecule that is furthest from the surface, which is favored in the sandwich configuration. In the second line, E_{tot}^{ref} is the total energy of an isolated graphene sheet plus that of the optimized dimer in its T, PD or SE form, respectively. The gained energies for the non-dissociating dimers (T and SE) are similar (∼0.46–0.49 eV for DFT and DFTB values), which is only very slightly above the binding energy of a single benzene with graphene (0.45 eV for DFT and DFTB values). In the third line, E_{tot}^{ref} is the energy of a benzene deposited on a graphene sheet plus that of an isolated benzene. It differs from the isolated dimers by 0.022 eV for the T-shaped structure and 0.036 eV for the SE structure at the DFTB level and 0.018 eV and 0.028 eV at the DFT level. The larger value obtained for the SE dimer can be, again, related to the expected larger interaction energy between the graphene sheet and the second further benzene unit in the SE configuration. The last line compares the binding energies with the one of a system where two benzene molecules would be deposited in a_4 configurations without interaction between them. This configuration appears to be more stable than those corresponding to the deposition of a T-shaped or SE dimer. The negative sign obtained for the $a_4 - cd$ structure shows that the latter is the most stable investigated configuration as it maximizes the interaction between each benzene molecule and the graphene surface while maintaining some stabilizing intermolecular interactions between the two benzene units. Again, this conclusion holds at both DFT and DFTB levels. It should also be noted that the values of the interaction energies are of similar order for these two levels of calculation.

6. Conclusions

In the present paper, we have reported a new implementation of periodic boundary conditions in the DFTB code deMonNano, as only the Γ-point approximation was available in the previous version of the code. An originality of our scheme is the inclusion of atomic charge corrections, which improves the description of intermolecular coulomb interactions. It allows the recovery of a reasonable description of molecular clusters, as shown in the particular case of benzene dimers in this work. Dispersion corrections are also mandatory for a proper description of such interactions, and we have benchmarked two empirical correction schemes. One of them produces the best C-C bond distance in graphene and graphite, whereas the second one provides the best interlayer distance in graphite according to previous reference calculations, experiments and new DFT calculations performed with the DFT-D3 dispersion-corrected functional.

Benzene monomer and dimers have been optimized at the DFTB and DFT levels, providing the following similar trends. For the deposition of a single benzene monomer on a graphene sheet, the adsorption of benzene centered on top of graphene carbon atom or C-C bond leads to almost degenerated structures, which are by far more stable than the superimposition of the benzene on top of a graphene aromatic cycle. The most stable one at the DFTB level has been selected to build initial conditions for benzene dimers deposition on graphene. The structural energy gap between the most stable T-shaped dimer and less stable Sandwich-like dimer is divided by two when the cluster is supported on graphene. The supported Parallel-Displaced structure appeared to be unstable at the DFTB level, resulting in a structure where the two benzene are deposited close to each other on the graphene surface. This structure is the most stable one of our calculations at DFT and DFTB levels and also more stable than the deposition of two benzene monomer at infinite distance, which is not the case of the deposited sandwich or T-shaped dimers.

As a conclusion, we have shown the ability of the new implementation to characterize properties of molecular clusters deposited on surfaces, opening the path to dynamical simulations that allow probing the role of deposition conditions (e.g., cluster/support collision energy) and temperature on the structure and stability. This work validates the implementation of the improved sampling of the reciprocal space (**k**-points) in deMonNano, which will allow to perform very precise calculations, as needed to study properties for which a very fine structure of the Brillouin zone is required, such as electronic transport, magnetism or topological states of matter [57].

Author Contributions: Conceptualization, M.R. and N.T.; methodology, M.R. and N.T.; software, M.R. and N.T.; validation, M.R. and N.T.; formal analysis, M.R. and N.T.; investigation, M.R. and N.T.; resources, M.R. and N.T.; data curation, M.R. and N.T.; writing—original draft preparation, M.R. and N.T.; writing—review and editing, M.R. and N.T.; visualization, M.R. and N.T.; supervision, M.R. and N.T.; project administration, M.R. and N.T. All authors have read and agreed to the published version of the manuscript.

Funding: This research received no external funding.

Institutional Review Board Statement: Not applicable.

Informed Consent Statement: Not applicable.

Data Availability Statement: Data is contained within the article.

Acknowledgments: This work was granted access to the HPC resources of CALMIP supercomputing center under allocation p18009.

Conflicts of Interest: The authors declare no conflict of interest.

References

1. Sadowska, M.; Cieśla, M.; Adamczyk, Z. Nanoparticle deposition on heterogeneous surfaces: Random sequential adsorption modeling and experiments. *Colloids Surf. A Physicochem. Eng. Asp.* **2021**, *617*, 126296. [CrossRef]
2. Li, L.; Plessow, P.N.; Rieger, M.; Sauer, S.; Sánchez-Carrera, R.S.; Schaefer, A.; Abild-Pedersen, F. Modeling the migration of platinum nanoparticles on surfaces using a kinetic Monte Carlo approach. *J. Phys. Chem. C* **2017**, *121*, 4261–4269. [CrossRef]
3. Rochefort, A.; Vernisse, L.; Gómez-Herrero, A.C.; Sánchez-Sánchez, C.; Martín-Gago, J.A.; Chérioux, F.; Clair, S.; Coraux, J.; Martínez, J.I. Role of the structure and reactivity of Cu and Ag surfaces in the formation of a 2D Metal–Hexahydroxytriphenylene network. *J. Phys. Chem. C* **2021**, *125*, 17333–17341. [CrossRef]
4. Bruix, A.; Margraf, J.; Andersen, M.; Reuter, K. First-principles-based multiscale modelling of heterogeneous catalysis. *Nat. Catal.* **2019**, *2*, 659–670. [CrossRef]
5. Abidi, N.; Lim, K.R.G.; Seh, Z.W.; Steinmann, S.N. Atomistic modeling of electrocatalysis: Are we there yet? *WIREs Comput. Mol. Sci.* **2021**, *11*, e1499. [CrossRef]
6. Zhang, C.; Chen, G.; Si, Y.; Liu, M. Surface modeling of photocatalytic materials for water splitting. *Phys. Chem. Chem. Phys.* **2022**, *24*, 1237–1261. [CrossRef] [PubMed]
7. Vilan, A.; Cahen, D. Chemical modification of semiconductor surfaces for molecular electronics. *Chem. Rev.* **2017**, *117*, 4624–4666. [CrossRef] [PubMed]
8. Shahmoradi, A.; Ahangari, M.G.; Jahanshahi, M.; Mirghoreishi, M.; Fathi, E.; Mashhadzadeh, A.H. Removal of methylmercaptan pollution using Ni and Pt-decorated graphene: An ab-initio DFT study. *J. Sulfur Chem.* **2020**, *41*, 593–604. [CrossRef]
9. Li, M.; Zhu, H.; Wei, G.; He, A.; Liu, Y. DFT calculation and analysis of the gas sensing mechanism of methoxy propanol on Ag decorated SnO_2 (110) surface. *RSC Adv.* **2019**, *9*, 35862–35871. [CrossRef]
10. Soini, T.M.; Rösch, N. Size-dependent properties of transition metal clusters: From molecules to crystals and surfaces–computational studies with the program ParaGauss. *Phys. Chem. Chem. Phys.* **2015**, *17*, 28463–28483. [CrossRef] [PubMed]
11. Porezag, D.; Frauenheim, T.; Köhler, T.; Seifert, G.; Kaschner, R. Construction of tight-binding-like potentials on the basis of density-functional theory: Application to carbon. *Phys. Rev. B* **1995**, *51*, 12947–12957. [CrossRef] [PubMed]
12. Seifert, G.; Porezag, D.; Frauenheim, T. Calculations of molecules, clusters, and solids with a simplified LCAO-DFT-LDA scheme. *Int. J. Quantum Chem.* **1996**, *58*, 185–192. [CrossRef]
13. Elstner, M.; Seifert, G. Density functional tight binding. *Philos. Trans. R. Soc. A Math. Phys. Eng. Sci.* **2014**, *372*, 20120483. [CrossRef] [PubMed]
14. Spiegelman, F.; Tarrat, N.; Cuny, J.; Dontot, L.; Posenitskiy, E.; Martí, C.; Simon, A.; Rapacioli, M. Density-functional tight-binding: Basic concepts and applications to molecules and clusters. *Adv. Phys. X* **2020**, *5*, 1710252. [CrossRef] [PubMed]
15. Aradi, B.; Hourahine, B.; Frauenheim, T. DFTB+, a sparse matrix-based implementation of the DFTB method. *J. Phys. Chem. A* **2007**, *111*, 5678–5684. [CrossRef] [PubMed]
16. Koskinen, P.; Makinen, V. Density-functional tight-binding for beginners. *Comput. Mater. Sci.* **2009**, *47*, 237–253. [CrossRef]
17. Te Velde, G.; Bickelhaupt, F.M.; Baerends, E.J.; Fonseca Guerra, C.; van Gisbergen, S.J.A.; Snijders, J.G.; Ziegler, T. Chemistry with ADF. *J. Comput. Chem.* **2001**, *22*, 931–967. [CrossRef]
18. Salomon-Ferrer, R.; Case, D.A.; Walker, R.C. An overview of the Amber biomolecular simulation package. *WIREs Comput. Mol. Sci.* **2013**, *3*, 198–210. [CrossRef]
19. Walker, R.C.; Crowley, M.F.; Case, D.A. The implementation of a fast and accurate QM/MM potential method in Amber. *J. Comput. Chem.* **2008**, *29*, 1019–1031. [CrossRef] [PubMed]
20. Berendsen, H.; van der Spoel, D.; van Drunen, R. GROMACS: A message-passing parallel molecular dynamics implementation. *Comput. Phys. Commun.* **1995**, *91*, 43–56. [CrossRef]

21. Hutter, J.; Iannuzzi, M.; Schiffmann, F.; VandeVondele, J. cp2k: Atomistic simulations of condensed matter systems. *WIREs Comput. Mol. Sci.* **2014**, *4*, 15–25. [CrossRef]
22. Heine, T.; Rapacioli, M.; Patchkovskii, S.; Frenzel, J.; Koster, A.; Calaminici, P.; Duarte, H.A.; Escalante, S.; Flores-Moreno, R.; Goursot, A.; et al. deMonNano. 2009. Available online: http://demon-nano.ups-tlse.fr (accessed on 9 March 2022).
23. Peeters, E.; Hony, S.; Van Kerckhoven, C.; Tielens, A.G.G.M.; Allamandola, L.J.; Hudgins, D.M.; Bauschlicher, C.W. The rich 6 to 9 μm spectrum of interstellar PAHs. *Astron. Astrophys.* **2002**, *390*, 1089–1113. [CrossRef]
24. Rapacioli, M.; Joblin, C.; Boissel, P. Spectroscopy of polycyclic aromatic hydrocarbons and very small grains in photodissociation regions *. *Astron. Astrophys.* **2005**, *429*, 193–204. [CrossRef]
25. Podeszwa, R.; Bukowski, R.; Szalewicz, K. Potential energy surface for the benzene dimer and perturbational analysis of pi-pi interactions. *J. Phys. Chem. A* **2006**, *110*, 10345–10354. [CrossRef] [PubMed]
26. Rapacioli, M.; Spiegelman, F.; Talbi, D.; Mineva, T.; Goursot, A.; Heine, T.; Seifert, G. Correction for dispersion and coulombic interactions in molecular clusters with density functional derived methods: Application to polycyclic aromatic hydrocarbon clusters. *J. Chem. Phys.* **2009**, *130*, 244304. [CrossRef] [PubMed]
27. Elstner, M.; Porezag, D.; Jungnickel, G.; Elsner, J.; Haugk, M.; Frauenheim, T.; Suhai, S.; Seifert, G. Self-consistent-charge density-functional tight-binding method for simulations of complex materials properties. *Phys. Rev. B* **1998**, *58*, 7260. [CrossRef]
28. Yang, Y.; Yu, H.; York, D.; Cui, Q.; Elstner, M. Extension of the self-consistent-charge density-functional tight-binding method: Third-order expansion of the density functional theory total energy and introduction of a modified effective coulomb interaction. *J. Phys. Chem. A* **2007**, *111*, 10861–10873. [CrossRef] [PubMed]
29. Li, J.; Zhu, T.; Cramer, C.; Truhlar, D. New class IV charge model for extracting accurate partial charges from wave functions. *J. Phys. Chem. A* **1998**, *102*, 1820–1831. [CrossRef]
30. Michoulier, E.; Ben Amor, N.; Rapacioli, M.; Noble, J.A.; Mascetti, J.; Toubin, C.; Simon, A. Theoretical determination of adsorption and ionisation energies of polycyclic aromatic hydrocarbons on water ice. *Phys. Chem. Chem. Phys.* **2018**, *20*, 11941–11953. [CrossRef]
31. Dontot, L.; Spiegelman, F.; Zamith, S.; Rapacioli, M. Dependence upon charge of the vibrational spectra of small Polycyclic Aromatic Hydrocarbon clusters: The example of pyrene. *Eur. Phys. J. D* **2020**, *74*, 216. [CrossRef]
32. Simon, A.; Rapacioli, M.; Michoulier, E.; Zheng, L.; Korchagina, K.; Cuny, J. Contribution of the density-functional-based tight-binding scheme to the description of water clusters: Methods, applications and extension to bulk systems. *Mol. Simul.* **2019**, *45*, 249–268. [CrossRef]
33. Slater, J.C.; Koster, G.F. Simplified LCAO method for the meriodic motential mroblem. *Phys. Rev.* **1954**, *94*, 1498–1524. [CrossRef]
34. Zhechkov, L.; Heine, T.; Patchkovskii, S.; Seifert, G.; Duarte, H.A. An efficient a posteriori treatment for dispersion interaction in density-functional-based tight binding. *J. Chem. Theory Comput.* **2005**, *1*, 841–847. [CrossRef] [PubMed]
35. Kresse, G.; Hafner, J. Norm-conserving and ultrasoft pseudopotentials for first-row and transition elements. *J. Phys. Rev. B* **1993**, *47*, 558–561. [CrossRef] [PubMed]
36. Kresse, G.; Furthmüller, J. Efficiency of ab-initio total energy calculations for metals and semiconductors using a plane-wave basis set. *Comput. Mater. Sci.* **1996**, *6*, 15–50. [CrossRef]
37. Kresse, G.; Furthmüller, J. Efficient iterative schemes for Ab Initio Total-Energy Calc. Using A Plane-Wave Basis Set. *Phys. Rev. B* **1996**, *54*, 11169–11186. [CrossRef] [PubMed]
38. Blöchl, P.E. Projector augmented-wave method. *Phys. Rev. B* **1994**, *50*, 17953–17979. [CrossRef] [PubMed]
39. Kresse, G.; Joubert, D. From ultrasoft pseudopotentials to the projector augmented-wave method. *Phys. Rev. B* **1999**, *59*, 1758–1775. [CrossRef]
40. Grimme, S.; Antony, J.; Ehrlich, S.; Krieg, H. A consistent and accurate ab initio parametrization of density functional dispersion correction (DFT-D) for the 94 elements H-Pu. *J. Chem. Phys.* **2010**, *132*, 154104. [CrossRef]
41. Lebedeva, I.V.; Lebedev, A.V.; Popov, A.M.; Knizhnik, A.A. Comparison of performance of van der Waals-corrected exchange-correlation functionals for interlayer interaction in graphene and hexagonal boron nitride. *Comput. Mater. Sci.* **2017**, *128*, 45–58. [CrossRef]
42. Methfessel, M.; Paxton, A.T. High-precision sampling for Brillouin-zone integration in metals. *Phys. Rev. B* **1989**, *40*, 3616–3621. [CrossRef] [PubMed]
43. Yu, Y.X. Can all nitrogen-doped defects improve the performance of graphene anode materials for lithium-ion batteries? *Phys. Chem. Chem. Phys.* **2013**, *15*, 16819–16827. [CrossRef] [PubMed]
44. Lebègue, S.; Harl, J.; Gould, T.; Ángyán, J.G.; Kresse, G.; Dobson, J.F. Cohesive properties and asymptotics of the dispersion interaction in graphite by the random phase approximation. *Phys. Rev. Lett.* **2010**, *105*, 196401.. [CrossRef] [PubMed]
45. Spanu, L.; Sorella, S.; Galli, G. Nature and strength of interlayer binding in graphite. *Phys. Rev. Lett.* **2009**, *103*, 196401. [CrossRef] [PubMed]
46. Cooper, D.R.; D'Anjou, B.; Ghattamaneni, N.; Harack, B.; Hilke, M.; Horth, A.; Majlis, N.; Massicotte, M.; Vandsburger, L.; Whiteway, E.; et al. Experimental review of graphene. *ISRN Condens. Matter Phys.* **2012**, *2012*, 501686. [CrossRef]
47. Bosak, A.; Krisch, M.; Mohr, M.; Maultzsch, J.; Thomsen, C. Elasticity of single-crystalline graphite: Inelastic x-ray scattering study. *Phys. Rev. B* **2007**, *75*, 153408. [CrossRef]
48. Trucano, P.; Chen, R. Structure of graphite by neutron diffraction. *Nature* **1975**, *258*, 136–137. [CrossRef]

49. Zhang, Y.H.; Zhou, K.G.; Xie, K.F.; Zeng, J.; Zhang, H.L.; Peng, Y. Tuning the electronic structure and transport properties of graphene by noncovalent functionalization: Effects of organic donor, acceptor and metal atoms. *Nanotechnology* **2010**, *21*, 065201. [CrossRef]
50. Berland, K.; Hyldgaard, P. Analysis of van der Waals density functional components: Binding and corrugation of benzene and C_{60} on boron nitride and graphene. *Phys. Rev. B* **2013**, *87*, 205421. [CrossRef]
51. Ershova, O.V.; Lillestolen, T.C.; Bichoutskaia, E. Study of polycyclic aromatic hydrocarbons adsorbed on graphene using density functional theory with empirical dispersion correction. *Phys. Chem. Chem. Phys.* **2010**, *12*, 6483–6491. [CrossRef]
52. Otyepková, E.; Lazar, P.; Čépe, K.; Tomanec, O.; Otyepka, M. Organic adsorbates have higher affinities to fluorographene than to graphene. *Appl. Mater. Today* **2016**, *5*, 142–149. [CrossRef]
53. Zacharia, R.; Ulbricht, H.; Hertel, T. Interlayer cohesive energy of graphite from thermal desorption of polyaromatic hydrocarbons. *Phys. Rev. B* **2004**, *69*, 155406. [CrossRef]
54. Rajesh, C.; Majumder, C.; Mizuseki, H.; Kawazoe, Y. A theoretical study on the interaction of aromatic amino acids with graphene and single walled carbon nanotube. *J. Chem. Phys.* **2009**, *130*, 124911. [CrossRef] [PubMed]
55. Yu, Y.X. A dispersion-corrected DFT study on adsorption of battery active materials anthraquinone and its derivatives on monolayer graphene and h-BN. *J. Mater. Chem. A* **2014**, *2*, 8910–8917. [CrossRef]
56. Yu, Y.X. Binding energy and work function of organic electrode materials phenanthraquinone, pyromellitic dianhydride and their derivatives adsorbed on graphene. *ACS Appl. Mater. Interfaces* **2014**, *6*, 16267–16275. [CrossRef] [PubMed]
57. Kratzer, P.; Neugebauer, J. The basics of electronic structure theory for periodic systems. *Front. Chem.* **2019**, *7*, 106. [CrossRef] [PubMed]

computation

MDPI

Article

Local Potential Functional Embedding Theory: A Self-Consistent Flavor of Density Functional Theory for Lattices without Density Functionals

Sajanthan Sekaran [1,*], Matthieu Saubanère [2] and Emmanuel Fromager [1]

1 Laboratoire de Chimie Quantique, Institut de Chimie, CNRS/Université de Strasbourg, 4 Rue Blaise Pascal, 67000 Strasbourg, France; fromagere@unistra.fr

2 ICGM, Université de Montpellier, CNRS, ENSCM, 34000 Montpellier, France; matthieu.saubanere@umontpellier.fr

* Correspondence: s.sekaran@unistra.fr

Abstract: Quantum embedding is a divide and conquer strategy that aims at solving the electronic Schrödinger equation of sizeable molecules or extended systems. We establish in the present work a clearer and in-principle-exact connection between density matrix embedding theory (DMET) and density-functional theory (DFT) within the simple but nontrivial one-dimensional Hubbard model. For that purpose, we use our recent reformulation of single-impurity DMET as a Householder transformed density-matrix functional embedding theory (Ht-DMFET). On the basis of well-identified density-functional approximations, a self-consistent local potential functional embedding theory (LPFET) is formulated and implemented. Combining both LPFET and DMET numerical results with our formally exact density-functional embedding theory reveals that a single statically embedded impurity can in principle describe the density-driven Mott–Hubbard transition, provided that a complementary density-functional correlation potential (which is neglected in both DMET and LPFET) exhibits a derivative discontinuity (DD) at half filling. The extension of LPFET to multiple impurities (which would enable to circumvent the modeling of DDs) and its generalization to quantum chemical Hamiltonians are left for future work.

Keywords: density matrix functional embedding; density-functional theory; householder transformation

Citation: Sekaran, S.; Saubanère, M.; Fromager, E. Local Potential Functional Embedding Theory: A Self-Consistent Flavor of Density Functional Theory for Lattices without Density Functionals. *Computation* **2022**, *10*, 45. https://doi.org/10.3390/computation10030045

Academic Editor: Henry Chermette

Received: 15 February 2022
Accepted: 14 March 2022
Published: 18 March 2022

1. Introduction

Kohn–Sham density-functional theory (KS-DFT) [1] has become over the last two decades the method of choice for computational chemistry and physics studies, essentially because it often provides a relatively accurate description of the electronic structure of large molecular or extended systems at a low computational cost. The major simplification of the electronic structure problem in KS-DFT lies in the fact that the ground-state energy is evaluated, in principle exactly, from a non-interacting single-configuration wave function, which is simply referred to as the KS determinant. The latter is obviously not the exact solution to the Schrödinger equation. However, its density matches the exact interacting ground-state density, so that the Hartree-exchange-correlation (Hxc) energy of the physical system, which is induced by the electronic repulsion, can be recovered from an appropriate (in principle exact and universal) Hxc density functional. Despite the success of KS-DFT, standard density-functional approximations still fail in describing strongly correlated electrons. To overcome this issue, various strategies have been explored and improved over the years, both in condensed matter physics [2–7] and quantum chemistry [8]. Note that, in the latter case, in-principle-exact multi-determinantal extensions of DFT based on the adiabatic connection formalism have been developed [9–12]. In these approaches, the KS system is only referred to in the design of density-functional approximations. In practice, a single (partially-interacting) many-body wave function is calculated self-consistently and the complement to the partial interaction energy is described with an appropriate density functional (which differs from the conventional xc one). In other words, there is no KS construction in the actual calculation. Some of these concepts have been reused in the study of model lattice Hamiltonians [12,13]. A similar strategy will be adopted in the present work, with an important difference though. The *reduced-in-size* correlated density-functional many-body wave function that we will introduce will be extracted from a quantum embedding theory where the KS determinant of the full system is a key ingredient that must be evaluated explicitly.

Quantum embedding theory [14] is at first sight a completely different approach to the strong electron correlation problem. Interestingly, some of its implementations, like the *density matrix embedding theory* (DMET) [15–23], rely on a reference Slater determinant that is computed for the full system. This is also the case in practical embedding calculations based on the exact factorization formalism [24,25]. Unlike the well-established *dynamical mean-field theory* (DMFT) [26–30], which relies on the one-electron Green's function, DMET is a static theory of ground electronic states. Most importantly, the bath, in which a fragment of the original system (referred to as impurity when it is a single localized orbital) is embedded, is drastically reduced in size in DMET. As a result, the "impurity+bath" embedding cluster can be accurately (if not exactly) described with wave function-based quantum chemical methods. The authors have shown recently that the Schmidt decomposition of the reference Slater determinant, which is central in DMET, can be recast into a (one-electron reduced) density-matrix functional Householder transformation [31], which is much simpler to implement. This approach, in which the bath orbitals can in principle be correlated directly through the density matrix [31], is referred to as *Householder transformed density matrix functional embedding theory* (Ht-DMFET). Since the seminal work of Knizia and Chan on DMET [15], various connections with DMFT and related approaches have been established [32–37]. Connections with DFT have been less explored, and only at the approximate level of theory. We can refer to the *density embedding theory* (DET) of Bulik et al. [38], which is a simplified version of DMET where only the diagonal elements of the embedded density matrix are mapped onto the reference Slater determinant of the full system. More recently, Senjean [39] combined DFT for lattices [40,41] with DMET, and Mordovina et al. [42] (see also Ref. [43]) proposed a *self-consistent density-functional embedding* (SDE), where the KS determinant is explicitly used as the reference wave function in the DMET algorithm.

In the present work, an in-principle-exact combination of KS-DFT with DMET is derived for the one-dimensional (1D) Hubbard lattice, as a proof of concept. For that purpose, we use the density-matrix functional Householder transformation introduced recently by the authors [31]. On the basis of well-identified density-functional approximations, we propose and implement a *local potential functional embedding theory* (LPFET) where the Hxc potential is evaluated self-consistently in the lattice by "learning" from the embedding cluster at each iteration of the optimization process. LPFET can be seen as a flavor of KS-DFT where no density functional is actually used.

The paper is organized as follows. After a short introduction to the 1D Hubbard model in Section 2.1, a detailed review of Ht-DMFET is presented in Section 2.2, for clarity and completeness. An exact density-functional reformulation of the theory is then proposed in Section 2.3. The resulting approximate LPFET and its comparison with SDE are detailed in Sections 2.4 and 2.5, respectively. The LPFET algorithm is summarized in Section 3. Results obtained for a 1000-site Hubbard ring are presented and discussed in Section 4. The conclusion and perspectives are finally given in Section 5.

2. Theory

2.1. One-Dimensional Hubbard Lattice

By analogy with Ref. [31], various quantum embedding strategies will be discussed in the following within the simple but nontrivial uniform 1D Hubbard model. The corresponding lattice Hamiltonian (for a L-site ring) reads as

$$\hat{H} = \hat{T} + \hat{U} + v_{\text{ext}}\hat{N}, \tag{1}$$

where the hopping operator (written in second quantization),

$$\hat{T} = -t \sum_{i=0}^{L-1} \sum_{\sigma=\uparrow,\downarrow} \left(\hat{c}^{\dagger}_{i\sigma}\hat{c}_{(i+1)\sigma} + \hat{c}^{\dagger}_{(i+1)\sigma}\hat{c}_{i\sigma} \right), \tag{2}$$

with parameter t, is the analog for lattices of the kinetic energy operator. For convenience, we will systematically use periodic boundary conditions, i.e., $\hat{c}^{\dagger}_{L\sigma} \equiv \hat{c}^{\dagger}_{0\sigma}$. On-site repulsions are only taken into account in the two-electron repulsion operator $\hat{U}$, i.e.,

$$\hat{U} = \sum_{i=0}^{L-1} \hat{U}_i, \tag{3}$$

where $\hat{U}_i = U\hat{n}_{i\uparrow}\hat{n}_{i\downarrow}$, U is the parameter that controls the strength of the interaction, and $\hat{n}_{i\sigma} = \hat{c}^{\dagger}_{i\sigma}\hat{c}_{i\sigma}$ is a site occupation operator for spin σ. Since the lattice is uniform, the local external potential (which would correspond to the nuclear potential in a conventional quantum chemical calculation) operator is proportional to the electron counting operator (see the last term on the right-hand side of Equation (1)),

$$\hat{N} = \sum_{i=0}^{L-1} \sum_{\sigma=\uparrow,\downarrow} \hat{n}_{i\sigma}. \tag{4}$$

The uniform value of the external potential can be rewritten as

$$v_{\text{ext}} = -\mu, \tag{5}$$

where the chemical potential μ controls the number of electrons N or, equivalently, the uniform density $n = N/L$ in the lattice. In this case, $\hat{H}$ is actually a (zero-temperature) grand canonical Hamiltonian. For convenience, we rewrite the hopping operator as follows,

$$\hat{T} \equiv \sum_{i,j=0}^{L-1} \sum_{\sigma=\uparrow,\downarrow} t_{ij}\hat{c}^{\dagger}_{i\sigma}\hat{c}_{j\sigma}, \tag{6}$$

where

$$t_{ij} = -t\left(\delta_{j(i+1)} + \delta_{i(j+1)}\right), \tag{7}$$

and $t_{(L-1)0} = t_{0(L-1)} = -t$. From now on the bounds in the summations over the full lattice will be dropped, for simplicity:

$$\sum_i \equiv \sum_{i=0}^{L-1}. \tag{8}$$

Note that the quantum embedding strategies discussed in the present work can be extended to more general (quantum chemical, in particular) Hamiltonians [20]. For that purpose, the true *ab initio* Hamiltonian should be written in a localized molecular orbital basis, thus leading to the more general Hamiltonian expression,

$$\hat{H} = \sum_{\sigma}\sum_{ij} h_{ij}\hat{c}^{\dagger}_{i\sigma}\hat{c}_{j\sigma} + \frac{1}{2}\sum_{\sigma,\sigma'}\sum_{ijkl}\langle ij|kl\rangle \hat{c}^{\dagger}_{i\sigma}\hat{c}^{\dagger}_{j\sigma'}\hat{c}_{l\sigma'}\hat{c}_{k\sigma}, \tag{9}$$

where h_{ij} and $\langle ij|kl\rangle$ are the (kinetic and nuclear attraction) one-electron and two-electron repulsion integrals, respectively. Using a localized orbital basis allows for the decomposition of the molecule under study into fragments that can be embedded afterward [20]. In the following, we will work with the simpler Hamiltonian of Equation (1), as a proof of concept.

2.2. *Review of Ht-DMFET*

For the sake of clarity and completeness, a review of Ht-DMFET [31] is presented in the following subsections. Various ingredients (operators and reduced quantities) that will be used later on in Section 2.3 in the derivation of a formally exact density-functional embedding theory (which is the main outcome of this work) are introduced. Real algebra will be used. For simplicity, we focus on the embedding of a single impurity. A multiple-impurity extension of the theory can be obtained from a block Householder transformation [31,44]. Unlike in the exact reformulation of the theory which is proposed in the following Section 2.3 and where the chemical potential μ controls the density of the uniform lattice, the total number of electrons will be *fixed* to the value N in the present section. In other words, the uniform density is set to $n = N/L$ and μ is an arbitrary constant (that could be set to zero).

2.2.1. Exact Non-Interacting Embedding

Let us first consider the particular case of a non-interacting ($U = 0$) lattice for which Ht-DMFET is exact [31]. As it will be applied later on (in Section 2.3) to the auxiliary KS lattice, it is important to highlight the key features of the non-interacting embedding. Following Ref. [31], we label as $i = 0$ one of the localized (lattice site in the present case) spin-orbital $|\chi^{\sigma}_0\rangle \equiv \hat{c}^{\dagger}_{0\sigma}|\text{vac}\rangle$ [we denote $|\text{vac}\rangle$ the vacuum state of second quantization] that, ultimately, will become the so-called *embedded impurity*. The ingredient that is central in Ht-DMFET is the (one-electron reduced) density matrix of the full system in the lattice representation, i.e.,

$$\gamma^{\uparrow} = \gamma^{\downarrow} = \gamma \equiv \gamma_{ij} = \langle\Phi|\hat{c}^{\dagger}_{i\sigma}\hat{c}_{j\sigma}|\Phi\rangle, \tag{10}$$

where we restrict ourselves to closed-shell singlet ground states $|\Phi\rangle$, for simplicity. Note that

$$\gamma_{00} = \frac{n}{2} = \frac{N}{2L} \tag{11}$$

is the uniform lattice filling per spin. Since the full lattice will always be described with a single Slater determinant in the following, the density matrix γ will always be *idempotent*. The latter is used to construct the Householder unitary transformation which, once it has been applied to the one-electron lattice space, defines the so-called *bath* spin-orbital with which the impurity will ultimately be exclusively entangled. More explicitly, the Householder transformation matrix

$$\mathbf{P} = \mathbf{I} - 2\mathbf{v}\mathbf{v}^{\dagger} \equiv P_{ij} = \delta_{ij} - 2\text{v}_i\text{v}_j, \tag{12}$$

where $\mathbf{I}$ is the identity matrix, is a functional of the density matrix, i.e.,

$$\mathbf{P} \equiv \mathbf{P}[\gamma], \tag{13}$$

where the density-matrix-functional Householder vector components read as [31]

$$\mathrm{v}_0 = 0, \tag{14}$$

$$\mathrm{v}_1 = \frac{\gamma_{10} - \tilde{\gamma}_{10}}{\sqrt{2\tilde{\gamma}_{10}(\tilde{\gamma}_{10} - \gamma_{10})}}, \tag{15}$$

$$\mathrm{v}_i \underset{i\geq 2}{=} \frac{\gamma_{i0}}{\sqrt{2\tilde{\gamma}_{10}(\tilde{\gamma}_{10} - \gamma_{10})}}, \tag{16}$$

with

$$\tilde{\gamma}_{10} = -\mathrm{sgn}(\gamma_{10})\sqrt{\sum_{j>0} \gamma_{j0}^2}, \tag{17}$$

and

$$\mathbf{v}^\dagger \mathbf{v} = \sum_{i\geq 1} \mathrm{v}_i^2 = 1. \tag{18}$$

Note that, in the extreme case of a two-site lattice, the denominator in Equations (15) and (16) is still well defined and it does not vanish. Indeed, by construction (see Equation (17)),

$$\tilde{\gamma}_{10} \underset{\left\{\gamma_{j0} \overset{j>1}{=} 0\right\}}{=} -\mathrm{sgn}(\gamma_{10})|\gamma_{10}| = -\gamma_{10} \tag{19}$$

in this case, thus leading to $\tilde{\gamma}_{10}(\tilde{\gamma}_{10} - \gamma_{10}) = 2\gamma_{10}^2 > 0$. Note also that $\boldsymbol{P}$ is hermitian and unitary, i.e., $\boldsymbol{P} = \boldsymbol{P}^\dagger$ and

$$\boldsymbol{P}^2 = \boldsymbol{P}\boldsymbol{P}^\dagger = \boldsymbol{P}^\dagger \boldsymbol{P} = \boldsymbol{I}. \tag{20}$$

The bath spin-orbital $|\varphi^\sigma_{\mathrm{bath}}\rangle$ is then constructed as follows in second quantization,

$$|\varphi^\sigma_{\mathrm{bath}}\rangle := \hat{d}^\dagger_{1\sigma}|\mathrm{vac}\rangle, \tag{21}$$

where, according to Equations (12) and (14),

$$\begin{aligned} \hat{d}^\dagger_{1\sigma} &= \sum_k P_{1k}\hat{c}^\dagger_{k\sigma} \\ &= \hat{c}^\dagger_{1\sigma} - 2\mathrm{v}_1 \sum_{k\geq 1} \mathrm{v}_k \hat{c}^\dagger_{k\sigma}. \end{aligned} \tag{22}$$

More generally, the entire lattice space can be Householder-transformed as follows,

$$\hat{d}^\dagger_{i\sigma} \underset{0\leq i\leq L-1}{=} \sum_k P_{ik}\hat{c}^\dagger_{k\sigma}, \tag{23}$$

and the back transformation simply reads as

$$\sum_i P_{li}\hat{d}^\dagger_{i\sigma} = \sum_{ik} P_{li}P_{ik}\hat{c}^\dagger_{k\sigma} = \sum_k \left[\boldsymbol{P}^2\right]_{lk}\hat{c}^\dagger_{k\sigma} = \hat{c}^\dagger_{l\sigma}. \tag{24}$$

We stress that the impurity is invariant under the Householder transformation, i.e.,

$$\hat{d}^\dagger_{0\sigma} = \hat{c}^\dagger_{0\sigma}, \tag{25}$$

and, according to the Appendix A, the Householder-transformed density matrix elements involving the impurity can be simplified as follows,

$$\langle\Phi|\hat{d}^{\dagger}_{j\sigma}\hat{d}_{0\sigma}|\Phi\rangle = \gamma_{j0} - \mathrm{v}_j\sqrt{2\tilde{\gamma}_{10}(\tilde{\gamma}_{10} - \gamma_{10})}. \tag{26}$$

As readily seen from Equations (15) and (26), the matrix element $\tilde{\gamma}_{10}$ introduced in Equation (17) is in fact the bath-impurity element of the density matrix in the Householder representation:

$$\langle\Phi|\hat{d}^{\dagger}_{1\sigma}\hat{d}_{0\sigma}|\Phi\rangle = \tilde{\gamma}_{10}. \tag{27}$$

If we denote

$$\tilde{\gamma} \equiv \tilde{\gamma}_{ij} = \langle\Phi|\hat{d}^{\dagger}_{i\sigma}\hat{d}_{j\sigma}|\Phi\rangle = \sum_{kl} P_{ik}\gamma_{kl}P_{lj} \equiv \boldsymbol{P}\gamma\boldsymbol{P} \tag{28}$$

the full Householder-transformed density matrix, we do readily see from Equations (16) and (26) that the impurity is exclusively entangled with the bath, i.e.,

$$\tilde{\gamma}_{i0} \underset{i\geq 2}{=} 0, \tag{29}$$

by construction [31]. As $\tilde{\gamma}$ inherits the idempotency of γ through the unitary Householder transformation, we deduce from Equation (29) that

$$\tilde{\gamma}_{i0} = \left[\tilde{\gamma}^2\right]_{i0} = \sum_j \tilde{\gamma}_{ij}\tilde{\gamma}_{j0} = \tilde{\gamma}_{i0}\tilde{\gamma}_{00} + \tilde{\gamma}_{i1}\tilde{\gamma}_{10}, \tag{30}$$

or, equivalently,

$$\tilde{\gamma}_{i1} = \frac{\tilde{\gamma}_{i0}(1 - \tilde{\gamma}_{00})}{\tilde{\gamma}_{10}}, \tag{31}$$

thus leading to (see Equation (29))

$$\tilde{\gamma}_{i1} \underset{i\geq 2}{=} 0, \tag{32}$$

and

$$\tilde{\gamma}_{00} + \tilde{\gamma}_{11} = 1. \tag{33}$$

Equations (32) and (33) simply indicate that, by construction [31], the bath is itself entangled exclusively with the impurity, and the Householder "impurity+bath" cluster, which is disconnected from its environment, contains exactly two electrons (one per spin). Therefore, the Householder cluster sector of the density matrix can be described exactly by a *two-electron* Slater determinant $\Phi^{\mathcal{C}}$:

$$\tilde{\gamma}_{ij} \underset{0\leq i,j\leq 1}{=} \left\langle \Phi^{\mathcal{C}}\middle|\hat{d}^{\dagger}_{i\sigma}\hat{d}_{j\sigma}\middle|\Phi^{\mathcal{C}}\right\rangle. \tag{34}$$

Note that, in the Householder representation, the lattice ground-state determinant reads as $\Phi \equiv \Phi^{\mathcal{C}}\Phi_{\text{core}}$, where the cluster's determinant $\Phi^{\mathcal{C}}$ is disentangled from the core one Φ_{core}. Once the cluster's block of the density matrix has been diagonalized, we obtain the sole occupied orbital that overlaps with the impurity, exactly like in DMET [20]. In other words, for non-interacting (or mean-field-like descriptions of) electrons, the Ht-DMFET construction of the bath is equivalent (although simpler) to that of DMET. We refer the reader to Ref. [31] for a more detailed comparison of the two approaches.

2.2.2. Non-Interacting Embedding Hamiltonian

As the Householder cluster is strictly disconnected from its environment in the non-interacting case, it is exactly described by the two-electron ground state $\left|\Phi^{\mathcal{C}}\right\rangle$ of the Householder-transformed hopping operator (that we refer to as kinetic energy operator from now on, like in DFT for lattices [13,41]) on projected onto the cluster [31], i.e.,

$$\hat{\mathcal{T}}^{\mathcal{C}}\left|\Phi^{\mathcal{C}}\right\rangle = \mathcal{E}_{\rm s}^{\mathcal{C}}\left|\Phi^{\mathcal{C}}\right\rangle, \tag{35}$$

where, according to Equations (6) and (24),

$$\hat{\mathcal{T}}^{\mathcal{C}} = \sum_{ij}\sum_{\sigma=\uparrow,\downarrow} t_{ij}\sum_{k,l=0}^{1} P_{ik}P_{jl}\hat{d}^{\dagger}_{k\sigma}\hat{d}_{l\sigma}. \tag{36}$$

For convenience, we will separate in $\hat{\mathcal{T}}^{\mathcal{C}}$ the physical per-site kinetic energy operator (see Equation (2)),

$$\hat{t}_{01} = -t\sum_{\sigma=\uparrow,\downarrow}\left(\hat{c}^{\dagger}_{0\sigma}\hat{c}_{1\sigma} + \hat{c}^{\dagger}_{1\sigma}\hat{c}_{0\sigma}\right), \tag{37}$$

from the correction induced (within the cluster) by the Householder transformation:

$$\hat{\tau}^{\mathcal{C}} = \hat{\mathcal{T}}^{\mathcal{C}} - \hat{t}_{01}. \tag{38}$$

Note that, since $t_{00} = 0$, $\hat{\tau}^{\mathcal{C}}$ can be expressed more explicitly as follows,

$$\begin{aligned}
\hat{\tau}^{\mathcal{C}} &= \sum_{\sigma=\uparrow,\downarrow}\left(\sum_{ij} P_{i1}P_{j0}t_{ij}\right)\left[\hat{d}^{\dagger}_{0\sigma}\hat{d}_{1\sigma} + \hat{d}^{\dagger}_{1\sigma}\hat{d}_{0\sigma}\right] \\
&\quad + \sum_{\sigma=\uparrow,\downarrow}\left(\sum_{ij} P_{i1}P_{j1}t_{ij}\right)\hat{d}^{\dagger}_{1\sigma}\hat{d}_{1\sigma} - \hat{t}_{01} \\
&= \sum_{\sigma=\uparrow,\downarrow}\left(\sum_{i} P_{i1}t_{i0}\right)\left[\hat{c}^{\dagger}_{0\sigma}\hat{d}_{1\sigma} + \hat{d}^{\dagger}_{1\sigma}\hat{c}_{0\sigma}\right] \\
&\quad + \sum_{\sigma=\uparrow,\downarrow}\left(\sum_{ij} P_{i1}P_{j1}t_{ij}\right)\hat{d}^{\dagger}_{1\sigma}\hat{d}_{1\sigma} - \hat{t}_{01} \\
&= \sum_{\sigma=\uparrow,\downarrow} t_{10}\left[\hat{c}^{\dagger}_{0\sigma}\hat{d}_{1\sigma} + \hat{d}^{\dagger}_{1\sigma}\hat{c}_{0\sigma}\right] \\
&\quad - 2\mathrm{v}_1\sum_{\sigma=\uparrow,\downarrow}\left(\sum_{i}\mathrm{v}_i t_{i0}\right)\left[\hat{c}^{\dagger}_{0\sigma}\hat{d}_{1\sigma} + \hat{d}^{\dagger}_{1\sigma}\hat{c}_{0\sigma}\right] \\
&\quad + \sum_{\sigma=\uparrow,\downarrow}\left(\sum_{ij} P_{i1}P_{j1}t_{ij}\right)\hat{d}^{\dagger}_{1\sigma}\hat{d}_{1\sigma} - \hat{t}_{01},
\end{aligned} \tag{39}$$

thus leading to

$$\begin{aligned}
\hat{\tau}^{\mathcal{C}} &= 2t\mathrm{v}_1\sum_{\sigma=\uparrow,\downarrow}\sum_{k\geq 1}\mathrm{v}_k\left[\hat{c}^{\dagger}_{0\sigma}\hat{c}_{k\sigma} + \hat{c}^{\dagger}_{k\sigma}\hat{c}_{0\sigma}\right] \\
&\quad - 2\mathrm{v}_1\sum_{\sigma=\uparrow,\downarrow}\left(\sum_{i}\mathrm{v}_i t_{i0}\right)\left[\hat{c}^{\dagger}_{0\sigma}\hat{d}_{1\sigma} + \hat{d}^{\dagger}_{1\sigma}\hat{c}_{0\sigma}\right] \\
&\quad + 4\left(\sum_{ij}\mathrm{v}_i\mathrm{v}_j\left(\mathrm{v}_1^2 - \delta_{j1}\right)t_{ij}\right)\sum_{\sigma=\uparrow,\downarrow}\hat{d}^{\dagger}_{1\sigma}\hat{d}_{1\sigma},
\end{aligned} \tag{40}$$

where we used Equations (12) and (22), as well as the fact that $t_{11} = 0$ and $t_{10} = -t$. Note that, when no Householder transformation is performed (i.e., when $\mathrm{v}_i = 0$ for $0 \leq i \leq L-1$), the bath site simply corresponds to the nearest neighbor ($i = 1$) of the impurity in the lattice (see Equation (22)) and, as readily seen from Equations (38) and (40), the non-interacting cluster's Hamiltonian $\hat{\mathcal{T}}^{\mathcal{C}}$ reduces to $\hat{t}_{01}$.

Unlike in the interacting case, which is discussed in Section 2.2.3, it is unnecessary to introduce an additional potential on the embedded impurity in order to ensure that it reproduces the correct lattice filling. Indeed, according to Equations (11), (14), (26), (28) and (34),

$$\left\langle \Phi^{\mathcal{C}} \middle| \hat{c}^{\dagger}_{0\sigma}\hat{c}_{0\sigma} \middle| \Phi^{\mathcal{C}} \right\rangle = \left\langle \Phi^{\mathcal{C}} \middle| \hat{d}^{\dagger}_{0\sigma}\hat{d}_{0\sigma} \middle| \Phi^{\mathcal{C}} \right\rangle = n/2. \tag{41}$$

This constraint is automatically fulfilled when Householder transforming the kinetic energy operator $\hat{T}$ of the full lattice, thanks to the local potential contribution on the bath (see the last term on the right-hand side of Equation (40)). Interestingly, the true (non-interacting in this case) per-site energy of the lattice can be determined solely from $\Phi^{\mathcal{C}}$. Indeed, according to Equation (10), the per-site kinetic energy can be evaluated from the lattice ground-state wave function Φ as follows,

$$\langle \Phi | \hat{t}_{01} | \Phi \rangle = -4t\gamma_{10}. \tag{42}$$

When rewritten in the Householder representation, Equation (42) gives (see Equations (24), (29) and (34))

$$\begin{aligned} \langle \Phi | \hat{t}_{01} | \Phi \rangle &= -4t \sum_i P_{1i}\tilde{\gamma}_{i0} \\ &= -4t \sum_{0 \leq i \leq 1} P_{1i}\tilde{\gamma}_{i0} \\ &= -4t \sum_{0 \leq i \leq 1} P_{1i} \left\langle \Phi^{\mathcal{C}} \middle| \hat{d}^{\dagger}_{i\sigma}\hat{d}_{0\sigma} \middle| \Phi^{\mathcal{C}} \right\rangle \\ &= -4t \sum_i P_{1i} \left\langle \Phi^{\mathcal{C}} \middle| \hat{d}^{\dagger}_{i\sigma}\hat{c}_{0\sigma} \middle| \Phi^{\mathcal{C}} \right\rangle, \end{aligned} \tag{43}$$

where we used Equation (25) and the fact that $\hat{d}_{i\sigma} \left| \Phi^{\mathcal{C}} \right\rangle \overset{i \geq 1}{=} 0$, since $\Phi^{\mathcal{C}}$ is constructed within the cluster. We finally recover from Equation (43) the following equality [31],

$$\begin{aligned} \langle \Phi | \hat{t}_{01} | \Phi \rangle &= -4t \left\langle \Phi^{\mathcal{C}} \middle| \hat{c}^{\dagger}_{1\sigma}\hat{c}_{0\sigma} \middle| \Phi^{\mathcal{C}} \right\rangle \\ &= \left\langle \Phi^{\mathcal{C}} \middle| \hat{t}_{01} \middle| \Phi^{\mathcal{C}} \right\rangle, \end{aligned} \tag{44}$$

which drastically (and exactly) simplifies the evaluation of non-interacting energies for lattices.

2.2.3. Approximate Interacting Embedding

The simplest (approximate) extension of Ht-DMFET to interacting electrons consists in introducing the on-impurity-site two-electron repulsion operator $\hat{U}_0$ into the non-interacting Householder cluster's Hamiltonian of Equation (35), by analogy with DMET [15,31]. In such a (standard) scheme, the interaction is treated *on top* of the non-interacting embedding. Unlike in the non-interacting case, it is necessary to introduce a chemical potential $\tilde{\mu}^{\mathrm{imp}}$ on the embedded impurity in order to ensure that it reproduces the correct lattice filling N/L [31], i.e.,

$$\langle \hat{n}_0 \rangle_{\Psi^{\mathcal{C}}} = N/L, \tag{45}$$

where the two-electron cluster's ground-state wave function $\Psi^{\mathcal{C}}$ fulfills the following interacting Schrödinger equation:

$$\left(\hat{\mathcal{T}}^{\mathcal{C}} + \hat{U}_0 - \tilde{\mu}^{\text{imp}}\hat{n}_0\right)\left|\Psi^{\mathcal{C}}\right\rangle = \mathcal{E}^{\mathcal{C}}\left|\Psi^{\mathcal{C}}\right\rangle. \tag{46}$$

The physical per-site energy (from which we remove the chemical potential contribution) is then evaluated as follows:

$$(E + \mu N)/L \underset{\text{Ht-DMFET}}{\approx} \left\langle\Psi^{\mathcal{C}}\middle|\hat{t}_{01} + \hat{U}_0\middle|\Psi^{\mathcal{C}}\right\rangle. \tag{47}$$

Let us stress that, in Ht-DMFET, the cluster is designed from a single determinantal (non-interacting in the present case) lattice wave function, like in regular DMET calculations [20]. In other words, the Householder transformation is constructed from an idempotent density matrix. Moreover, the interacting cluster is described as a *closed* (two-electron) subsystem. As shown for small Hubbard rings, the exact interacting cluster is in principle an open subsystem [31]. It rigorously contains two electrons only at half filling, as a consequence of the hole-particle symmetry of the Hubbard lattice Hamiltonian [31].

Note finally that, if we Householder transform the two-electron repulsion operator $\hat{U}$ of the full lattice, one can in principle take into account its complete projection onto the cluster. It means that the interaction on the bath site could be added to the Hamiltonian in Equation (46). For simplicity, we will focus in the following on the (so-called) non-interacting bath formulation of the theory, which is described by Equation (46). Let us finally mention that, in the present single-impurity embedding, DMET, DET, and Ht-DMFET are equivalent [31].

2.3. *Exact Density-Functional Embedding*

We will show in the following that, once it has been merged with KS-DFT, Ht-DMFET can be made formally exact. For clarity, we start with reviewing briefly KS-DFT for lattice Hamiltonians in Section 2.3.1. A multi-determinantal extension of the theory based on the interacting Householder cluster's wave function is then proposed in Section 2.3.2.

2.3.1. KS-DFT for Uniform Lattices

According to the Hohenberg–Kohn (HK) variational principle [45], which is applied in this work to lattice Hamiltonians [41], the ground-state energy of the full lattice can be determined as follows,

$$E = \min_{n}\{F(n) + v_{\text{ext}}nL\}, \tag{48}$$

where the HK density functional reads as

$$F(n) = \langle\Psi(n)|\hat{T} + \hat{U}|\Psi(n)\rangle, \tag{49}$$

and $|\Psi(n)\rangle$ is the lattice ground state with uniform density profile $n \overset{0\leq i<L}{=} \langle\Psi(n)|\hat{n}_i|\Psi(n)\rangle$. Strictly speaking, $F(n)$ is a function of the site occupation n, hence the name *site occupation functional theory* often given to DFT for lattices [13,41]. Note that the ground-state energy E is in fact a (zero-temperature) grand canonical energy since a change in uniform density n induces a change in the number $N = nL$ of electrons. In the thermodynamic $N \to +\infty$ and $L \to +\infty$ limit, with N/L fixed to n, one can in principle describe *continuous* variations in n with a pure-state wave function $\Psi(n)$. The derivations that follow will be based on this assumption. If we introduce the per-site analog of the HK functional,

$$f(n) = F(n)/L = \langle\Psi(n)|\hat{t}_{01} + \hat{U}_0|\Psi(n)\rangle, \tag{50}$$

and use the notation of Equation (5), then Equation (48) becomes

$$E/L \equiv E(\mu)/L = \min_n\{f(n) - \mu n\}, \tag{51}$$

and the minimizing density $n(\mu)$ fulfills the following stationarity condition:

$$\mu = \left.\frac{\partial f(n)}{\partial n}\right|_{n=n(\mu)}. \tag{52}$$

In the conventional KS formulation of DFT, the per-site HK functional is decomposed as follows,

$$f(n) = t_s(n) + e_{\rm Hxc}(n), \tag{53}$$

where

$$t_s(n) = \langle\Phi(n)|\hat{t}_{01}|\Phi(n)\rangle = \frac{1}{L}\langle\Phi(n)|\hat{T}|\Phi(n)\rangle \tag{54}$$

is the (per-site) analog for lattices of the non-interacting kinetic energy functional, and the Hxc density functional reads as [41]

$$e_{\rm Hxc}(n) = \frac{U}{4}n^2 + e_c(n), \tag{55}$$

where $e_c(n)$ is the exact (per-site) correlation energy functional of the interacting lattice. The (normalized) density-functional lattice KS determinant $\Phi(n)$ fulfills the (non-interacting) KS equation

$$(\hat{T} - \mu_s(n)\hat{N})|\Phi(n)\rangle = \mathcal{E}_s(n)|\Phi(n)\rangle, \tag{56}$$

so that (see Equation (54))

$$\begin{aligned}\frac{\partial t_s(n)}{\partial n} &= \frac{2}{L}\left\langle\frac{\partial\Phi(n)}{\partial n}\middle|\hat{T}\middle|\Phi(n)\right\rangle\\ &= \frac{2\mu_s(n)}{L}\left\langle\frac{\partial\Phi(n)}{\partial n}\middle|\hat{N}\middle|\Phi(n)\right\rangle\\ &= \frac{\mu_s(n)}{L}\frac{\partial(nL)}{\partial n}\\ &= \mu_s(n),\end{aligned} \tag{57}$$

since $\langle\Phi(n)|\hat{N}|\Phi(n)\rangle = N = nL$. Thus, we recover from Equations (52) and (53) the well-known relation between the physical and KS chemical potentials:

$$\mu_s(n(\mu)) \equiv \mu_s = \mu - v_{\rm Hxc}, \tag{58}$$

where the density-functional Hxc potential reads as $v_{\rm Hxc} = v_{\rm Hxc}(n(\mu))$ with

$$v_{\rm Hxc}(n) = \frac{\partial e_{\rm Hxc}(n)}{\partial n}. \tag{59}$$

Note that the exact non-interacting density-functional chemical potential can be expressed analytically as follows [40]:

$$\mu_s(n) = -2t\cos\left(\frac{\pi}{2}n\right). \tag{60}$$

Capelle and coworkers [40,41] have designed a local density approximation (LDA) to $e_{\rm Hxc}(n)$ on the basis of exact Bethe Ansatz (BA) solutions [46] (the functional is usually referred to as BALDA).

Unlike in conventional *ab initio* DFT, the Hxc functional of lattice Hamiltonians is not truly universal in the sense that it is universal for a given choice of (hopping) one-electron and two-electron repulsion operators. In other words, the Hxc functional does not depend on the (possibly non-uniform) one-electron local potential operator $\sum_i v_{{\rm ext},i}\hat{n}_i$, which is the analog for lattices of the nuclear potential in molecules, but it is t- and U-dependent and, in the present case, it should be designed specifically for the 1D Hubbard model. Even though BALDA can be extended to higher dimensions [47], there is no general strategy for constructing (localized) orbital-occupation functional approximations, thus preventing direct applications to quantum chemistry [12], for example. Turning ultimately to a potential-functional theory, as proposed in Section 2.4, is appealing in this respect. With this change of paradigm, which is the second key result of the paper, the Hxc energy and potential become implicit functionals of the density, and they can be evaluated from a (few-electron) correlated wave function through a quantum embedding procedure.

2.3.2. Density-Functional Interacting Cluster

We propose in this section an alternative formulation of DFT based on the interacting Householder cluster introduced in Section 2.2.3. For that purpose, we consider the following *exact* decomposition,

$$f(n) = f^{\mathcal{C}}(n) + \overline{e}_{\rm c}(n), \tag{61}$$

where the Householder cluster HK functional

$$f^{\mathcal{C}}(n) = \left\langle \Psi^{\mathcal{C}}(n) \middle| \hat{t}_{01} + \hat{U}_0 \middle| \Psi^{\mathcal{C}}(n) \right\rangle \tag{62}$$

is evaluated from the two-electron cluster density-functional wave function $\Psi^{\mathcal{C}}(n)$, and $\overline{e}_{\rm c}(n)$ is the complementary correlation density functional that describes the missing correlation effects of the interacting bath and the Householder cluster's environment on the embedded impurity [31]. Note that, according to Section 2.2.3, $|\Psi^{\mathcal{C}}(n)\rangle$ fulfills the following Schrödinger-like equation,

$$\hat{\mathcal{H}}^{\mathcal{C}}(n)\left|\Psi^{\mathcal{C}}(n)\right\rangle = \mathcal{E}^{\mathcal{C}}(n)\left|\Psi^{\mathcal{C}}(n)\right\rangle, \tag{63}$$

where (we use the same notations as in Section 2.2.3)

$$\hat{\mathcal{H}}^{\mathcal{C}}(n) \equiv \hat{\mathcal{T}}^{\mathcal{C}}(n) + \hat{U}_0 - \tilde{\mu}^{\rm imp}(n)\,\hat{n}_0 \tag{64}$$

and

$$\hat{\mathcal{T}}^{\mathcal{C}}(n) \equiv \hat{t}_{01} + \hat{\tau}^{\mathcal{C}}(n). \tag{65}$$

The dependence in n of the (projected-onto-the-cluster) Householder-transformed kinetic energy operator $\hat{\mathcal{T}}^{\mathcal{C}}(n)$ comes from the fact that the KS lattice density matrix $\gamma(n) \equiv \langle\Phi(n)|\hat{c}^\dagger_{i\sigma}\hat{c}_{j\sigma}|\Phi(n)\rangle$ (on which the Householder transformation is based) is, like the KS determinant $\Phi(n) \equiv \Phi^{\mathcal{C}}(n)\Phi_{\rm core}(n)$ of the lattice, a functional of the uniform density n. On the other hand, for a given uniform lattice density n, the local potential $-\tilde{\mu}^{\rm imp}(n)$ is adjusted on the embedded impurity such that the interacting cluster reproduces n, i.e.,

$$\left\langle \Psi^{\mathcal{C}}(n) \middle| \hat{n}_0 \middle| \Psi^{\mathcal{C}}(n) \right\rangle = n. \tag{66}$$

Interestingly, on the basis of the two decompositions in Equations (53), (61) and (62), we can relate the exact Hxc functional to the density-functional Householder cluster as follows,

$$e_{\text{Hxc}}(n) \quad = \langle \Psi^{\mathcal{C}}(n)|\hat{t}_{01} + \hat{U}_0|\Psi^{\mathcal{C}}(n)\rangle - t_{\text{s}}(n) + \bar{e}_{\text{c}}(n), \tag{67}$$

where, as shown in Equation (44), the per-site non-interacting kinetic energy can be determined exactly from the two-electron cluster's part $\Phi^{\mathcal{C}}(n)$ of the KS lattice determinant $\Phi(n)$, i.e.,

$$t_{\text{s}}(n) = \left\langle \Phi^{\mathcal{C}}(n) \middle| \hat{t}_{01} \middle| \Phi^{\mathcal{C}}(n) \right\rangle, \tag{68}$$

thus leading to the final expression

$$e_{\text{Hxc}}(n) = \left\langle \Psi^{\mathcal{C}}(n) \middle| \hat{t}_{01} + \hat{U}_0 \middle| \Psi^{\mathcal{C}}(n) \right\rangle - \left\langle \Phi^{\mathcal{C}}(n) \middle| \hat{t}_{01} \middle| \Phi^{\mathcal{C}}(n) \right\rangle + \bar{e}_{\text{c}}(n). \tag{69}$$

Note that, according to Equations (35) and (38), $\Phi^{\mathcal{C}}(n)$ fulfills the KS-like equation

$$\left(\hat{t}_{01} + \hat{\tau}^{\mathcal{C}}(n)\right)\left|\Phi^{\mathcal{C}}(n)\right\rangle = \mathcal{E}_{\text{s}}^{\mathcal{C}}(n)\left|\Phi^{\mathcal{C}}(n)\right\rangle, \tag{70}$$

where the Householder transformation ensures that $\langle \Phi^{\mathcal{C}}(n)|\hat{n}_0|\Phi^{\mathcal{C}}(n)\rangle \;=\; n$ (see Equation (41)).

We will now establish a clearer connection between the KS lattice system and the Householder cluster *via* the evaluation of the Hxc density-functional potential in the lattice. According to Equations (59) and (69), the latter can be expressed as follows,

$$\begin{aligned} v_{\text{Hxc}}(n) = {} & 2\left\langle \frac{\partial \Psi^{\mathcal{C}}(n)}{\partial n} \middle| \hat{t}_{01} + \hat{U}_0 \middle| \Psi^{\mathcal{C}}(n) \right\rangle \\ & - 2\left\langle \frac{\partial \Phi^{\mathcal{C}}(n)}{\partial n} \middle| \hat{t}_{01} \middle| \Phi^{\mathcal{C}}(n) \right\rangle + \frac{\partial \bar{e}_{\text{c}}(n)}{\partial n}, \end{aligned} \tag{71}$$

or, equivalently (see Equations (63), (66) and (70)),

$$\begin{aligned} v_{\text{Hxc}}(n) = {} & \tilde{\mu}^{\text{imp}}(n) - 2\left\langle \frac{\partial \Psi^{\mathcal{C}}(n)}{\partial n} \middle| \hat{\tau}^{\mathcal{C}}(n) \middle| \Psi^{\mathcal{C}}(n) \right\rangle \\ & + 2\left\langle \frac{\partial \Phi^{\mathcal{C}}(n)}{\partial n} \middle| \hat{\tau}^{\mathcal{C}}(n) \middle| \Phi^{\mathcal{C}}(n) \right\rangle + \frac{\partial \bar{e}_{\text{c}}(n)}{\partial n}. \end{aligned} \tag{72}$$

If we introduce the following bi-functional of the density,

$$\tau_{\text{c}}^{\mathcal{C}}(n,\nu) = \left\langle \Psi^{\mathcal{C}}(\nu) \middle| \hat{\tau}^{\mathcal{C}}(n) \middle| \Psi^{\mathcal{C}}(\nu) \right\rangle - \left\langle \Phi^{\mathcal{C}}(\nu) \middle| \hat{\tau}^{\mathcal{C}}(n) \middle| \Phi^{\mathcal{C}}(\nu) \right\rangle, \tag{73}$$

which can be interpreted as a kinetic correlation energy induced within the density-functional cluster by the Householder transformation and the interaction on the impurity, we obtain the final *exact* expression

$$v_{\text{Hxc}}(n) = \tilde{\mu}^{\text{imp}}(n) - \left.\frac{\partial \tau_{\text{c}}^{\mathcal{C}}(n,\nu)}{\partial \nu}\right|_{\nu=n} + \frac{\partial \bar{e}_{\text{c}}(n)}{\partial n}, \tag{74}$$

which is the first key result of this paper.

Before turning Equation (74) into a practical self-consistent embedding method (see Section 2.4), let us briefly discuss its physical meaning and connection with Ht-DMFET. As pointed out in Section 2.2.1, the (density-functional) operator $\hat{\tau}^{\mathcal{C}}(n)$ is an auxiliary correction to the true per-site kinetic energy operator $\hat{t}_{01}$ which originates from the Householder-transformation-based embedding of the impurity. It is not physical and its impact on the impurity chemical potential $\tilde{\mu}^{\rm imp}(n)$, which is determined in the presence of $\hat{\tau}^{\mathcal{C}}(n)$ in the cluster's Hamiltonian (see Equations (63)–(65)), should be removed when evaluating the Hxc potential of the true lattice, hence the minus sign in front of the second term on the right-hand side of Equation (74). Finally, the complementary correlation potential $\partial\bar{e}_{\rm c}(n)/\partial n$ is in charge of recovering the electron correlation effects that were lost when considering an (impurity-only) interacting cluster that is disconnected from its environment [31]. We should stress at this point that, in Ht-DMFET (which is equivalent to DMET or DET when a single impurity is embedded [31]), the following density-functional approximation is made:

$$\bar{e}_{\rm c}(n) \underset{\rm Ht-DMFET}{\approx} 0, \tag{75}$$

so that the physical density-functional chemical potential is evaluated as follows [31],

$$\mu(n) \underset{\rm Ht-DMFET}{\approx} \frac{\partial f^{\mathcal{C}}(n)}{\partial n}. \tag{76}$$

Interestingly, even though it is never computed explicitly in this context, the corresponding (approximate) Hxc potential simply reads as

$$v_{\rm Hxc}(n) \underset{\rm Ht-DMFET}{\approx} \frac{\partial\left(f^{\mathcal{C}}(n) - t_{\rm s}(n)\right)}{\partial n}, \tag{77}$$

or, equivalently (see Equations (74) and (75)),

$$v_{\rm Hxc}(n) \underset{\rm Ht-DMFET}{\approx} \tilde{\mu}^{\rm imp}(n) - \left.\frac{\partial\tau_{\rm c}^{\mathcal{C}}(n,\nu)}{\partial\nu}\right|_{\nu=n}. \tag{78}$$

Therefore, Ht-DMFET can be seen as an approximate formulation of KS-DFT where the Hxc potential is determined solely from the density-functional Householder cluster. As illustrated in Figure 9 of Ref. [31], the approximation of Equation (75) leads, for example, to a substantial underestimation of the per-site energy, except in the vicinity of half filling where the energy is overestimated. Describing the electron repulsion in the bath (not considered in the present work, for simplicity) lowers the energy even further, thus leading to accurate results only at half filling, because of error cancellations [31]. Most importantly, Equation (75) implies that Ht-DMFET neglects the fluctuations in the electron number within the Householder cluster (see Equations (61)–(63)). Consequently, as further discussed in Ref. [31] and Section 4, Ht-DMFET is unable to describe the opening of the gap at half filling.

2.4. Local Potential Functional Embedding Theory

Until now the Householder transformation has been described as a functional of the uniform density n or, more precisely, as a functional of the KS density matrix, which is itself a functional of the density. If we opt for a potential-functional reformulation of the theory, as suggested in the following, the Householder transformation becomes a functional of the KS chemical potential $\mu_{\rm s}$ instead, and, consequently, the Householder correction to the per-site kinetic energy operator within the cluster (see Equation (65)) is also a functional of $\mu_{\rm s}$:

$$\hat{\tau}^{\mathcal{C}}(n) \rightarrow \hat{\tau}^{\mathcal{C}}(\mu_{\rm s}). \tag{79}$$

Similarly, the interacting cluster's wave function becomes a bi-functional of the KS *and* interacting embedded impurity chemical potentials:

$$\Psi^{\mathcal{C}}(n) \rightarrow \Psi^{\mathcal{C}}\left(\mu_{\rm s}, \tilde{\mu}^{\rm imp}\right). \tag{80}$$

In the exact theory, for a given chemical potential value μ in the true interacting lattice, both the KS lattice and the embedded impurity reproduce the interacting lattice density $n(\mu)$, i.e.,

$$n(\mu) = n^{\rm KS}_{\rm lattice}(\mu - v_{\rm Hxc}) = n^{\mathcal{C}}\left(\mu - v_{\rm Hxc}, \tilde{\mu}^{\rm imp}\right), \tag{81}$$

where

$$n^{\rm KS}_{\rm lattice}(\mu_{\rm s}) \equiv \langle \hat{n}_0 \rangle_{\hat{T} - \mu_{\rm s}\hat{N}}, \tag{82}$$

and

$$\begin{aligned} n^{\mathcal{C}}\left(\mu_{\rm s}, \tilde{\mu}^{\rm imp}\right) &= \langle \hat{n}_0 \rangle_{\Psi^{\mathcal{C}}\left(\mu_{\rm s}, \tilde{\mu}^{\rm imp}\right)} \\ &\equiv \langle \hat{n}_0 \rangle_{\hat{t}_{01} + \hat{\tau}^{\mathcal{C}}(\mu_{\rm s}) + \hat{U}_0 - \tilde{\mu}^{\rm imp}\hat{n}_0}, \end{aligned} \tag{83}$$

with, according to Equation (74),

$$\begin{aligned} \tilde{\mu}^{\rm imp} &= \tilde{\mu}^{\rm imp}(n(\mu)) \\ &= v_{\rm Hxc} - \left[\frac{\partial \overline{e}_{\rm c}(\nu)}{\partial \nu} - \frac{\partial \tau^{\mathcal{C}}_{\rm c}(n(\mu), \nu)}{\partial \nu}\right]_{\nu = n(\mu)}. \end{aligned} \tag{84}$$

The density constraint of Equation (81) combined with Equation (84) allows for an in-principle-exact evaluation of the Hxc potential $v_{\rm Hxc}$. Most importantly, these two equations can be used for designing an alternative (and self-consistent) embedding strategy on the basis of well-identified density-functional approximations. Indeed, in Ht-DMFET, the second term on the right-hand side of Equation (84) is simply dropped, for simplicity (see Equation (75)). If, in addition, we neglect the Householder kinetic correlation density-bi-functional potential correction $\partial \tau^{\mathcal{C}}_{\rm c}(n, \nu)/\partial \nu$ [last term on the right-hand side of Equation (84)], we obtain from Equation (81) the following self-consistent equation,

$$n^{\rm KS}_{\rm lattice}(\mu - \tilde{v}_{\rm Hxc}) = n^{\mathcal{C}}(\mu - \tilde{v}_{\rm Hxc}, \tilde{v}_{\rm Hxc}), \tag{85}$$

from which an approximation $\tilde{v}_{\rm Hxc} \equiv \tilde{v}_{\rm Hxc}(\mu)$ to the Hxc potential can be determined. Equation (85) is the second main result of this paper. Since $\tilde{v}_{\rm Hxc}$ is now the to-be-optimized quantity on which the embedding fully relies, we refer to the approach as *local potential functional embedding theory* (LPFET), in which the key density-functional approximation that is made reads as

$$v_{\rm Hxc}(n) \underset{\rm LPFET}{\approx} \tilde{\mu}^{\rm imp}(n). \tag{86}$$

The approach is graphically summarized in Figure 1.

In order to verify that the first HK theorem [45] still holds at the LPFET level of approximation, let us assume that two chemical potentials μ and $\mu + \Delta\mu$ lead to the same density. If so, the converged Hxc potentials should differ by $\tilde{v}_{\rm Hxc}(\mu + \Delta\mu) - \tilde{v}_{\rm Hxc}(\mu) = \Delta\mu$, so that both calculations give the same KS chemical potential value (see Equation (58)). According to Equations (85) and (86), it would imply that two different values of the interacting embedded impurity chemical potential can give the same density, which is impossible [31,48]. Therefore, when convergence is reached in Equation (85), we can generate an approximate map

$$\mu \rightarrow n(\mu) \underset{\rm LPFET}{\approx} n^{\rm KS}_{\rm lattice}(\mu - \tilde{v}_{\rm Hxc}) = \langle \hat{n}_0 \rangle_{\Psi^{\mathcal{C}}(\mu - \tilde{v}_{\rm Hxc}, \tilde{v}_{\rm Hxc})}, \tag{87}$$

and compute approximate per-site energies as follows,

$$\frac{E(\mu)}{L} + \mu n(\mu) \underset{\text{LPFET}}{\approx} \langle \hat{t}_{01} + \hat{U}_0 \rangle_{\Psi^{\mathcal{C}}(\mu - \tilde{v}_{\text{Hxc}}, \tilde{v}_{\text{Hxc}})}, \tag{88}$$

since the approximation in Equation (75) is also used in LPFET, as discussed above.

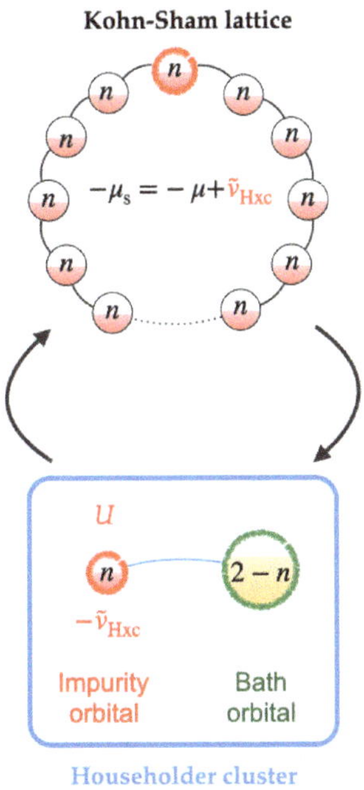

Figure 1. Graphical representation of the LPFET procedure. Note that the *same* Hxc potential $\tilde{v}_{\text{Hxc}}$ is used in the KS lattice and the embedding Householder cluster. It is optimized self-consistently in order to fulfill the density constraint of Equation (85). See text for further details.

Note that Ht-DMFET (which is equivalent to DMET in the present context) and LPFET use the same per-site energy expression (see Equation (47)), which is a functional of the interacting cluster's wave function. In both approaches, the latter and the non-interacting lattice share the same density. Therefore, if the per-site energy or the double occupation $\langle \hat{n}_{0\uparrow}\hat{n}_{0\downarrow} \rangle$ were plotted as functions of the (converged) lattice filling n, as it is usually done in the literature [15], both methods would give exactly the same results. The reason is that, at convergence of the LPFET algorithm, the density constraint of Equation (85) should be fulfilled, exactly like in Ht-DMFET (see Equations (45) and (46)). However, if properties were plotted as functions of the chemical potential value μ in the true interacting lattice, LPFET and Ht-DMFET would give different results, simply because the densities obtained (for a given μ value) with the two methods would be different. Indeed, as shown in Section 2.3.2, Ht-DMFET can be viewed as an approximation to KS-DFT where the Hxc density-functional potential of Equation (78) is employed. As readily seen from Equation (86), the LPFET and Ht-DMFET Hxc potentials differ by the Householder kinetic correlation potential (which is neglected in LPFET). If the corresponding KS densities were the same then the Hxc potential, the Householder transformation, and, therefore, the chemical potential on the interacting embedded impurity would be the same, which is impossible according to Equations (78) and (86). In summary, differences in properties between LPFET and Ht-DMFET are directly related to differences in density. This is the reason why, in order to compare the two methods, we will restrict ourselves to the computation of chemical-potential-density maps (see Section 4).

2.5. Comparison with SDE

At this point we should stress that LPFET is very similar to the SDE approach of Mordovina et al. [42]. The major difference between SDE and LPFET (in addition to the fact that LPFET has a clear connection with a formally exact density-functional embedding theory based on the Householder transformation) is that no KS construction is made within the cluster. Instead, the Hxc potential is directly updated in the KS lattice, on the basis of the correlated embedded impurity density. This becomes even more clear when rewriting Equation (85) as follows,

$$\tilde{v}_{\rm Hxc} = \mu - \left[n^{\rm KS}_{\rm lattice}\right]^{-1}\left(n^{\mathcal{C}}(\mu - \tilde{v}_{\rm Hxc}, \tilde{v}_{\rm Hxc})\right), \tag{89}$$

where $\left[n^{\rm KS}_{\rm lattice}\right]^{-1} : n \rightarrow \mu_{\rm s}(n)$ is the inverse of the non-interacting chemical-potential-density map. A practical advantage of such a procedure (which remains feasible since the full system is treated at the non-interacting KS level only) lies in the fact that the KS construction within the cluster is automatically (and exactly) generated by the Householder transformation, once the density has been updated in the KS lattice (see Equation (41) and the comment that follows). Most importantly, the density in the KS lattice and the density of the non-interacting KS embedded impurity (which, unlike the embedded *interacting* impurity, is not used in the actual calculation) will match *at each iteration* of the Hxc potential optimization process, as it should when convergence is reached. If, at a given iteration, the KS construction were made directly within the cluster, there would always be a "delay" in density between the KS lattice and the KS cluster, which would only disappear at convergence. Note that, when the latter is reached, the (approximate) Hxc potential of the lattice should match the one extracted from the cluster, which is defined in SDE as the difference between the KS cluster Hamiltonian and the one-electron part of the interacting cluster's Hamiltonian [42], both reproducing the density of the KS lattice. Therefore, according to Equations (64), (65) and (70), the converged Hxc potential will simply correspond to the chemical potential on the interacting embedded impurity, exactly like in LPFET (see Equation (86)).

Note finally that the simplest implementation of LPFET, as suggested by Equation (89), can be formally summarized as follows:

$$\begin{aligned} \tilde{v}^{(i+1)}_{\rm Hxc} &= \mu - \left[n^{\rm KS}_{\rm lattice}\right]^{-1}\left(n^{\mathcal{C}}\left(\mu - \tilde{v}^{(i)}_{\rm Hxc}, \tilde{v}^{(i)}_{\rm Hxc}\right)\right), \\ \tilde{v}^{(i=0)}_{\rm Hxc} &= 0. \end{aligned} \tag{90}$$

A complete description of the algorithm is given in the next section.

3. LPFET Algorithm

The LPFET approach introduced in Section 2.4 aims at computing the interacting chemical-potential-density $\mu \rightarrow n(\mu)$ map through the self-consistent optimization of the uniform Hxc potential. A schematics of the algorithm is provided in Figure 2. It can be summarized as follows.

1. We start by diagonalizing the one-electron Hamiltonian (i.e., the hopping in the present case) matrix $\boldsymbol{t} \equiv t_{ij}$ (see Equation (7)). Thus, we obtain the "molecular" spin-orbitals and their corresponding energies. We fix the chemical potential of the interacting lattice to some value μ and (arbitrarily) initialize the Hxc potential to $\tilde{v}_{\rm Hxc} = 0$. Therefore, at the zeroth iteration, the KS chemical potential $\mu_{\rm s}$ equals μ.
2. We occupy all the molecular spin-orbitals with energies below $\mu_{\rm s} = \mu - \tilde{v}_{\rm Hxc}$ and construct the corresponding density matrix (in the lattice representation). The latter

provides the uniform KS density (denoted $n^{\text{KS}}_{\text{lattice}}$ in Figure 2) and the embedding Householder cluster Hamiltonian (see Equation (46)) in which the impurity chemical potential is set to $\tilde{\mu}^{\text{imp}} = \tilde{v}_{\text{Hxc}}$ (see Equation (86)).

3. We solve the interacting Schrödinger equation for the two-electron Householder cluster and deduce the occupation of the embedded impurity (which is denoted $n^{\mathscr{C}}$ in Figure 2). This can be done analytically since the Householder cluster is an asymmetric Hubbard dimer [31].
4. We verify that the density in the KS lattice $n^{\text{KS}}_{\text{lattice}}$ and the occupation of the interacting embedded impurity $n^{\mathscr{C}}$ match (a convergence threshold has been set to 10^{-4}). If this is the case, the calculation has converged and $n^{\mathscr{C}}$ is interpreted as (an approximation to) the density $n(\mu)$ in the true interacting lattice. If the two densities do not match, the Hxc potential $\tilde{v}_{\text{Hxc}}$ is adjusted in the KS lattice such that the latter reproduces $n^{\mathscr{C}}$ (see Equation (90)) or, equivalently, such that the KS lattice contains $Ln^{\mathscr{C}}$ electrons. We then return to step 2.

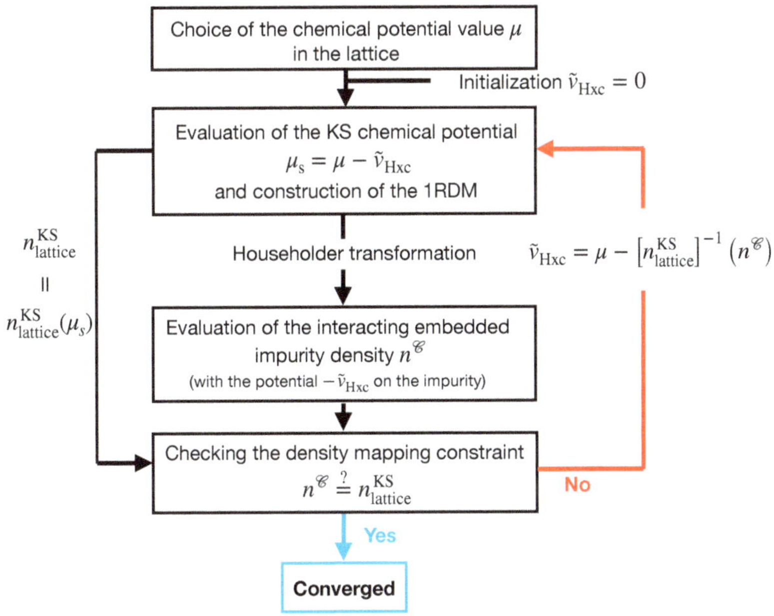

Figure 2. Schematics of the LPFET algorithm. The (one-electron reduced) density matrix of the KS lattice is referred to as the 1RDM. See text for further details.

4. Results and Discussion

In the following, LPFET is applied to a uniform Hubbard ring with a large $L = 1000$ number of sites in order to approach the thermodynamic limit. Periodic boundary conditions have been used. The hopping parameter is set to $t = 1$. As explained at the end of Section 2.4, plotting Ht-DMFET (which is equivalent to DMET or DET in the present single embedded impurity case) and LPFET properties such as the per-site energy or the double occupation as functions of the (converged) lattice filling n would give exactly the same results. We refer the reader to Ref. [31] for a detailed analysis of the Ht-DMFET scheme and its performance. On the other hand, the two methods are expected to give different chemical-potential-density $\mu \rightarrow n(\mu)$ maps since they rely on different density-functional approximations (see Equations (78) and (86)). We focus in the following on the self-consistent computation of this map at the LPFET level of theory. Comparison will be made with Ht-DMFET and the exact BA results.

As illustrated by the strongly correlated results of Figures 3 and 4, the LPFET self-consistency loop converges smoothly in few iterations. The same observation is made in weaker correlation regimes (not shown). The deviation in density between the KS lattice and the embedded impurity is drastically reduced after the first iteration (see Figure 3). This is also reflected in the large variation of the Hxc potential from the zeroth to the first iteration (see Figure 4). It originates from the fact that, at the zeroth iteration, the Hxc potential is set to zero in the lattice while, in the embedding Householder cluster, the interaction on the impurity site is "turned on". As shown in Figure 3, the occupation of the interacting embedded impurity is already at the zeroth iteration a good estimate of the self-consistently converged density. A few additional iterations are needed to refine the result.

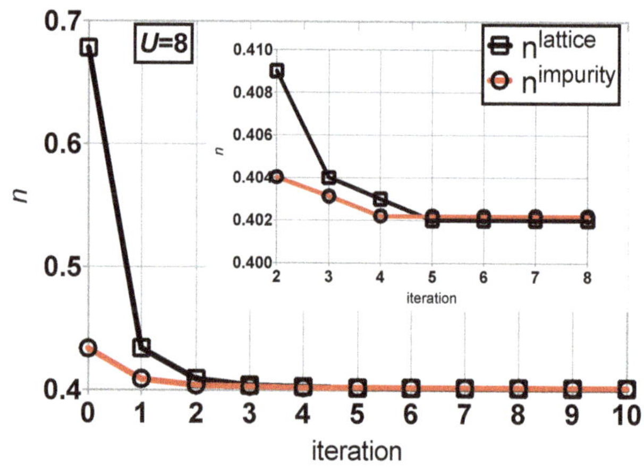

Figure 3. Comparison of the KS lattice and embedded impurity densities at each iteration of the LPFET calculation. The interaction strength and chemical potential values are set to $U/t = 8$ and $\mu/t = -0.97$, respectively. As shown in the inset, convergence is reached after five iterations.

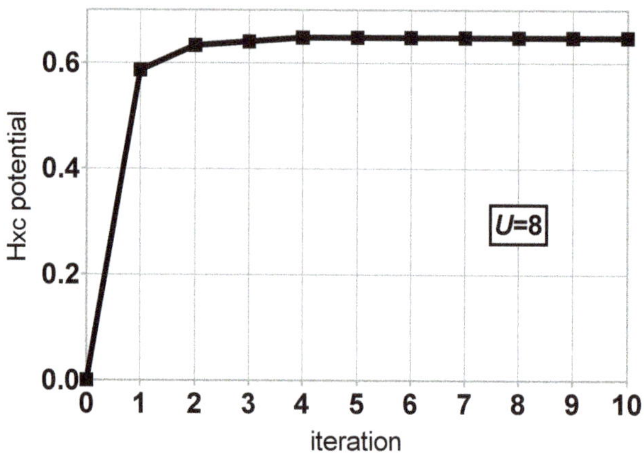

Figure 4. Convergence of the LPFET Hxc potential for $U/t = 8$ and $\mu/t = -0.97$.

The converged LPFET densities are plotted in Figure 5 as functions of the chemical potential μ in various correlation regimes. The non-interacting $U = 0$ curve describes the KS lattice at the zeroth iteration of the LPFET calculation. Thus, we can visualize, as U deviates from zero, how much the KS lattice learns from the interacting two-electron Householder cluster. LPFET is actually quite accurate (even more than Ht-DMFET, probably because of error cancellations) in the low filling regime. Even though LPFET deviates from Ht-DMFET when electron correlation is strong, as expected, their chemical-potential-density maps are quite similar. This is an indication that neglecting the Householder kinetic correlation potential contribution to the Hxc potential, as done in LPFET, is not a crude approximation, even in the strongly correlated regime. As expected [15,31], LPFET and Ht-DMFET poorly perform when approaching half filling. Like the well-established single-site DMFT (see Figure 7 of Ref. [5]), they are unable to describe the density-driven Mott–Hubbard transition (i.e., the opening of the gap). As discussed in Ref. [31], this might be related to the fact that, in the exact theory, the Householder cluster is not disconnected from its environment and it contains a fractional number of electrons, away from half filling, unlike in the (approximate) Ht-DMFET and LPFET schemes. In the language of KS-DFT, modeling the gap opening is equivalent to modeling the derivative discontinuity in the density-functional correlation potential $v_c(n) = \mu(n) - \mu_s(n) - \frac{U}{2}n$ at half filling. As clearly shown in Figure 6, Ht-DMFET and LPFET do not reproduce this feature. In the language of the exact density-functional embedding theory derived in Section 2.3, both Ht-DMFET and LPFET approximations neglect the complementary density-functional correlation energy $\bar{e}_c(n)$ that is induced by the interacting bath and the environment of the (closed) density-functional Householder cluster. As readily seen from Equation (74), it should be possible to describe the density-driven Mott–Hubbard transition with a single statically embedded impurity, provided that we can model the derivative discontinuity in $\partial\bar{e}_c(n)/\partial n$ at half filling. This is obviously a challenging task that is usually bypassed by embedding more impurities [15,31]. The implementation of a multiple-impurity LPFET as well as its generalization to higher-dimension lattice or quantum chemical Hamiltonians is left for future work.

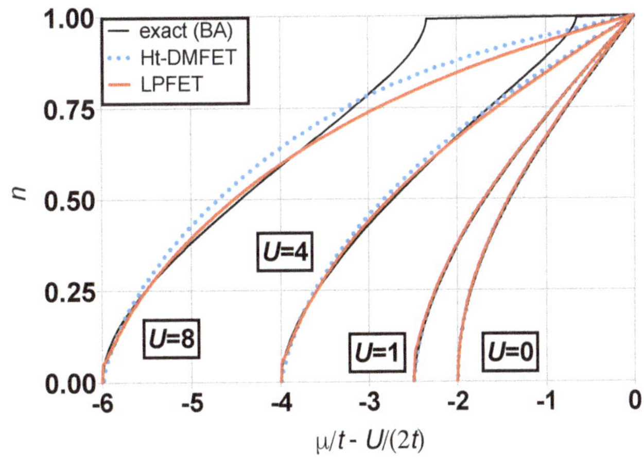

Figure 5. Converged LPFET densities (red solid lines) plotted as functions of the chemical potential μ in various correlation regimes. Comparison is made with the exact BA (black solid lines) and Ht-DMFET (blue dotted lines) results. In the latter case, the chemical potential is evaluated *via* the numerical differentiation of the density-functional Ht-DMFET per-site energy (see Equations (62) and (76)). The non-interacting ($U = 0$) chemical-potential-density map (see Equation (60)) is shown for analysis purposes.

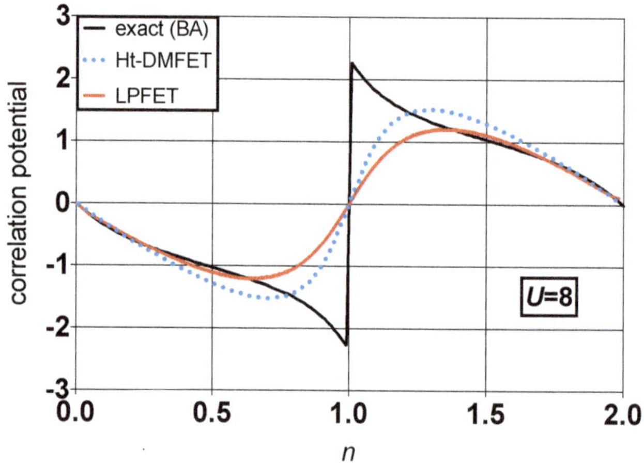

Figure 6. Correlation potential $v_c(n) = \mu(n) - \mu_s(n) - \frac{U}{2}n$ plotted as a function of the lattice filling n at the Ht-DMFET (blue dashed line) and LPFET (red solid line) levels of approximation for $U/t = 8$. Comparison is made with the exact BA correlation potential (black solid line).

5. Conclusions and Perspectives

An in-principle-exact density-functional reformulation of the recently proposed *Householder transformed density matrix functional embedding theory* (Ht-DMFET) [31] has been derived for the uniform 1D Hubbard Hamiltonian with a single embedded impurity. On that basis, an approximate *local potential functional embedding theory* (LPFET) has been proposed and implemented. Ht-DMFET, which is equivalent to DMET or DET in the particular case of a single impurity, is reinterpreted in this context as an approximation to DFT where the complementary density-functional correlation energy $\bar{e}_c(n)$ induced by the interacting bath and the environment of the (closed) embedding "impurity+bath" cluster is neglected. LPFET neglects, in addition, the kinetic correlation effects induced by the Householder transformation on the impurity chemical potential. We have shown that combining the two approximations is equivalent to approximating the latter potential with the Hxc potential of the full lattice. Thus, an approximate Hxc potential can be determined *self-consistently* for a given choice of external (chemical in the present case) potential in the true interacting lattice. The self-consistency loop, which does not exist in regular single-impurity DMET or DET [36], emerges naturally in LPFET from the exact density constraint, i.e., by forcing the KS lattice and interacting embedded impurity densities to match. In this context, the energy becomes a functional of the Hxc potential. In this respect, LPFET can be seen as a flavor of KS-DFT where no density functional is used. LPFET is very similar to SDE [42]. The two approaches essentially differ in the optimization of the potential. In LPFET, no KS construction is made within the embedding cluster, unlike in SDE. Instead, the Hxc potential is directly updated in the lattice. As a result, the KS cluster (which is not used in the actual calculation) can be automatically generated with the correct density by applying the Householder transformation to the KS lattice Hamiltonian.

LPFET and Ht-DMFET chemical-potential-density maps have been computed for a 1000-site Hubbard ring. Noticeable differences appear in the strongly correlated regime. LPFET is more accurate than Ht-DMFET in the low filling regime, probably because of error cancellations. As expected from previous works [15,31], their performance deteriorates as we approach half filling. It appears that, in the language of density-functional embedding theory, it should be possible to describe the density-driven Mott–Hubbard transition (i.e., the opening of the gap), provided that the complementary correlation potential $\partial\bar{e}_c(n)/\partial n$ exhibits a derivative discontinuity at half filling. Since the latter is neglected in both methods, the gap opening is not reproduced. The missing correlation effects might be recovered by applying a multi-reference Görling–Levy-type perturbation theory on top of the correlated cluster calculation [31]. Extending LPFET to multiple impurities by means of a block Householder transformation is another viable strategy [31]. Work is currently in progress in these directions. Note that, like DMET or SDE, LPFET is in principle applicable to quantum chemical Hamiltonians written in a localized molecular orbital basis. A general computational implementation of the theory will be presented in a forthcoming paper and the code will be made available on that occasion.

Author Contributions: Conceptualization, E.F. and M.S.; methodology, E.F.; software, S.S.; validation, E.F., S.S. and M.S.; formal analysis, E.F. and S.S.; investigation, E.F. and S.S.; resources, S.S.; data curation, S.S.; writing—original draft preparation, E.F. and S.S.; writing—review and editing, E.F., S.S. and M.S.; visualization, E.F. and S.S.; supervision, E.F.; project administration, E.F.; funding acquisition, E.F. and M.S. All authors have read and agreed to the published version of the manuscript.

Funding: LabEx CSC (grant number: ANR-10-LABX-0026-CSC) and ANR (grant numbers: ANR-19-CE29-0002 DESCARTES and ANR-19-CE07-0024-02 CoLab).

Institutional Review Board Statement: Not applicable.

Informed Consent Statement: Not applicable.

Data Availability Statement: Data is contained within the article.

Acknowledgments: The authors thank Saad Yalouz (for his comments on the manuscript and many fruitful discussions) as well as Thomas Baker and Martin Rafael Gulin for stimulating discussions.

Conflicts of Interest: The authors declare that they have no conflict of interest.

Appendix A. Simplification of Density Matrix Elements in the Householder Representation

Starting from the expression in Equation (23) of the creation operators in the Householder representation and Equations (12), (14)–(16) and (18), we can simplify step by step the expression of the density matrix elements that involve the impurity as follows,

$$
\begin{aligned}
\langle\Phi|\hat{d}^{\dagger}_{j\sigma}\hat{d}_{0\sigma}|\Phi\rangle &= \sum_i P_{ji}\gamma_{i0} \\
&= \gamma_{j0} - 2\mathrm{v}_j \sum_{i\geq 1} \mathrm{v}_i\gamma_{i0} \\
&= \gamma_{j0} - 2\mathrm{v}_j\mathrm{v}_1\gamma_{10} - 2\mathrm{v}_j\sqrt{2\tilde{\gamma}_{10}(\tilde{\gamma}_{10}-\gamma_{10})}\sum_{i\geq 2}\mathrm{v}_i^2 \\
&= \gamma_{j0} - 2\mathrm{v}_j\mathrm{v}_1\gamma_{10} - 2\mathrm{v}_j\sqrt{2\tilde{\gamma}_{10}(\tilde{\gamma}_{10}-\gamma_{10})}\left(1-\mathrm{v}_1^2\right) \\
&= \gamma_{j0} - 2\mathrm{v}_j\mathrm{v}_1\gamma_{10} - 2\mathrm{v}_j\sqrt{2\tilde{\gamma}_{10}(\tilde{\gamma}_{10}-\gamma_{10})} + 2\mathrm{v}_j\mathrm{v}_1^2\sqrt{2\tilde{\gamma}_{10}(\tilde{\gamma}_{10}-\gamma_{10})} \qquad \text{(A1)} \\
&= \gamma_{j0} - 2\mathrm{v}_j\mathrm{v}_1\gamma_{10} - 2\mathrm{v}_j\sqrt{2\tilde{\gamma}_{10}(\tilde{\gamma}_{10}-\gamma_{10})} + 2\mathrm{v}_j\mathrm{v}_1(\gamma_{10}-\tilde{\gamma}_{10}) \\
&= \gamma_{j0} - 2\mathrm{v}_j\left(\mathrm{v}_1\tilde{\gamma}_{10} + \sqrt{2\tilde{\gamma}_{10}(\tilde{\gamma}_{10}-\gamma_{10})}\right) \\
&= \gamma_{j0} - 2\mathrm{v}_j\sqrt{2\tilde{\gamma}_{10}(\tilde{\gamma}_{10}-\gamma_{10})}\left(1+\frac{\tilde{\gamma}_{10}(\gamma_{10}-\tilde{\gamma}_{10})}{2\tilde{\gamma}_{10}(\tilde{\gamma}_{10}-\gamma_{10})}\right) \\
&= \gamma_{j0} - \mathrm{v}_j\sqrt{2\tilde{\gamma}_{10}(\tilde{\gamma}_{10}-\gamma_{10})}.
\end{aligned}
$$

References

1. Kohn, W.; Sham, L.J. Self-Consistent Equations Including Exchange and Correlation Effects. *Phys. Rev.* **1965**, *140*, A1133–A1138. [CrossRef]
2. Anisimov, V.I.; Aryasetiawan, F.; Lichtenstein, A.I. First-principles calculations of the electronic structure and spectra of strongly correlated systems: The LDA +U method. *J. Phys. Condens. Matter* **1997**, *9*, 767–808. [CrossRef]

3. Anisimov, V.I.; Poteryaev, A.I.; Korotin, M.A.; Anokhin, A.O.; Kotliar, G. First-principles calculations of the electronic structure and spectra of strongly correlated systems: Dynamical mean-field theory. *J. Phys. Condens. Matter* **1997**, *9*, 7359–7367. [CrossRef]
4. Lichtenstein, A.I.; Katsnelson, M.I. Ab initio calculations of quasiparticle band structure in correlated systems: LDA++ approach. *Phys. Rev. B* **1998**, *57*, 6884–6895. [CrossRef]
5. Kotliar, G.; Savrasov, S.Y.; Haule, K.; Oudovenko, V.S.; Parcollet, O.; Marianetti, C.A. Electronic structure calculations with dynamical mean-field theory. *Rev. Mod. Phys.* **2006**, *78*, 865–951. [CrossRef]
6. Haule, K. Exact Double Counting in Combining the Dynamical Mean Field Theory and the Density Functional Theory. *Phys. Rev. Lett.* **2015**, *115*, 196403. [CrossRef]
7. Requist, R.; Gross, E.K.U. Model Hamiltonian for strongly correlated systems: Systematic, self-consistent, and unique construction. *Phys. Rev. B* **2019**, *99*, 125114. [CrossRef]
8. Ghosh, S.; Verma, P.; Cramer, C.J.; Gagliardi, L.; Truhlar, D.G. Combining Wave Function Methods with Density Functional Theory for Excited States. *Chem. Rev.* **2018**, *118*, 7249–7292. [CrossRef]
9. Savin, A. *Recent Developments and Applications of Modern Density Functional Theory*; Elsevier: Amsterdam, The Netherlands, 1996; p. 327.
10. Toulouse, J.; Colonna, F.; Savin, A. Long-range–short-range separation of the electron-electron interaction in density-functional theory. *Phys. Rev. A* **2004**, *70*, 062505. [CrossRef]
11. Sharkas, K.; Toulouse, J.; Savin, A. Double-hybrid density-functional theory made rigorous. *J. Chem. Phys.* **2011**, *134*, 064113. [CrossRef]
12. Fromager, E. On the exact formulation of multi-configuration density-functional theory: Electron density versus orbitals occupation. *Mol. Phys.* **2015**, *113*, 419–434. [CrossRef]
13. Senjean, B.; Nakatani, N.; Tsuchiizu, M.; Fromager, E. Site-occupation embedding theory using Bethe ansatz local density approximations. *Phys. Rev. B* **2018**, *97*, 235105. [CrossRef]
14. Wasserman, A.; Pavanello, M. Quantum embedding electronic structure methods. *Int. J. Quantum Chem.* **2020**, *120*, e26495. [CrossRef]
15. Knizia, G.; Chan, G.K.L. Density Matrix Embedding: A Simple Alternative to Dynamical Mean-Field Theory. *Phys. Rev. Lett.* **2012**, *109*, 186404. [CrossRef] [PubMed]
16. Knizia, G.; Chan, G. Density matrix embedding: A strong-coupling quantum embedding theory. *J. Chem. Theory Comput.* **2013**, *9*, 1428–1432. [CrossRef] [PubMed]
17. Tsuchimochi, T.; Welborn, M.; Van Voorhis, T. Density matrix embedding in an antisymmetrized geminal power bath. *J. Chem. Phys.* **2015**, *143*, 024107. [CrossRef] [PubMed]
18. Welborn, M.; Tsuchimochi, T.; Van Voorhis, T. Bootstrap embedding: An internally consistent fragment-based method. *J. Chem. Phys.* **2016**, *145*, 074102. [CrossRef]
19. Sun, Q.; Chan, G.K.L. Quantum Embedding Theories. *Accounts Chem. Res.* **2016**, *49*, 2705–2712. [CrossRef]
20. Wouters, S.; Jiménez-Hoyos, C.A.; Sun, Q.; Chan, G.K.L. A Practical Guide to Density Matrix Embedding Theory in Quantum Chemistry. *J. Chem. Theory Comput.* **2016**, *12*, 2706–2719. [CrossRef]
21. Wu, X.; Cui, Z.H.; Tong, Y.; Lindsey, M.; Chan, G.K.L.; Lin, L. Projected density matrix embedding theory with applications to the two-dimensional Hubbard model. *J. Chem. Phys.* **2019**, *151*, 064108. [CrossRef]
22. Cui, Z.H.; Zhu, T.; Chan, G.K.L. Efficient Implementation of Ab Initio Quantum Embedding in Periodic Systems: Density Matrix Embedding Theory. *J. Chem. Theory Comput.* **2020**, *16*, 119–129. [CrossRef] [PubMed]
23. Faulstich, F.M.; Kim, R.; Cui, Z.H.; Wen, Z.; Kin-Lic Chan, G.; Lin, L. Pure State v-Representability of Density Matrix Embedding Theory. *J. Chem. Theory Comput.* **2022**, *18*, 851–864. [CrossRef] [PubMed]
24. Lacombe, L.; Maitra, N.T. Embedding via the Exact Factorization Approach. *Phys. Rev. Lett.* **2020**, *124*, 206401. [CrossRef]
25. Requist, R.; Gross, E.K.U. Fock-Space Embedding Theory: Application to Strongly Correlated Topological Phases. *Phys. Rev. Lett.* **2021**, *127*, 116401. [CrossRef] [PubMed]
26. Georges, A.; Kotliar, G. Hubbard model in infinite dimensions. *Phys. Rev. B* **1992**, *45*, 6479–6483. [CrossRef]
27. Georges, A.; Kotliar, G.; Krauth, W.; Rozenberg, M.J. Dynamical mean-field theory of strongly correlated fermion systems and the limit of infinite dimensions. *Rev. Mod. Phys.* **1996**, *68*, 13–125. [CrossRef]
28. Kotliar, G.; Vollhardt, D. Strongly Correlated Materials: Insights From Dynamical Mean-Field Theory. *Phys. Today* **2004**, *57*, 53–59. [CrossRef]
29. Held, K. Electronic structure calculations using dynamical mean field theory. *Adv. Phys.* **2007**, *56*, 829–926. [CrossRef]
30. Zgid, D.; Chan, G.K.L. Dynamical mean-field theory from a quantum chemical perspective. *J. Chem. Phys.* **2011**, *134*, 094115. [CrossRef]
31. Sekaran, S.; Tsuchiizu, M.; Saubanère, M.; Fromager, E. Householder-transformed density matrix functional embedding theory. *Phys. Rev. B* **2021**, *104*, 035121. [CrossRef]
32. Ayral, T.; Lee, T.H.; Kotliar, G. Dynamical mean-field theory, density-matrix embedding theory, and rotationally invariant slave bosons: A unified perspective. *Phys. Rev. B* **2017**, *96*, 235139. [CrossRef]
33. Lee, T.H.; Ayral, T.; Yao, Y.X.; Lanata, N.; Kotliar, G. Rotationally invariant slave-boson and density matrix embedding theory: Unified framework and comparative study on the one-dimensional and two-dimensional Hubbard model. *Phys. Rev. B* **2019**, *99*, 115129. [CrossRef]

34. Fertitta, E.; Booth, G.H. Rigorous wave function embedding with dynamical fluctuations. *Phys. Rev. B* **2018**, *98*, 235132. [CrossRef]
35. Fertitta, E.; Booth, G.H. Energy-weighted density matrix embedding of open correlated chemical fragments. *J. Chem. Phys.* **2019**, *151*, 014115. [CrossRef] [PubMed]
36. Sriluckshmy, P.V.; Nusspickel, M.; Fertitta, E.; Booth, G.H. Fully algebraic and self-consistent effective dynamics in a static quantum embedding. *Phys. Rev. B* **2021**, *103*, 085131. [CrossRef]
37. Lee, T.H.; Lanatà, N.; Kim, M.; Kotliar, G. Efficient Slave-Boson Approach for Multiorbital Two-Particle Response Functions and Superconductivity. *Phys. Rev. X* **2021**, *11*, 041040. [CrossRef]
38. Bulik, I.W.; Scuseria, G.E.; Dukelsky, J. Density matrix embedding from broken symmetry lattice mean fields. *Phys. Rev. B* **2014**, *89*, 035140. [CrossRef]
39. Senjean, B. Projected site-occupation embedding theory. *Phys. Rev. B* **2019**, *100*, 035136. [CrossRef]
40. Lima, N.A.; Silva, M.F.; Oliveira, L.N.; Capelle, K. Density Functionals Not Based on the Electron Gas: Local-Density Approximation for a Luttinger Liquid. *Phys. Rev. Lett.* **2003**, *90*, 146402. [CrossRef]
41. Capelle, K.; Campo, V.L. Density functionals and model Hamiltonians: Pillars of many-particle physics. *Phys. Rep.* **2013**, *528*, 91–159. [CrossRef]
42. Mordovina, U.; Reinhard, T.E.; Theophilou, I.; Appel, H.; Rubio, A. Self-Consistent Density-Functional Embedding: A Novel Approach for Density-Functional Approximations. *J. Chem. Theory Comput.* **2019**, *15*, 5209–5220. [CrossRef] [PubMed]
43. Theophilou, I.; Reinhard, T.E.; Rubio, A.; Ruggenthaler, M. Approximations based on density-matrix embedding theory for density-functional theories. *Electron. Struct.* **2021**, *3*, 035001. [CrossRef]
44. Rotella, F.; Zambettakis, I. Block Householder transformation for parallel QR factorization. *Appl. Math. Lett.* **1999**, *12*, 29–34. [CrossRef]
45. Hohenberg, P.; Kohn, W. Inhomogeneous Electron Gas. *Phys. Rev.* **1964**, *136*, B864–B871. [CrossRef]
46. Lieb, E.H.; Wu, F.Y. Absence of Mott Transition in an Exact Solution of the Short-Range, One-Band Model in One Dimension. *Phys. Rev. Lett.* **1968**, *20*, 1445–1448. [CrossRef]
47. Vilela, L.N.P.; Capelle, K.; Oliveira, L.N.; Campo, V.L. Approximate expression for the ground-state energy of the two- and three-dimensional Hubbard model at arbitrary filling obtained from dimensional scaling. *J. Phys. Condens. Matter* **2019**, *31*, 455601. [CrossRef]
48. Senjean, B.; Tsuchiizu, M.; Robert, V.; Fromager, E. Local density approximation in site-occupation embedding theory. *Mol. Phys.* **2017**, *115*, 48–62. [CrossRef]

computation

MDPI

Article

Generation of Basis Sets for Accurate Molecular Calculations: Application to Helium Atom and Dimer

Ignacio Ema, Guillermo Ramírez, Rafael López and José Manuel García de la Vega *

Departamento de Química Física Aplicada, Facultad de Ciencias, Universidad Autónoma de Madrid, 28049 Madrid, Spain; nacho.ema@uam.es (I.E.); guillermo.ramirez@uam.es (G.R.); rafael.lopez@uam.es (R.L.)
* Correspondence: garcia.delavega@uam.es

Abstract: A new approach for basis set generation is reported and tested in helium atom and dimer. The basis sets thus computed, named sigma, range from DZ to 5Z and consist of the same composition as Dunning basis sets but with a different treatment of contractions. The performance of the sigma sets is analyzed for energy and other properties of He atom and He dimer, and the results are compared with those obtained with Dunning and ANO basis sets. The sigma basis sets and their extended versions up to triple augmented provide better energy values than Dunning basis sets of the same composition, and similar values to those attained with the currently available ANO. Extrapolation to complete basis set of correlation energy is compared between the sigma basis sets and those of Dunning, showing the better performance of the former in this respect.

Keywords: He atomic basis sets; helium dimer; He_2 potential well; correlation energy; complete basis set; sigma basis set

Citation: Ema, I.; Ramírez, G.; López, R.; García de la Vega, J.M. Generation of Basis Sets for Accurate Molecular Calculations: Application to Helium Atom and Dimer. *Computation* **2022**, *10*, 65. https://doi.org/10.3390/computation10050065

Academic Editor: Henry Chermette

Received: 30 March 2022
Accepted: 18 April 2022
Published: 20 April 2022

1. Introduction

The study of weak van der Waals (vdW) interactions has always been one of the most challenging applications of theoretical calculations of electron structure. Thus, methods based on Kohn–Sham (KS) density functional theory (DFT) have shown limitations for including weak and long-range interactions in the exchange-correlation term of the KS equation. In particular, standard functionals of DFT fail to explain these interactions because the stabilization is determined by dispersion interactions, and is not explained by these functionals. Exchange-correlation potentials, derived from local and semi-local models, often exhibit artifacts when applied to systems with large non-local correlation effects. Nevertheless, there is a continuous effort to include vdW interactions within the framework of KS theory [1], and remarkable progress on corrections to this fact has been made, such as the exchange-hole dipole moment (XDM) model [2,3].

In the long range, vdW interactions are dominated by dispersion and permanent multipole moment interaction and include superposition and exchange contributions. The behavior of an exchange functional in the region of small density and large density gradients plays a very important role. From a topological point of view, the presence of vdW interactions can be identified by reduced density gradient (RDG) analysis. In this respect, a novel procedure for studying van der Waals interactions has been developed as an extension of Bader's QTAIM in combination with RDG analysis in which a volumetric source function is used for describing the atomic composition of vdW interactions [4].

The above considerations reinforce that the well known fact is that to reliably study vdW interactions, not only large basis sets (BS) are needed, but also fine electron correlation treatments are required to reproduce experimental data. In particular, noble gases have been the focus of much research as a starting point for the study of rare gas dimer interactions [5], a criterion for the performance of basis sets and correlation methods in vdW studies.

Among the systems linked by vdW dispersion interactions, helium occupies an outstanding position. The properties of gaseous helium are close to those of the ideal gas because the interactions between helium atoms are extremely weak, and its behavior at very low temperatures makes helium a paramount system on its own, with unusual physico-chemical properties under these conditions. Furthermore, vdW interaction in helium dimer, He_2, is a touchstone for testing the capabilities of the most precise theoretical procedures available [6], as its existence is due to purely electronic correlation effects.

The accurate computation of the He_2 potential energy curve is a big challenge, and a great amount of work has been devoted to this task. Ab initio studies on He_2 interactions have been published by Van Mourik et al., who report dimer calculations using different methods and computational levels (MPn, CCSD(T), FCI, ...) with large basis sets, including polarization [7–9]. In the late 20th and 21st centuries, many other authors have published works on He_2, using various sets of a consistent basis for correlation and a high level of theory: SAPT [10], CCSD(T) [11], r12-MR-ACPF [12–14], MRCI [15], CCSD(T) [16], Monte Carlo [17], Compton profiles [18] and Gaussian geminal theory [19]. The most accurate results were obtained with BS supplemented with an additional set of bond functions [20], obtaining better results when compared with larger BS without bond functions.

Because of the high accuracy intended and the feeble interaction involved, the question about the full elimination of basis set superposition error (BSSE) in the calculations of He_2 energy has been a matter of discussion [21], which still remains today. This problem comes from the difficulty of saturating the dispersion energy in calculations with conventional basis sets, and it is especially relevant in weakly interacting species. A good representation of the dispersion energy requires polarization functions with small exponents, and extrapolation methods from the raw energies without Counterpoise (CP) correction have been proposed to reduce the BSSE [22–24]. In particular, extrapolation to complete basis set (CBS) limit for the helium dimer was studied by A. Varandas [22,25], who carried out calculations with size-consistent methods such as Hartree–Fock (HF) and full configuration interaction (FCI) using Dunning BS.

As a consequence of this extraordinary effort, the accuracy achieved is certainly impressive and, according to the best estimations, the potential energy curve of He_2 has a minimum at a distance $R_e = 2.9676$ Å, with an energy of $-34.82\ \mu E_h$ with respect to the limit of separated atoms [20,26]. This minimum in the potential energy curve is remarkably shallow and has been found to admit only one vibrational state, the mean distance between nuclei in this state being ca. ten times the equilibrium distance [27].

In this work, we report a way of constructing new basis sets for molecular calculations which overcome the variational performance of the existing ones of equivalent composition. Furthermore, we apply the procedure to the development of basis sets, hereafter named sigma (σBS), for helium atoms and use them for the study of helium atom and dimer.

The article is organized as follows. In Section 2, the procedure for developing sigma BS is explained, and the contraction scheme and composition are described. Supplementary Materials are included, which contain the link for the sigma basis sets for He. In Section 3, a brief summary of computational procedures, methods, bases and programs is given. The results on the He atom and the He_2 dimer are reported in Section 4, in which the the precision of the total and dissociation energies as well as the equilibrium distances is discussed. Results on the CBS extrapolation of correlation energies are included in a subsection of Section 4. Finally, conclusions are drawn from these results in Section 5.

2. Criteria for Basis Set Optimization and Contraction Scheme: Size and Composition

The σBS consists of linear combinations (contractions) of radial primitive Gaussians aimed at providing highly accurate energy values at different computational levels. In the construction of σBS, we have exploited our previous experience in the development of exponential type BS [28–31]. Guided by this experience, we have decided to design σBS with the characteristic that, if a given primitive contains a spherical harmonic of quantum number $l = L$, all the primitives with the same exponent and $l < L$ are also present in the set. For instance, if there is a d function with exponent α_i, then s and p functions with this exponent will appear in the BS. This characteristic was thought to reduce the computation cost of integrals involving primitives, although most standard packages are usually not prepared to profit from it. Another feature of σBS is that all primitives in a given shell, i.e., with the same angular part, participate in all contractions of the same shell. The combination of both features makes it possible for the number of primitives in the contractions can be increased, and the quality of the BS functions is thus improved, without penalizing the computational cost. In summary, contractions in σBS are built from the same set of exponentials combined with different angular functions. Furthermore, whereas polarization functions of Dunning BS consist of single Gaussian functions (one primitive per function without contraction), in sigma basis sets they are true contractions, yielding significant improvements in the results on energy.

As a rule of thumb, the number of primitives included in each shell of polarization functions is equal to the number of contractions in the shell plus two. As mentioned before, the radial parts of the primitives used in the polarization functions are also present in the functions of the core shells. The choice of primitives for polarization functions, among those of core shells functions, is not obvious and must be accomplished by optimizing the exponents together for both types of functions. It is also noticeable that, albeit not specifically intended, the exponents of the primitives thus obtained almost follow an even tempered sequence, with slight variations and covering a wide range of values.

To simplify the notation, XZ will be used in the following as an abbreviation of cc-pVXZ, and aXZ of aug-cc-pcVXZ families. The equivalent Atomic Natural Orbitals (ANO) and σBS will be denoted as anoXZ, aanoXZ, σXZ and aσXZ, respectively. In each of the six families, we have considered basis sets ranging from X = 2 (DZ) up to 5 (5Z).

The optimization of helium σBS follows the general lines of Dunning's procedure which relies on the minimization of CISD energy in He atoms. Starting with a given set of exponents in the primitives, the contractions are constructed in a stepwise way. In the case of He, the (1s) contraction of the σDZ is determined by minimizing the HF energy for the ground state. Next, the CISD is used to add a new shell and one more contraction per shell, but keeping unchanged the contraction previously optimized. These two steps yield the σDZ basis set. This procedure is repeated, changing the primitive exponents until their optimum values are obtained.

To build the σTZ, we proceed as in the case of the σDZ. After optimizing the 2(s) + 1(p) contractions, a new shell and a new contraction per shell, 1(s) + 1(p) + 1(d), are added and optimized with CISD of the He atom. The scheme for all σBS follows the procedure described but repeating the steps of contraction/optimization as many times as required according to the BS level. Proceeding in this way, and taking into account that the number of primitives is increased according to the rule mentioned, σBS tend to saturate one-electron space per shell, yielding energies as close as possible to the best attainable values according to their size.

In general, these σBS give energy values for He atoms lower than those of Dunning BS and similar to those of ANO BS, as we will see the next section. Unlike in Dunning and ANO BS, in the case of the augmented σBS (aσBS), the CISD energy of the dimer at the best available equilibrium distance (BED) $R_e = 2.9676$ Å [26] is considered for the optimization of σBS exponents.

3. Computational Details

As it is well known, basis sets augmented with polarization and diffuse functions can adequately reproduce weak dispersion interactions [9,25,32]. Bearing this in mind, to study He_2 interactions in a systematic way, we have carried out calculations using atom-centered basis sets. In particular, in this work we use correlation consistent basis sets developed by Dunning et al. [32–35] as a reference for testing the performance of σBS reported herein. Dunning BS are widely used, and they are especially well suited for our purposes because they incorporate polarization functions and, in the case of augmented versions, diffuse functions. In addition, it has been shown that correlation-consistent basis sets doubly augmented with diffuse functions can be used to nearly saturate the radial contribution to the dispersion energy in rare gas dimers [8,9,36]. Furthermore, these basis sets are grouped in families whose members are ranked in size (and quality), and this facilitates the extrapolation of results to the CBS limit.

For testing the accuracy of energy values attained with the σBS, we have also used the ANO BS, which have proved to be able to exhaust the capabilities of the underlying Gaussian expansion basis (to minimize the contraction error) and provide a highly accurate reference [37,38].

Electronic structure calculations of He and He_2 have been carried out at HF, CISD and FCI levels, using MOLPRO suite [39].

4. Results and Discussion

4.1. Sigma Basis Sets vs. Dunning and ANO Basis Sets

The composition of the basis sets for the He atom is detailed in Table 1, in which the numbers of exponentials, primitives and contractions are quoted. Notice that the composition is the same for the three families, the only difference being the number of primitives, smaller in Dunning BS and similar in ANO and σBS. On the other hand, the number of exponentials is smaller in σBS than in the other two.

Table 1. Composition of Dunning, ANO and sigma basis sets.

Basis Sets	# Exponentials	#	Primitives	#	Contracted
DZ	5	7	(4s, 1p)	5	[2s, 1p]
TZ	8	16	(5s, 2p, 1d)	14	[3s, 2p, 1d]
QZ	12	32	(6s, 3p, 2d, 1f)	30	[4s, 3p, 2d, 1f]
5Z	17	57	(7s, 4p, 3d, 2f, 1g)	55	[5s, 4p, 3d, 2f, 1g]
anoDZ	15	25	(10s, 5p)	5	[2s, 1p]
anoTZ	19	45	(10s, 5p, 4d)	14	[3s, 2p, 1d]
anoQZ	23	66	(10s, 5p, 4d, 3f)	30	[4s, 3p, 2d, 1f]
ano5Z	25	84	(10s, 5p, 4d, 3f, 2g)	55	[5s, 4p, 3d, 2f, 1g]
σDZ	10	19	(10s, 3p)	5	[2s, 1p]
σTZ	10	37	(10s, 4p, 3d)	14	[3s, 2p, 1d]
σQZ	10	66	(10s, 5p, 4d, 3f)	30	[4s, 3p, 2d, 1f]
σ5Z	10	108	(10s, 6p, 5d, 4f, 3g)	55	[5s, 4p, 3d, 2f, 1g]
aDZ	7	11	(5s, 2p)	9	[3s, 2p]
aTZ	11	25	(6s, 3p, 2d)	23	[4s, 3p, 2d]
aQZ	16	48	(7s, 4p, 3d, 2f)	46	[5s, 4p, 3d, 2f]
a5Z	22	82	(8s, 5p, 4d, 3f, 2g)	80	[6s, 5p, 4d, 3f, 2g]
aanoDZ	15	25	(10s, 5p)	9	[3s, 2p]
aanoTZ	19	45	(10s, 5p, 4d)	23	[4s, 3p, 2d]
aanoQZ	23	66	(10s, 5p, 4d, 3f)	46	[5s, 4p, 3d, 2f]
aano5Z	25	84	(10s, 5p, 4d, 3f, 2g)	80	[6s, 5p, 4d, 3f, 2g]
aσDZ	11	23	(11s, 4p)	9	[3s, 2p]
aσTZ	11	46	(11s, 5p, 4d)	23	[4s, 3p, 2d]
aσQZ	11	82	(11s, 6p, 5d, 4f)	46	[5s, 4p, 3d, 2f]
aσ5Z	11	133	(11s, 7p, 6d, 5f, 4g)	80	[6s, 5p, 4d, 3f, 2g]

In Table 2, some properties computed at HF and FCI levels are reported. As can be appreciated, FCI energies computed with σBS are lower than those computed with their equivalent partners of the other two families. In the case of He_2, energies have been calculated at the equilibrium distance optimized at each computational level, and no result is displayed in the case of HF calculations with ANO or σBS because no minimum is found in these cases.

Table 2. He and He_2 energies (E_h), dissociation energies and equilibrium distances at HF and FCI levels.

	He Atom		He_2 Dimer		D_e(μH)		R_e(Å)	
	HF	FCI	HF	FCI	HF	FCI	HF	FCI
DZ	−2.85516048	−2.88759483	−5.71032241	−5.77519594	1.47	6.28	3.209	3.090
TZ	−2.86115334	−2.90023217	−5.72230756	−5.80047345	0.87	9.11	3.637	3.296
QZ	−2.86151423	−2.90241088	−5.72302886	−5.80483479	0.40	13.03	3.806	3.236
5Z	−2.86162483	−2.90315188	−5.72324982	−5.80632032	0.15	16.55	4.005	3.155
anoDZ	−2.86165583	−2.89748229	—	−5.79496717	—	2.60	—	3.623
anoTZ	−2.86166988	−2.90170267	—	−5.80341360	—	8.26	—	3.335
anoQZ	−2.86167139	−2.90282021	—	−5.80565442	—	13.99	—	3.194
ano5Z	−2.86167205	−2.90324192	—	−5.80650279	—	18.95	—	3.112
σDZ	−2.86166454	−2.89755779	—	−5.79511772	—	2.13	—	3.645
σTZ	−2.86166897	−2.90175275	—	−5.80351122	—	5.73	—	3.426
σQZ	−2.86167087	−2.90283439	—	−5.80567817	—	9.39	—	3.301
σ5Z	−2.86167106	−2.90324791	—	−5.80651300	—	17.17	—	3.141
aDZ	−2.85570467	−2.88954849	−5.71141075	−5.77914013	1.41	43.16	3.417	2.998
aTZ	−2.86118343	−2.90059792	−5.72236734	−5.80122808	0.49	32.22	3.806	3.007
aQZ	−2.86152200	−2.90253360	−5.72304429	−5.80510033	0.30	33.13	4.057	2.979
a5Z	−2.86162693	−2.90320053	−5.72325393	−5.80643456	0.07	34.96	4.131	2.981
aanoDZ	−2.86166988	−2.89996594	—	−5.79993919	—	7.31	—	3.378
aanoTZ	−2.86167139	−2.90245288	—	−5.80490575	—	13.88	—	3.197
aanoQZ	−2.86167205	−2.90312610	—	−5.80627113	—	18.93	—	3.113
aano5Z	−2.86167282	−2.90339507	—	−5.80681090	—	20.76	—	3.088
aσDZ	−2.86167614	−2.89997789	—	−5.79996081	—	5.02	—	3.464
aσTZ	−2.86167722	−2.90245411	—	−5.80491755	—	9.33	—	3.300
aσQZ	−2.86167716	−2.90313545	—	−5.80628469	—	13.84	—	3.197
aσ5Z	−2.86167707	−2.90340070	—	−5.80681741	—	16.02	—	3.143

Dissociation energies of He_2, D_e have been computed as the difference between the energies of the separated atoms and that in the minimum of the curve, R_e. In the case of Dunning BS, a shallow minimum is obtained at the HF level at a distance that increases with the BS size. As mentioned above, this minimum does not appear in HF calculations with the two other families, suggesting that the improvements in the core zone with respect to those of Dunning should prevent the presence of a minimum at the HF level.

The convergence of D_e and R_e towards the currently best available values ($D_e = 34.82\ \mu E_h$ and $R_e = 2.9676$ Å) [26] has been analyzed in FCI calculations, and the results are depicted in Figures 1 and 2. In Figure 1, D_e is plotted vs. the BS size for the three types of BS, including the augmented versions. It is observed that non-augmented BS have a slow convergence towards the exact result, and they are far away from it even at the 5Z level. Augmented ANO and σBS improve the results slightly, but the convergence is still far from the reference. On the other hand, Dunning augmented BS, although not optimized for CISD energy of He_2 at R_e, yield astonishing precise depth values of the well.

In the case of equilibrium distance, displayed in Figure 2, the behavior is quite similar to that of D_e, with Dunning augmented BS giving a very good agreement even with the smallest set (aDZ).

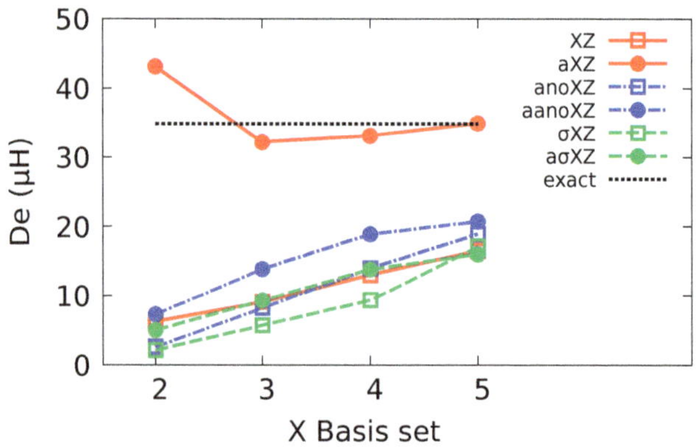

Figure 1. Dissociation energy, D_e (μE_h).

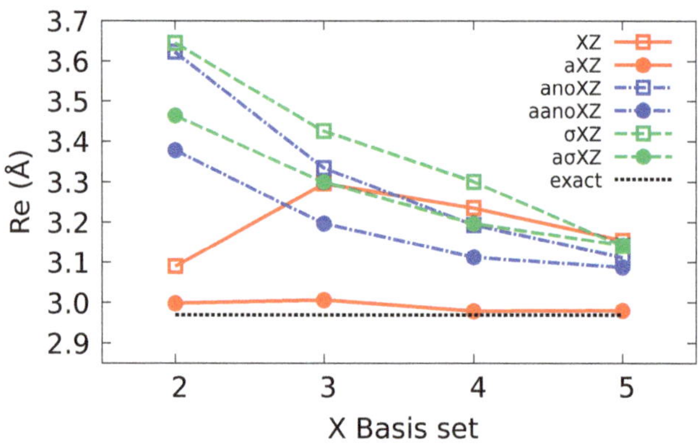

Figure 2. Equilibrium distance, R_e (angstrom).

4.2. Multiple Augmented Basis Sets

Given the excellent performance of Dunning BS with regard to the dissociation energy of He_2, and in an attempt to understand why this is not so in the case of σBS, we decided to explore the performance of multiple augmented basis sets in both families.

We developed double and triple augmented σBS but, unlike in Dunning BS, which were designed to improve the polarizability of the He atom, we followed the methodology based only on energy. Thus, we obtained double and triple augmented σBS (daσXZ and taσXZ) to compare with the corresponding Dunning BS (daXZ and taXZ, for short). These multiply augmented BS are described in Table 3. In analogy with Table 1, the number of exponentials, primitives and contracted functions are given, and the same comments on the composition, size and characteristics mentioned before also hold for the new multiple augmented σBS.

Table 3. Composition of double and triple augmented Dunning and σBS.

Basis Sets	# Exponentials	#	Primitives	#	Contracted
daDZ	9	15	(6s, 3p)	13	[4s, 3p]
daTZ	14	34	(7s, 4p, 3d)	32	[5s, 4p, 3d]
daQZ	20	64	(8s, 5p, 4d, 3f)	62	[6s, 5p, 4d, 3f]
da5Z	27	107	(9s, 6p, 5d, 4f, 3g)	105	[7s, 6p, 5d, 4f, 3g]
daσDZ	12	27	(12s, 5p)	13	[4s, 3p]
daσTZ	12	55	(12s, 6p, 5d)	32	[5s, 4p, 3d]
daσQZ	12	98	(12s, 7p, 6d, 5f)	62	[6s, 5p, 4d, 3f]
daσ5Z	12	158	(12s, 8p, 7d, 6f, 5g)	105	[7s, 6p, 5d, 4f, 3g]
taDZ	11	19	(7s, 4p)	17	[5s, 4p]
taTZ	17	43	(8s, 5p, 4d)	41	[6s, 5p, 4d]
taQZ	24	80	(9s, 6p, 5d, 4f)	78	[7s, 6p, 5d, 4f]
ta5Z	32	132	(10s, 7p, 6d, 5f, 4g)	130	[8s, 7p, 6d, 5f, 4g]
taσDZ	13	31	(13s, 6p)	17	[5s, 4p]
taσTZ	13	64	(13s, 7p, 6d)	41	[6s, 5p, 4d]
taσQZ	13	114	(13s, 8p, 7d, 6f)	78	[7s, 6p, 5d, 4f]
taσ5Z	13	183	(13s, 9p, 8d, 7f, 6g)	130	[8s, 7p, 6d, 5f, 4g]

Results obtained with multiple augmented BS are summarized in Table 4, whose structure is identical to that of Table 2. Regarding the atomic calculations at HF and FCI levels, it is observed again that energies computed with σBS are always better than those of the equivalent Dunning sets. In fact, taσXZ, albeit not directly intended, yields a good agreement with the HF limit (−2.861679995 E_h) [40,41] even for the smallest set of the series, with an error ca. 10^{-6} E_h in all cases.

Table 4. He and He_2 energies (E_h), dissociation energies and equilibrium distances at HF and FCI levels for double and triple augmented BS.

	He Atom		He_2 Dimer		D_e(μH)		R_e(Å)	
	HF	FCI	HF	FCI	HF	FCI	HF	FCI
daDZ	−2.85570939	−2.88959436	−5.71142686	−5.77924689	8.09	58.17	3.288	2.964
daTZ	−2.86118387	−2.90060812	−5.72237146	−5.80126097	3.72	44.72	3.837	2.995
daQZ	−2.86152234	−2.90253661	−5.72304596	−5.80511435	1.28	41.13	4.045	2.958
da5Z	−2.86162717	−2.90320194	−5.72325487	−5.80644187	0.52	37.98	3.882	2.965
daσDZ	−2.86167803	−2.90035561	—	−5.80072031	—	9.08	—	3.320
daσTZ	−2.86167821	−2.90265038	—	−5.80531813	—	17.37	—	3.136
daσQZ	−2.86167822	−2.90323388	—	−5.80648926	—	21.52	—	3.080
daσ5Z	−2.86167816	−2.90343963	—	−5.80690928	—	30.02	—	3.002
taDZ	−2.85571146	−2.88960188	−5.71143417	−5.77926374	11.22	59.92	3.351	2.984
taTZ	−2.86118406	−2.90061002	−5.72237488	−5.80128012	6.76	60.33	3.654	2.996
taQZ	−2.86152247	−2.90253721	−5.72304861	−5.80511936	3.67	44.93	3.973	2.957
ta5Z	−2.86162726	−2.90320225	−5.72325624	−5.80644560	1.72	41.10	3.850	2.960
taσDZ	−2.86167928	−2.90046500	—	−5.80094227	—	12.26	—	3.247
taσTZ	−2.86167885	−2.90272428	—	−5.80546977	—	21.21	—	3.087
taσQZ	−2.86167886	−2.90328593	—	−5.80659710	—	25.24	—	3.037
taσ5Z	−2.86167829	−2.90344482	—	−5.80692585	—	32.21	—	2.982

With respect to the dimer, Dunning BS (daXZ and taXZ) again give minima at the HF level, the well depth being greater than that obtained with the smaller XZ and aXZ, and the equilibrium distances range from 3.2 to 4.0 Å. Again, the σBS (daσXZ and taσXZ) give no minimum. In all cases, the FCI energies for the diatom tend to the exact value −5.80748357 E_h, estimated from the best available values for the He energy

(2.903724377 ... E$_h$) [42] and for the well depth (34.82 μE$_h$) [26]. Notice that energy values computed with daσXZ y taσXZ are significantly lower than those obtained with the equivalent Dunning BS.

Regarding the dissociation energy, Figure 3 shows that in Dunning BS, the inclusion of new polarization functions in the sequence aXZ, daXZ, taXZ tends to worsen the results, whereas in σBS, a regular improvement in the value of D_e is observed. In the case of equilibrium distance, R_e, Dunning BS produce results closer to the exact than those of σBS, but the results attained with the latter also exhibit a regular approach to the right value, as shown in Figure 4.

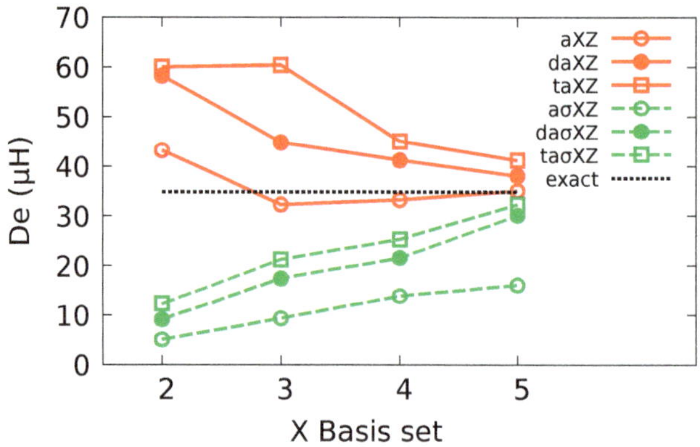

Figure 3. Dissociation energy, D_e (µE$_h$), multiple augmented BS.

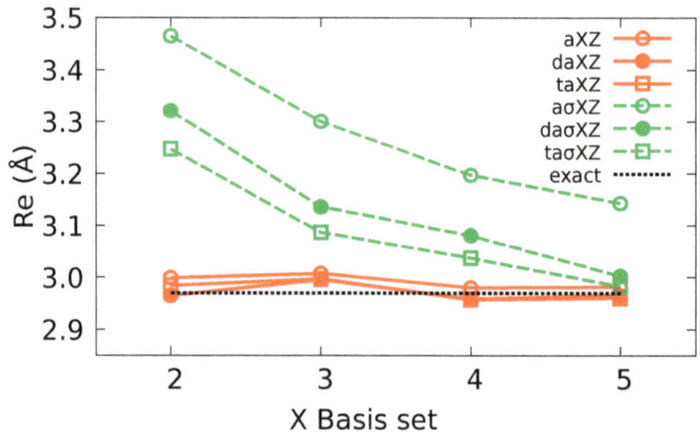

Figure 4. Equilibrium distance, R_e (angstrom), multiple augmented BS.

4.3. Extrapolation to CBS

The performance of the basis sets in extrapolations to CBS has been analyzed in the case of correlation energy both for the atom and for the dimer at BED. Although there are many extrapolation schemes, we have taken one due to Helgaker [43]:

$$E_{XZ} = E_{CBS} + \alpha / X^{\beta}.$$

The correlation energy of the atom has been calculated as the difference between the exact value [42] and the HF limit energy [40] and that of the dimer with the HF value at BED and adding the energy of the well (34.82 µE$_h$).

Correlation energy values thus computed plus the extrapolated CBS energies obtained for the Dunning and σBS are collected in Table 5 and depicted in Figures 5 and 6. The best available correlation energies appear at the top of Table 5, with values of 0.0420444 E_h and 0.0841528 E_h for the atom and the dimer, respectively. To facilitate comparisons, figures coincident with reference values are displayed in bold type.

Table 5. Correlation energy (E_h) for He and He_2 dimer at the equilibrium distance (2.97 Å).

	He Atom		**He_2 Dimer**	
Limit	**0.0420444**		**0.0841528**	
BS	**Dunning**	**Sigma**	**Dunning**	**Sigma**
DZ	−0.0324343	−0.0358932	−0.0648756	−0.0717958
TZ	−0.0390788	−0.0400838	−0.0781781	−0.0801879
QZ	−0.0408967	−0.0411635	−0.0818256	−0.0823559
5Z	−0.0415271	−0.0415768	−0.0830947	−0.0831955
CBS [a]	−**0.042**3430	−**0.0420**352	−**0.084**7403	−**0.0841**253
aDZ	−0.0338438	−0.0383018	−0.0677448	−0.0766221
aTZ	−0.0394145	−0.0407769	−0.0788879	−0.0815831
aQZ	−0.0410116	−0.0414583	−0.0820829	−0.0829540
a5Z	−0.0415736	−0.0417231	−0.0832083	−0.0834887
CBS [a]	−**0.042**3532	−**0.0420**496	−**0.084**7687	−**0.08415**10
daDZ	−0.0338850	−0.0386776	−0.0678307	−0.0773832
daTZ	−0.0394242	−0.0409722	−0.0789168	−0.0819873
daQZ	−0.0410143	−0.0415557	−0.0820937	−0.0831597
da5Z	−0.0415748	−0.0417615	−0.0832142	−0.0835811
CBS [a]	−**0.042**3531	−**0.0419**952	−**0.084**7654	−**0.0840**554
taDZ	−0.0338904	−0.0385857	−0.0678420	−0.0776054
taTZ	−0.0394260	−0.0410454	−0.0789302	−0.0821388
taQZ	−0.0410147	−0.0416071	−0.0820963	−0.0832671
ta5Z	−0.0415750	−0.0418099	−0.0832159	−0.0836765
CBS [a]	−**0.042**3526	−**0.0420**285	−**0.084**7545	−**0.0841**117

[a] Extrapolation to complete basis set with $E_{XZ} = E_{CBS} + \alpha / X^{\beta}$.

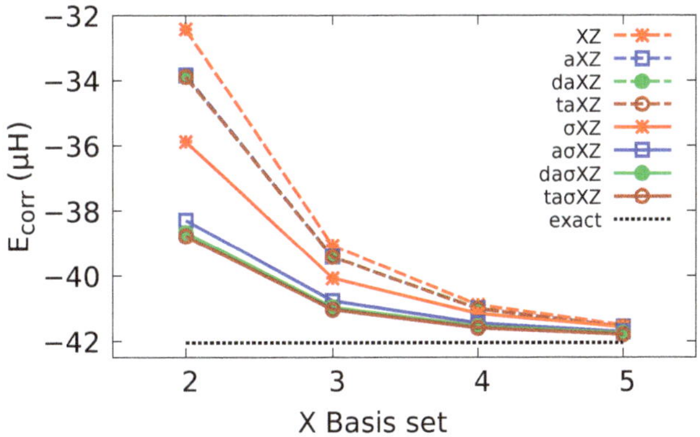

Figure 5. He: correlation energy.

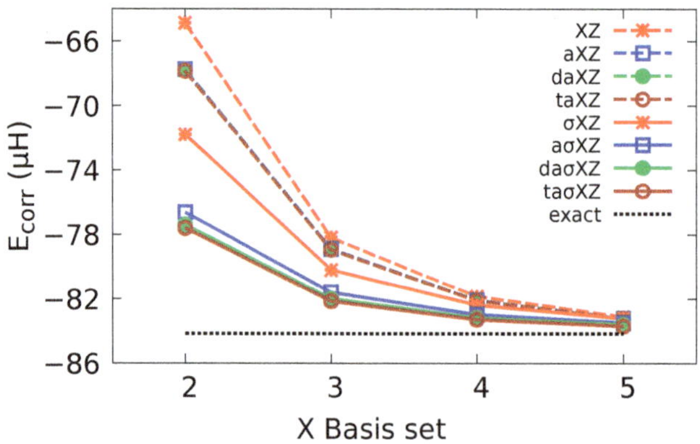

Figure 6. He_2: correlation energy.

The better performance of σBS over Dunning BS is evident. CBS extrapolation yields a gain of at least two figures in the former, while scarcely one figure is gained in the latter. The case of σaXZ is specially noticeable, as three figures are gained in the CBS extrapolation both in the atom and in the diatom. Furthermore, CBS extrapolation in Dunning BS lies below the exact value, whereas it converges to the right value in the case of σBS. This behavior is clearly visible in Figures 7 and 8.

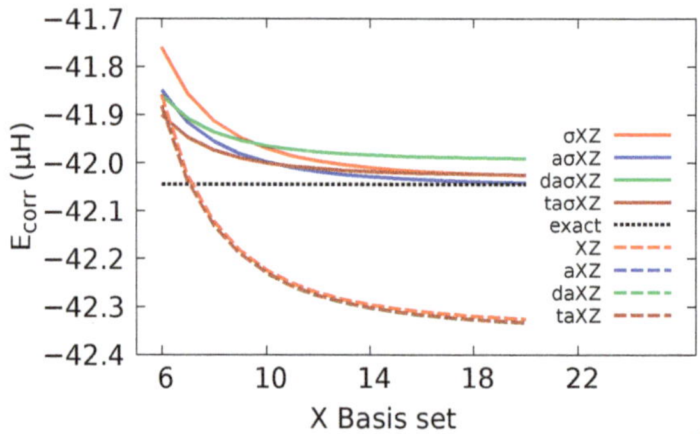

Figure 7. He: CBS extrapolation of correlation energy.

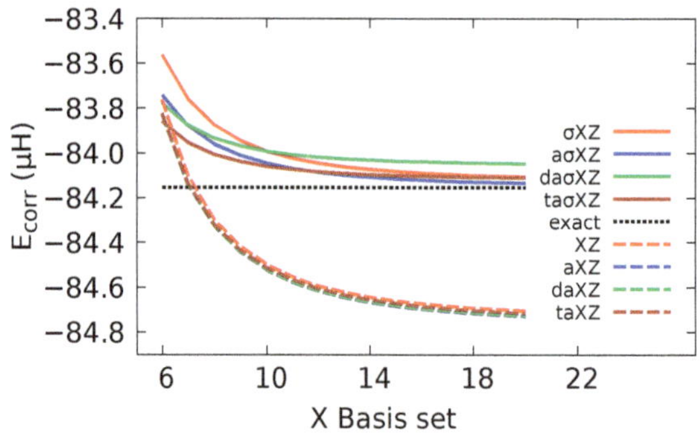

Figure 8. He_2: CBS extrapolation of correlation energy.

5. Conclusions

A new scheme for developing basis sets is reported and applied to helium atom and dimer. The σBS thus developed are configured by combining the same set of exponentials with different angular functions and saturating the corresponding one-electron spaces. The new family ranges from DZ to 5Z, with the same composition as Dunning BS, and also includes, simply, double and triple augmented versions. Energy values for helium atom and diatom computed with the σBS are lower than those obtained with their partners in Dunning BS, and similar to ANO BS, both at HF and FCI levels.

The analysis of the energy of He_2 reveals the presence of a minimum in the curve at the HF level in the case of Dunning BS that does not appear with σ or ANO BS. In the case of FCI, all calculations yield a well whose depth tends to the reference value as the BS quality improves. Augmented σ BS display a good convergence which, in the case of multiple augmented BS, is even more regular than in the corresponding Dunning BS.

CBS extrapolation of the correlation energy of He_2 at BED has also been examined, proving that the results with σBS are clearly superior to those attained with Dunning BS. Especially remarkable is the result with the aσXZ, which shows an agreement of five decimal figures with the best available result. Furthermore, it can be noticed that the CBS extrapolation with Dunning BS falls below the exact value, while the sigma converges to the right value.

Supplementary Materials: The σBS used in this work can be found online in MOLPRO format at https://www.mdpi.com/article/10.3390/computation10050065/s1.

Author Contributions: Conceptualization, I.E. and J.M.G.d.l.V.; methodology, I.E.; software, I.E., G.R., R.L. and J.M.G.d.l.V.; data curation, I.E. and J.M.G.d.l.V.; writing—original draft preparation, I.E., G.R., R.L. and J.M.G.d.l.V.; writing—review and editing, I.E., G.R., R.L. and J.M.G.d.l.V. All authors have read and agreed to the published version of the manuscript.

Funding: This research received no external funding

Institutional Review Board Statement: Not applicable

Informed Consent Statement: Not applicable

Data Availability Statement: Not applicable

Conflicts of Interest: The authors declare no conflict of interest.

Abbreviations

The following abbreviations are used in this manuscript:

ANO	Atomic Natural Orbitals
aANO	Augmented ANO
aXZ	aug-cc-pVXZ
aσBS	Augmented Sigma Basis Set
BED	Best Available Equilibrium Distance of He_2
BS	Basis set(s)
BSSE	Basis Set Superposition Error
CBS	Complete Basis Set
CC	Coupled Clusters
CCSD(T)	Coupled Clusters Singles Doubles (Triples)
CISD	Configuration Interactions Singles and Doubles
CP	Cunterpoise
daXZ	aug-aug-cc-pVXZ
daσBS	Double Augmented Sigma Basis Set
FCI	Full Configuration Interaction
HF	Hartree-Fock
MP	Möller-Plesset
MR-ACPF	Multi-Reference Averaged Coupled-Pair Functional
MRCI	Multi-Reference Configuration Interaction

SAPT	Symmetry-Adapted Intermolecular Perturbation Theory
σBS	Sigma Basis Set
taXZ	aug-aug-aug-cc-pVXZ
taσBS	Triple Augmented Sigma Basis Set
vdW	van der Waals
XZ	cc-pVXZ

References

1. Riley, K.E.; Pitoňák, M.; Jurečka, P.; Hobza, P. Stabilization and Structure Calculations for Noncovalent Interactions in Extended Molecular Systems Based on Wave Function and Density Functional Theories. *Chem. Rev.* **2010**, *110*, 5023–5063. [CrossRef] [PubMed]
2. Becke, A.D.; Johnson, E.R. Exchange-hole dipole moment and the dispersion interaction revisited. *J. Chem. Phys.* **2007**, *127*, 154108. [CrossRef] [PubMed]
3. Otero-de-la Roza, A.; Johnson, E.R. Van der Waals interactions in solids using the exchange-hole dipole moment model. *J. Chem. Phys.* **2012**, *136*, 174109. [CrossRef] [PubMed]
4. Tantardini, C.; Michalchuk, A.A.L.; Samtsevich, A.; Rota, C.; Kvashnin, A.G. The Volumetric Source Function: Looking Inside van der Waals Interactions. *Sci. Rep.* **2020**, *10*, 7816. [CrossRef] [PubMed]
5. Laschuk, E.F.; Martins, M.M.; Evangelisti, S. Ab initio potentials for weakly interacting systems: Homonuclear rare gas dimers. *Int. J. Quantum Chem.* **2003**, *95*, 303–312. [CrossRef]
6. Sheng, X.; Toennies, J.P.; Tang, K.T. Conformal Analytical Potential for All the Rare Gas Dimers over the Full Range of Internuclear Distances. *Phys. Rev. Lett.* **2020**, *125*, 253402. [CrossRef] [PubMed]
7. Van Mourik, T.; van Lenthe, J.H. Benchmark full configuration interaction calculations on the helium dimer. *J. Chem. Phys.* **1995**, *102*, 7479–7483. [CrossRef]
8. Van Mourik, T.; Dunning, T.H., Jr. A new ab initio potential energy curve for the helium dimer. *J. Chem. Phys.* **1999**, *111*, 9248–9258. [CrossRef]
9. Van Mourik, T.; Wilson, A.K.; Dunning, T.H., Jr. Benchmark calculations with correlated molecular wavefunctions. XIII. Potential energy curves for He_2, Ne_2 and Ar_2 using correlation consistent basis sets through augmented sextuple zeta. *Mol. Phys.* **1999**, *96*, 529–547. [CrossRef]
10. Korona, T.; Williams, H.L.; Bukowski, R.; Jeziorski, B.; Szalewicz, K. Helium dimer potential from symmetry-adapted perturbation theory calculations using large Gaussian geminal and orbital basis sets. *J. Chem. Phys.* **1997**, *106*, 5109–5122. [CrossRef]
11. Cybulski, S.M.; Toczyłowski, R.R. Ground state potential energy curves for He_2, Ne_2, Ar_2, He–Ne, He–Ar, and Ne–Ar: A coupled-cluster study. *J. Chem. Phys.* **1999**, *111*, 10520–10528. [CrossRef]
12. Gdanitz, R.J. Accurately solving the electronic Schrödinger equation of atoms and molecules using explicitly correlated (r_{12}-)MR-CI IV. The helium dimer (He_2). *Mol. Phys.* **1999**, *96*, 1423–1434.
13. Gdanitz, R.J. Accurately solving the electronic Schrödinger equation of atoms and molecules by extrapolating to the basis set limit. I. The helium dimer (He_2). *J. Chem. Phys.* **2000**, *113*, 5145–5153. [CrossRef]
14. Gdanitz, R.J. Accurately solving the electronic Schrödinger equation of atoms and molecules using explicitly correlated (r_{12}-) MR-CI. VI. The helium dimer (He_2) revisited. *Mol. Phys.* **2001**, *99*, 923–930. [CrossRef]
15. Van de Bovenkamp, J.; van Duijneveldt, F.B. MRCI calculations on the helium dimer employing an interaction optimized basis set. *J. Chem. Phys.* **1999**, *110*, 11141–11151. [CrossRef]
16. Klopper, W. A critical note on extrapolated helium pair potentials. *J. Chem. Phys.* **2001**, *115*, 761–765. [CrossRef]
17. Anderson, J.B. Comment on "An exact quantum Monte Carlo calculation of the helium–helium intermolecular potential" [J. Chem. Phys. 115, 4546 (2001)]. *J. Chem. Phys.* **2004**, *120*, 9886–9887. [CrossRef]
18. Komasa, J. Exponentially correlated Gaussian functions in variational calculations. Momentum space properties of the ground state helium dimer. *J. Chem. Phys.* **2001**, *115*, 158–165. [CrossRef]
19. Cencek, W.; Jeziorska, M.; Bukowski, R.; Jaszuński, M.; Jeziorski, B.; Szalewicz, K. Helium Dimer Interaction Energies from Gaussian Geminal and Orbital Calculations. *J. Phys. Chem. A* **2004**, *108*, 3211–3224. [CrossRef]
20. Patkowski, K.; Cencek, W.; Jeziorska, M.; Jeziorski, B.; Szalewicz, K. Accurate pair interaction energies for helium from supermolecular Gaussian geminal calculations. *J. Phys. Chem. A* **2007**, *111*, 7611–7623. [CrossRef]
21. Gutowski, M.; Lenthe, J.V.; Verbeek, J.; Van Duijneveldt, F.; Chałasinski, G. The basis set superposition error in correlated electronic structure calculations. *Chem. Phys. Lett.* **1986**, *124*, 370–375. [CrossRef]
22. Varandas, A. Can extrapolation to the basis set limit be an alternative to the counterpoise correction? A study on the helium dimer. *Theor. Chem. Acc.* **2008**, *119*, 511–521. [CrossRef]
23. Alvarez-Idaboy, J.R.; Galano, A. Counterpoise corrected interaction energies are not systematically better than uncorrected ones: Comparison with CCSD(T) CBS extrapolated values. *Theor. Chem. Acc.* **2010**, *126*, 75–85. [CrossRef]
24. Sheng, X.W.; Mentel, L.; Gritsenko, O.V.; Baerends, E.J. Counterpoise correction is not useful for short and Van der Waals distances but may be useful at long range. *J. Comput. Chem.* **2011**, *32*, 2896–901. [CrossRef] [PubMed]
25. Varandas, A.J.C. Extrapolation to the Complete Basis Set Limit without Counterpoise. The Pair Potential of Helium Revisited. *J. Phys. Chem. A* **2010**, *114*, 8505–8516. [CrossRef] [PubMed]

26. Cencek, W.; Przybytek, M.; Komasa, J.; Mehl, J.B.; Jeziorski, B.; Szalewicz, K. Effects of adiabatic, relativistic, and quantum electrodynamics interactions on the pair potential and thermophysical properties of helium. *J. Chem. Phys.* **2012**, *136*, 224303. [CrossRef]
27. Luo, F.; Giese, C.F.; Gentry, W.R. Direct measurement of the size of the helium dimer. *J. Chem. Phys.* **1996**, *104*, 1151–1154. [CrossRef]
28. García de la Vega, J.M.; Miguel, B.; Ramírez, G. Single-exponent Slater function expansions for lithium to neon atoms. *J. Phys. B* **1996**, *29*, 5273–5282. [CrossRef]
29. Koga, T.; García de la Vega, J.M.; Miguel, B. Double-zeta Slater-type basis sets with noninteger principal quantum numbers and common exponents. *Chem. Phys. Lett.* **1998**, *283*, 97–101. [CrossRef]
30. Miguel, B.; Koga, T.; García de la Vega, J.M. Double- and triple-zeta Slater-type basis sets with common exponents. *Theor. Chem. Acc.* **2000**, *104*, 167–171. [CrossRef]
31. Ema, I.; García de la Vega, J.M.; Ramírez, G.; López, R.; Rico, J.F.; Meissner, H.; Paldus, J. Polarized basis sets of Slater-type orbitals: H to Ne atoms. *J. Comput. Chem.* **2003**, *24*, 859–868. [CrossRef] [PubMed]
32. Kendall, R.A.; Dunning, T.H., Jr.; Harrison, R.J. Electron affinities of the first-row atoms revisited. Systematic basis sets and wave functions. *J. Chem. Phys.* **1992**, *96*, 6796–6806. [CrossRef]
33. Dunning, T.H., Jr. Gaussian basis sets for use in correlated molecular calculations. I. The atoms boron through neon and hydrogen. *J. Chem. Phys.* **1989**, *90*, 1007–1023. [CrossRef]
34. Woon, D.E.; Dunning, T.H., Jr. Gaussian basis sets for use in correlated molecular calculations. III. The atoms aluminum through argon. *J. Chem. Phys.* **1993**, *98*, 1358–1371. [CrossRef]
35. Woon, D.E.; Dunning, T.H., Jr. Gaussian basis sets for use in correlated molecular calculations. IV. Calculation of static electrical response properties. *J. Chem. Phys.* **1994**, *100*, 2975–2988. [CrossRef]
36. Woon, D.E. Benchmark calculations with correlated molecular wave functions. V. The determination of accurate ab initio intermolecular potentials for He_2, Ne_2, and Ar_2. *J. Chem. Phys.* **1994**, *100*, 2838–2850. [CrossRef]
37. Almlöf, J.; Taylor, P.R. Atomic natural orbital (ANO) basis sets for quantum chemical calculations. In *Advances in Quantum Chemistry*; Elsevier: Amsterdam, The Netherlands, 1991; Volume 22, pp. 301–373.
38. Neese, F.; Valeev, E.F. Revisiting the atomic natural orbital approach for basis sets: Robust systematic basis sets for explicitly correlated and conventional correlated ab initio methods? *J. Chem. Theory Comput.* **2011**, *7*, 33–43. [CrossRef]
39. Werner, H.J.; Knowles, P.J.; Knizia, G.; Manby, F.R.; Schütz, M.; Celani, P.; Korona, T.; Lindh, R.; Mitrushenkov, A.; Rauhut, G.; et al. MOLPRO. A Package of Ab Initio Programs. Version 2012.1. 2012. Available online: https://www.molpro.net/ (accessed on 19 April 2022).
40. Bunge, C.F.; Barrientos, J.A.; Bunge, A.V.; Cogordan, J.A. Hartree-Fock and Roothaan-Hartree-Fock energies for the ground states of He through Xe. *Phys. Rev. A* **1992**, *46*, 3691–3696. [CrossRef]
41. Bunge, C.; Barrientos, J.; Bunge, A. Roothaan-Hartree-Fock Ground-State Atomic Wave Functions: Slater-Type Orbital Expansions and Expectation Values for Z = 2-54. *Atom. Data Nucl. Data Tables* **1993**, *53*, 113–162. [CrossRef]
42. Nakashima, H.; Nakatsuji, H. Solving the Schrödinger equation for helium atom and its isoelectronic ions with the free iterative complement interaction (ICI) method. *J. Chem. Phys.* **2007**, *127*, 224104. [CrossRef]
43. Helgaker, T.; Klopper, W.; Koch, H.; Noga, J. Basis-set convergence of correlated calculations on water. *J. Chem. Phys.* **1997**, *106*, 9639–9646. [CrossRef]

computation

MDPI

Article

LFDFT—A Practical Tool for Coordination Chemistry

Harry Ramanantoanina

Department Chemie, Johannes Gutenberg-Universität Mainz, Fritz-Strassmann Weg 2, 55128 Mainz, Germany; haramana@uni-mainz.de

Abstract: The electronic structure of coordination compounds with lanthanide ions is studied by means of density functional theory (DFT) calculations. This work deals with the electronic structure and properties of open-shell systems based on the calculation of multiplet structure and ligand-field interaction, within the framework of the Ligand–Field Density-Functional Theory (LFDFT) method. Using effective Hamiltonian in conjunction with the DFT, we are able to reasonably calculate the low-lying excited states of the molecular [Eu(NO$_3$)$_3$(phenanthroline)$_2$] complex, subjected to the Eu^{3+} configuration $4f^6$. The results are compared with available experimental data, revealing relative uncertainties of less than 5% for many energy levels. We also demonstrate the ability of the LFDFT method to simulate absorption spectrum, considering cerocene as an example. Ce $M_{4,5}$ X-ray absorption spectra are simulated for the complexes [Ce($\eta^8-C_8H_8$)$_2$] and [Ce($\eta^8-C_8H_8$)$_2$][Li(tetrahydrofurane)$_4$], which are approximated by the Ce oxidation states 4+ and 3+, respectively. The results showed a very good agreement with the experimental data for the Ce^{3+} compound, unlike for the Ce^{4+} one, where charge transfer electronic structure is still missing in the theoretical model. Therefore this presentation reports the benefits of having a theoretical method that is primarily dedicated to coordination chemistry, but it also outlines limitations and places the ongoing developmental efforts in the broader context of treating complex molecular systems.

Keywords: atomic multiplet theory; crystal/ligand-field theory; coordination compounds; electronic structure; density functional theory

Citation: Ramanantoanina, H. LFDFT—A Practical Tool for Coordination Chemistry. *Computation* **2022**, *10*, 70. https://doi.org/10.3390/computation10050070

Received: 31 March 2022
Accepted: 28 April 2022
Published: 2 May 2022

1. Introduction

Coordination compounds play an important role in modern chemistry, as they are involved in many fields of research: for example in catalysis [1,2], optics [3,4], magnetism [5–7], etc. The electronic structures of coordination compounds of metal ions (including transition metals, lanthanide, and actinide elements) exhibit open-shell species and near-degeneracy correlation [8–10]. Low-lying excited states are often very challenging to identify from the experiments, and to calculate from theoretical modeling. In particular, the latter is not possible unless a proper treatment of the multi-electronic system is taken into consideration [11–13]. Post Hartree–Fock methods, including many-body treatment of electron correlation effects, have been extensively developed to deal with coordination compounds. For instance, complete active space self-consistent field (CASSCF) and related methodologies [14–16] are currently enjoying wide popularity in the community of computational chemists. However, because the configuration interaction expansion increases exponentially with the number of active orbitals, the calculations of large systems becoming difficult.

Kohn–Sham Density Functional Theory (DFT), on the other hand, is generally applied to ground state electronic structure [17–19]. Its scope includes calculation of large size molecules as well as condensed matter [20]. In DFT, excited states are often approached *via* linear response theory as it is implemented in the time-dependent DFT (TDDFT) formalism [21–23]. The disadvantage of TDDFT, however, is that it lacks computational protocols for addressing highly correlated electrons, and multiplet structures, which interactions are very relevant in coordination chemistry [24,25]. Therefore, new developments in DFT encompass a methodology to incorporate many-body corrections and configuration interaction models [11,13,26], in order to solve open-shell electronic structures and strongly correlated materials.

We developed LFDFT, Ligand–Field Density Functional Theory [27–30], to bring a methodological concept for calculating multiplet structures and properties of coordination compounds with metal ions across the periodic table of elements. In this work, we demonstrate how LFDFT can be applied to solve electronic structure problems and to provide rapid estimation of spectroscopic properties at low computational cost. We use selective applications that consist in: (1) calculating the ground and low-lying excited states of the molecular [Eu(NO_3)$_3$(phenanthroline)$_2$] complex in order to understand the luminescence properties of the Eu^{3+} $4f^6 \longrightarrow 4f^6$ transitions; and (2) calculating the core-electron excitation in cerocene in order to simulate the X-ray absorption spectral profiles of the Ce $M_{4,5}$-edge that corresponds to the $4f^n \longrightarrow 3d^9 4f^{n+1}$ transitions, with n = 0, and 1.

2. Theory

2.1. General

The concept of ligand-field and its theoretical foundation have been extensively described and can be found elsewhere [31–34]. Hereafter, we are giving a more practical view of the LFDFT methodology with recent calculation possibilities and technical development. LFDFT is now available in the Amsterdam Density Functional (ADF) code that is part of the Amsterdam Modeling Suite (AMS2021 onwards) [35,36]. In LFDFT, near-degeneracy correlation is explicitly treated using ad hoc full-configuration interaction algorithm within an active subspace of the Kohn–Sham molecular orbitals [27–30]. Kohn–Sham molecular orbitals are occupied with fractional electrons to build a statistically averaged electron density that is isomorphic with the basis of a model Hamiltonian for a configuration system with open-shell d or f electrons. The model Hamiltonian is defined so that the most relevant quantum–chemical interactions are taken into consideration. These include inter-electron repulsion, relativistic spin-orbit coupling and ligand-field potential. LFDFT uses a parameterization scheme, but it does not rely upon empiricism [30]. In practice, the parameters (Slater–Condon integrals, spin-orbit coupling constants and ligand-field potential) are derived from the DFT calculation[30], therefore LFDFT has a good predictive power.

2.2. Computational Details

The main results reported in this presentation have been carried out by means of the AMS2021 code [35,36]. To perform geometries and vibrational analysis, we used DFT functional based on the generalized gradient approximation (GGA) Perdew–Burke–Ernzerhof (PBE) [37]. To calculate the electronic structure, we used DFT functional based on the GGA PBE [37], as well as hybrid functional following the B3LYP, [38], PBE0 [39,40] and KMLYP parameterization [41]. Molecular orbitals were expanded by means of the Slater-type Orbital (STO) functions for all elements at the triple-zeta plus polarization extra functions (TZ2P) level [42]. Relativistic corrections were added by using the Zeroth-Order Regular Approximation (ZORA) of the Dirac-equation method [43–45]. All electronic structures were done at the scalar ZORA relativistic level of theory, and spin-orbit coupling interaction was included by using the spin-orbit ZORA method. The self-consistent field (SCF) was set up to take into account all electrons.

2.3. Methodology

First of all, the definition of the structural inputs is described as follows. We have defined three structures for this presentation: [Eu(NO_3)$_3$(phenanthroline)$_2$] as well as cerocene [Ce($\eta^8-C_8H_8$)$_2$] and [Ce($\eta^8-C_8H_8$)$_2$][Li(tetrahydrofurane)$_4$]. For simplicity, we will denominate the two cerocene molecules by the following: [Ce(COT)$_2$] and [Ce(COT)$_2$]$^-$, respectively. We have built the molecular complexes with the help of the graphical user-interface "AMSINPUT" of the AMS2021 program [35,36], where prototypes for possible coordination compounds were available. Then we relaxed the molecular structures with

DFT by using the GGA PBE functional [37]: the total energies were minimized and the symmetry was restricted to the point groups that represented the experimental structures (i.e., C_2 [46], D_{8h} [47] and C_1 [47] for [Eu(NO_3)$_3$(phenanthroline)$_2$], [Ce(COT)$_2$] and [Ce(COT)$_2$]$^-$, respectively). We note that the [Ce(COT)$_2$]$^-$ unit had intrinsically high symmetry D_{8h}, but the descent in symmetry to C_1 resulted from the presence of the counterion [Li(tetrahydrofurane)$_4$]$^+$. The optimized structures were confirmed by vibrational analysis, and no imaginary frequencies have been computed. Schematic representations of the structures are given in Figure 1. For [Eu(NO_3)$_3$(phenanthroline)$_2$], the average Eu-N and Eu-O optimized bond lengths were 2.588 Å, and 2.559 Å, respectively, close to the experimental data (2.566 Å, and 2.510 Å) [46,48]. For [Ce(COT)$_2$], the average Ce-C bond lengths was 2.703 Å, in agreement with the experimental data (2.675 Å) [47]. For [Ce(COT)$_2$]$^-$, the average Ce-C bond lengths was 2.733 Å, also in agreement with the experimental data (2.741 Å) [47].

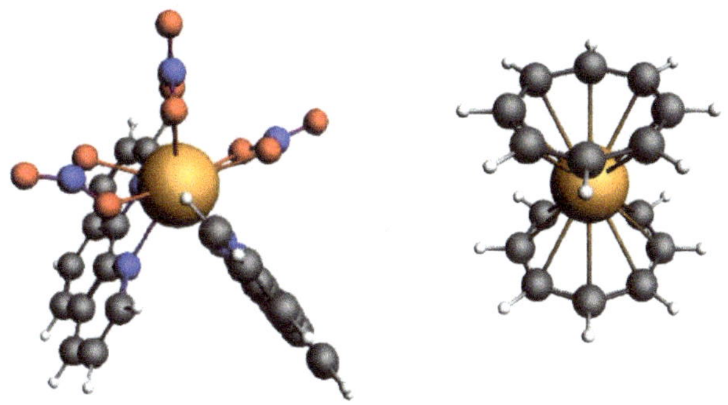

Figure 1. Ball-and-stick molecular model of the structures of [Eu(NO_3)$_3$(phenanthroline)$_2$] (left-hand-side) and cerocene (right-hand-side). Color code: grey (carbon), red (oygen), white (hydrogen), blue (nitrogen), and Orange (Europium and Cerium).

The procedural steps for computing the electronic structure are described as follows. Based on the geometrical data given by the previous step, we performed single-point DFT calculations. We used the keyword "IRREPOCCUPATIONS" in ADF [35,36] to set fractional electron occupations of the molecular orbitals. For [Eu(NO_3)$_3$ (phenanthroline)$_2$], seven molecular orbitals were occupied with fractional 6/7 electrons. These molecular orbitals were identified with large atomic $4f$ characters, and therefore constituted the active subspace of the ligand-field calculation. Figure 2 shows a section of the ADF output file [35,36] depicting this active subspace of the Kohn–Sham orbitals that were used to calculate the multiplet structures of Eu configuration $4f^6$. For the cerocene molecules, the calculations were done in two steps. First, we calculated the system with Ce^{3+} configuration $4f^1$ (and subsequently Ce^{4+} $4f^0$), which represented the ground state multiplet structure of the systems. Seven molecular orbitals with large atomic $4f$ parentage are populated with fractional electrons following similar procedure as above. Then, we calculated the systems with a core-hole, i.e., Ce^{3+} configuration $3d^9 4f^2$ (and subsequently Ce^{4+} configuration $3d^9 4f^1$), which represented the XAS electronic state. For that, three core-orbitals were occupied with fractional 9/5 electrons. These orbitals were identified with 100% atomic $3d$ characters. Additionally, as previously, seven orbitals that have larger atomic $4f$ characters were occupied with fractional electrons.

Finally, the ligand-field analyses are performed based on the single-point DFT calculation. We use the ADF keyword "LFDFT" [35,36] to set up the calculation of the multiplet energies. In the output of the LFDFT calculation, we obtained the parameters including the Slater–Condon integrals, the spin-orbit coupling constants and the matrix elements of the ligand-field potential without empirical corrections [30]. We also obtained the calculated multiplet energies and projection analysis of all the energy levels on to the atomic configuration. For the X-ray Absorption calculations, we use the keyword "LFDFT_TDM" to compute the matrix elements of the transition dipole moment that correspond to the $4f^n \longrightarrow 3d^9 4f^{n+1}$, with $n = 0, 1$.

```
-4.955  0.86   170 A        44.90%    1 F:z2y       -2.330  1.00    1 Eu
                            44.69%    1 F:z2x       -2.330  1.00    1 Eu
                             2.58%    1 F:x         -2.330  1.00    1 Eu
                             1.30%    1 P:z         -9.955  1.33    4 O
                             1.30%    1 P:z         -9.955  1.33    5 O
-4.918  0.86   171 A        43.18%    1 F:z3        -2.330  1.00    1 Eu
                            39.26%    1 F:z         -2.330  1.00    1 Eu
                            15.18%    1 F:xyz       -2.330  1.00    1 Eu
-4.909  0.86   172 A        34.89%    1 F:z2y       -2.330  1.00    1 Eu
                            26.33%    1 F:x         -2.330  1.00    1 Eu
                            17.60%    1 F:z2x       -2.330  1.00    1 Eu
                            16.97%    1 F:y         -2.330  1.00    1 Eu
-4.874  0.86   173 A        80.49%    1 F:xyz       -2.330  1.00    1 Eu
                            11.13%    1 F:z         -2.330  1.00    1 Eu
                             3.80%    1 F:z3        -2.330  1.00    1 Eu
-4.874  0.86   174 A        46.68%    1 F:y         -2.330  1.00    1 Eu
                            27.33%    1 F:z2x       -2.330  1.00    1 Eu
                            12.12%    1 F:z2y       -2.330  1.00    1 Eu
                             6.09%    1 F:x         -2.330  1.00    1 Eu
                             1.25%    1 P:y         -9.955  1.33    7 O
                             1.25%    1 P:y         -9.955  1.33    6 O
-4.855  0.86   175 A        50.16%    1 F:z3        -2.330  1.00    1 Eu
                            47.31%    1 F:z         -2.330  1.00    1 Eu
-4.848  0.86   176 A        61.20%    1 F:x         -2.330  1.00    1 Eu
                            29.78%    1 F:y         -2.330  1.00    1 Eu
                             6.36%    1 F:z2x       -2.330  1.00    1 Eu
                             1.25%    1 F:z2y       -2.330  1.00    1 Eu
```

Figure 2. Selective section of the ADF output for the calculation of [Eu(NO_3)$_3$(phenanthroline)$_2$] by using DFT with the hybrid PBE0 functional, showing the energy of the seven MOs in eV (in blue) that have larger atomic Eu $4f$ parentage together with the electron occupation number (in yellow) that are assigned with the MOs. The right-hand-side of the figure shows the percentage characters of the MOs based on Mulliken population analysis.

3. Results and Discussion

3.1. Low-Lying Excited States of [Eu(NO$_3$)$_3$(phenanthroline)$_2$]

Eu^{3+} compounds are often used in trichromatic phosphors for lighting purposes [49,50], where they are known for red-color emission. The red emission results from the $4f^6 \longrightarrow 4f^6$ transitions involving ground states ^7F_J (with J designating spin-orbit components, i.e., 0, 1, 2, …, 6), and low-lying excited states $^5D_0$ [51]. In order to understand the mechanism of the electron transition process, it is necessary to calculate these energy levels, and to assess the effect of the ligand-field interaction on to the atomic multiplets. The luminescence properties of [Eu(NO_3)$_3$(phenanthroline)$_2$] have been experimentally reported [48,52], together with analysis of the electron transition process, including the energy values at the ligand-field level [52]. We therefore choose this system as testbed for the theoretical method.

Table 1 shows the tabulated energy levels of the Eu^{3+} configuration $4f^6$ in [Eu(NO_3)$_3$ (phenanthroline)$_2$], which are obtained from the LFDFT calculations at different levels of DFT function. Results are shown for the calculated vertical excitation energies from the ground state (zero of energy) for 28 low-lying excited states that arise from the ^7F and ^5D atomic spectral terms We note that there is no influence of geometrical changes in the energy levels since the four sets of theoretical calculations in Table 1 come from the same atomic configuration of [Eu(NO_3)$_3$(phenanthroline)$_2$] (see the Methodology section). The energy levels are also compared with the experiments [52], which correspond to photophysical measurement at low temperature. The calculated energies agree within a few hundreds of cm^{-1} with the experimental data [52] (see Table 1), although larger discrepancies are more likely observed for certain levels. To assess the accuracy of theoretical results, we also list in Table 1 the calculated percent error for each level with respect to the reference energies. The percent error is calculated as 100 $*$ (theoretical values $-$ reference values)/reference values: a positive value indicating overestimation of the energy levels, and a negative value indicating underestimation. In terms of absolute value, the percent errors decrease from the GGA results to the hybrid ones, which can be attributed to the self-interaction error in DFT. Self-interaction error can, in part, be corrected by the inclusion of Hartree–Fock exchange as it is in the formulation of the DFT hybrid functional [53–55]. More particularly, the higher the percentage of the Hartree–Fock exchange, the more accurate are the predicted energy levels, as per the default values of the Hartree–Fock exchange in B3LYP [38] PBE0 [39,40], and KMLYP [41] equal 20.0%, 25.0% and 55.7%, respectively. We could not modulate the hybrid functional to include larger percentage of Hartree–Fock exchange, since we observe that although this helps improve higher-energy excited states, this also makes the prediction of the lower-energy ones poorer (see Table 1).

Table 1. Selective energy values of the multiplet states of Eu^{2+} configuration $4f^6$ (in cm^{-1}) in the system [$Eu(NO_3)_3$(phenanthroline)$_2$] obtained from LFDFT using the PBE (1) [37], B3LYP (2) [38], PBE0 (3) [39,40] and KMLYP (4) [41] functional, together with the calculated Percent Error (in%) with respect to the experimental data (Exp.)

Levels		**LFDFT**				**Exp.** [a]	**Percent Error**			
State	***I***	**(1)**	**(2)**	**(3)**	**(4)**		**(1)**	**(2)**	**(3)**	**(4)**
$^7F_0$	A	0	0	0	0	0	-	-	-	-
$^7F_1$	A	214	261	273	349	295	−27.46	−11.53	−7.46	18.31
	B	393	396	392	368	367	7.08	7.90	6.81	0.27
	B	700	523	498	389	444	57.66	17.79	12.16	−12.39
$^7F_2$	B	946	971	974	986	947	−0.11	2.53	2.85	4.12
	B	1020	975	980	1022	981	3.98	−0.61	-0.10	4.18
	A	1047	1117	1108	1023	1016	3.05	9.94	9.06	0.69
	A	1288	1134	1109	1027	1080	19.26	5.00	2.69	−4.91
	A	1323	1135	1112	1039	1111	19.08	2.16	0.09	−6.48
$^7F_3$	B	1882	1884	1874	1839		-	-	-	-
	A	1909	1894	1882	1852		-	-	-	-
	B	1952	1919	1905	1859	1808	7.96	6.14	5.37	2.82
	A	2011	1926	1910	1865	1846	8.94	4.33	3.47	1.03
	B	2027	1932	1913	1870	1857	9.15	4.04	3.02	0.70
	B	2032	1932	1921	1873	1893	7.34	2.06	1.48	−1.06
	A	2137	1985	1962	1874		-	-	-	-
$^7F_4$	B	2244	2742	2764	2771	2587	−13.26	5.99	6.84	7.11
	A	2473	2812	2818	2780	2603	−4.99	8.03	8.26	6.80
	A	2698	2834	2834	2799	2633	2.47	7.63	7.63	6.30
	B	2790	2890	2876	2801	2648	5.36	9.14	8.61	5.78
	A	2866	2897	2885	2812	2735	4.79	5.92	5.48	2.82
	A	2945	2913	2888	2838	2872	2.54	1.43	0.56	−1.18
	A	3072	2915	2898	2843	2946	4.28	−1.05	−1.63	−3.50
	B	3179	2983	2945	2850	2967	7.15	0.54	−0.74	−3.94
	B	3245	2987	2950	2886	3086	5.15	−3.21	−4.41	−6.48
$^5D_0$	A	16,081	16,517	16,535	16,874	17,241	−6.73	−4.20	−4.09	−2.13
$^5D_1$	A	17,705	18,128	18,143	18,485	18,945	−6.55	−4.31	−4.23	−2.43
	B	17,716	18,164	18,176	18,488		-	-	-	-
	B	17,806	18,199	18,206	18,493		-	-	-	-

[a] taken from ref. [52].

At this point, it is worth stressing the following: (1) the energy levels are reasonably predicted by the LFDFT calculations, the uncertainties vis-à-vis the experiments are relatively small independent of the choice of the DFT functional. (2) We primarily observe overestimation of the energy levels (many numbers in the percent error columns of Table 1 have positive sign) that can be attributed to the self-consistent error in DFT. (3) The inconsistency with experiments may be removed by using hybrid functional that reduces largely the percent error for many levels. (4) The inconsistency with experiments can also be removed by different starting geometries of the molecular complex, by using, for instance, the experimental structure from X-ray diffraction or other techniques as input.

Table 2 lists the values for the energy parameters obtained for Eu^{3+} ion configuration $4f^6$ in $[Eu(NO_3)_3(phenanthroline)_2]$. These parameters refer to the three Slater–Condon $F^k(4f,4f)$ integrals, with k = 2, 4 and 6, the spin-orbit coupling constant $\zeta(4f)$ [30], and the ligand-field potential in the form of the Wybourne-normalized crystal-field parameters [56,57]. For comparison, the reference data in Table 2 are drawn from earlier reports of the average free ion parameters [51] and theoretical CASSCF/NEVPT2 computation of the free ion [58]. In Table 2, the parameters values that are reported in the experimental work [52] are also listed for comparison. We note that changes in the parameters values from free ions to molecular complexes are expected, in terms of a reduction of the values as results of covalence and the nephelauxetic effect [59–61]. Therefore the comparison of the values in Table 2 is qualitative. For the spin-orbit coupling constant, the agreement with reference values is almost perfect. For the Slater–Condon integrals, on the other hand, our $F^2(4f,4f)$ parameters are overestimated, whereas $F^4(4f,4f)$ and $F^6(4f,4f)$ are underestimated (see Table 2). In Table 2, the Wybourne parameters [56,57] are only part of the multipole expansion of the ligand-field potential, but allow us to compare the theoretical values with the only two values reported in the experiments. Overall the ligand-field potential varies strongly upon the choice of the DFT functional. However, we also see that the LFDFT calculation with hybrid functional yields relatively adequate values.

Table 2. Calculated parameters of the ligand-field Hamiltonian: Slater–Condon integrals, spin-orbit coupling constant and the ligand-field potential in terms of the Wybourne-normalized crystal-field parameters (in eV) obtained from LFDFT using the PBE (1) [37], B3LYP (2) [38], PBE0 (3) [39,40] and KMLYP (4) [41] functional, compared with reference data taken from the literature: experimental parameters (a), average free-ion parameter (a) and other theoretical model (c).

	LFDFT				Reference		
	(1)	**(2)**	**(3)**	**(4)**	**(a)** [a]	**(b)** [b]	**(c)** [c]
$F^2(4f,4f)$	11.4244	11.7216	11.7334	11.9464	8.7164	10.2648	10.7841
$F^4(4f,4f)$	7.1200	7.3052	7.3126	7.4444		7.3652	7.4879
$F^6(4f,4f)$	5.1085	5.2413	5.2466	5.3410		5.2875	5.6741
$\zeta(4f)$	0.1604	0.1610	0.1607	0.1600		0.1652	0.1731
$B^2_2(4f,4f)$	−0.0519	−0.0245	−0.0188	0.0172	−0.0196		
$B^2_0(4f,4f)$	−0.2515	−0.1332	−0.1147	−0.0217	−0.0471		

[a] taken from ref. [52]; [b] taken from ref. [51]; [c] these values are derived from the Racah parameters in ref. [58] using conversion factor in ref. [62].

3.2. X-ray Absorption Spectra of Cerocene

Core-electron excitation has been exploited for decades to understand the properties and chemistry of materials with various techniques: X-ray absorption, X-ray emission, X-ray magnetic circular dichroism, electron energy loss spectroscopy, resonant inelastic scattering, etc. [63,64]. X-ray Absorption Spectroscopy (XAS) has many advantages, most importantly, its element specificity and local electronic and atomic structures probing [63,64]. Lanthanide compounds are often studied at the $M_{4,5}$-edge XAS [65], which correspond to the process in which incident photons are absorbed by promoting one electron from the core $3d$ orbitals to the valence $4f$. In the absorption spectra, strong features appeared representing the $4f^n \longrightarrow 3d^9 4f^{n+1}$ transitions governed by the electric-dipole mechanism [30].

Cerocene has been extensively studied in terms of the molecular orbital diagram and ground state electronic structure [47,66–68]. The Ce $4f$ orbitals split in energy into four molecular orbitals within the approximate D_{8h} symmetry. The molecular orbital with a_{1u} representation (with predominant $4f_{z^3}$) has the lowest energy, followed by the two-fold degenerate e_{3u} ($4f_{x(x^2-3y^2)}$ and $4f_{y(3x^2-y^2)}$), e_{1u} ($4f_{z^2x}$ and $4f_{z^2y}$) and e_{2u} ($4f_{z(x^2-y^2)}$ and $4f_{xyz}$). Mulliken population analysis of the $[Ce(COT)_2]$ shows that the a_{1u}, e_{3u} and e_{1u} are principally metallic orbitals with $4f$ parentage coefficients greater than 95%. e_{2u} has stronger interaction with the C $2p$ orbitals, with a reduced $4f$ parentage of 74% only. The

Mulliken population analysis of $[Ce(COT)_2]^-$ shows similar behavior, except that all the molecular orbitals are now relatively localized with $4f$ parentage coefficients greater than 90%. The molecular orbital diagrams are very similar to other theoretical results [47,66–68], so that we use these orbitals as the active subspace of the LFDFT calculation.

Table 3 lists the theoretical values for the energy parameters obtained for Ce ions with configurations $3d^9 4f^1$ and $3d^9 4f^2$ in $[Ce(COT)_2]$ and $[Ce(COT)_2]^-$, respectively. These parameters refer to the the Slater–Condon integrals: $F^k(4f,4f)$, with k = 2, 4 and 6; $F^k(3d,4f)$, with k = 2, and 4; and $G^k(3d,4f)$, with k = 1, 3 and 5; the spin-orbit coupling constant $\zeta(3d)$, and $\zeta(4f)$; and the ligand-field potential in the form of Wybourne parameters [56,57]. For comparison, the reference data in Table 3 correspond to the free ion parameter values for La^{3+}, which is iso-electronic to Ce^{4+}, and Ce^{3+} [65]. In case of the Slater–Condon integrals and spin-orbit coupling constants, the parameters are reduced vis-à-vis the reference free ion values, which shows the decrease of the electron density on the central metal ions *via* the nephelauxetic effect [59–61]. The ligand-field parameters also shows that GGA functional slight overestimates the metal–ligand interaction, as it is also obtained for the Eu^{3+} complex (see above).

Table 3. Calculated parameters of the ligand-field Hamiltonian for the $[Ce(COT)_2]$ and $[Ce(COT)_2]^-$ systems: Slater–Condon integrals, spin-orbit coupling constant and the ligand-field potential in terms of the Wybourne-normalized crystal-field parameters (in eV) obtained from LFDFT using the PBE (1) [37], and PBE0 (2) [39,40] functional, compared with reference data taken from the litterature: average free-ion parameter for La^{3+} (iso-electronic to Ce^{4+}) (a) and Ce^{3+} (b).

	LFDFT				Reference	
	$[Ce(COT)_2]$		$[Ce(COT)_2]^-$			
	(1)	(2)	(1)	(2)	(a) [a]	(b) [b]
$F^2(4f,4f)$	-	-	9.0259	8.5968	-	10.01
$F^4(4f,4f)$	-	-	5.6126	5.3400	-	6.35
$F^6(4f,4f)$	-	-	4.0234	3.8265	-	4.57
$G^1(3d,4f)$	3.4783	3.2930	3.7890	3.6082	3.78	4.06
$G^3(3d,4f)$	2.0595	1.9495	2.2432	2.1359	2.21	2.37
$G^5(3d,4f)$	1.4287	1.3524	1.5560	1.4816	1.52	1.64
$F^2(3d,4f)$	5.4411	5.1962	5.9528	5.7140	5.65	5.99
$F^4(3d,4f)$	2.4219	2.3010	2.6421	2.5243	2.53	2.71
$\zeta(3d)$	7.5344	7.5331	7.5357	7.5343	6.80	7.45
$\zeta(4f)$	0.0781	0.0742	0.0852	0.0814	0.086	0.106
$B_0^2(4f,4f)$	−0.0523	−0.0460	−0.0350	−0.0058		
$B_0^4(4f,4f)$	−3.0090	0.9311	−1.5888	−0.3282		
$B_0^6(4f,4f)$	0.3738	−0.9137	0.1651	−0.2111		

[a] taken from ref. [65]; [b] taken from ref. [65].

Figure 3 shows the calculated spectral profiles of $[Ce(COT)_2]$ and $[Ce(COT)_2]^-$ obtained from the LFDFT calculations by using the PBE [37] and PBE0 DFT functional [39,40]. The calculated oscillator strengths of the electric-dipole $4f^n \longrightarrow 3d^9 4f^{n+1}$ transitions, with n = 0, and 1 are represented in bar diagrams. The colorful curves represent the broadening of all the oscillator strengths with a Lorentzian function with a constant half-width-at-half-maximum parameter of 0.25 eV to mimic the core-hole lifetime [30]. The calculated spectra of $[Ce(COT)_2]$ are relatively simple, with two sharp peaks and fine structures, resulting from the large spin-orbit coupling of the $3d$ electrons (see Table 3. The spectra in (Figure 3a) and in (Figure 3c) exhibit similar profiles, except for the excitation energies, where the hybrid functional shifts the energy to higher values, which is already observed in earlier studies [26,30]. To validate the results, we use the experimental spectrum of $[Ce(COT)_2]$ in ref. [68] The two sharp peaks are also present in the experiment [68], but additional features appeared also in the form of satellites indicating the mixing between pure metallic Ce^{4+} $4f^0$

and ligand-to-metal charge transfer (LMCT) C $2p \longrightarrow$ Ce $4f$ electronic states. Besides, it has been demonstrated that the ground state of [Ce(COT)$_2$] is, in fact, multiconfigurational [69], limiting then the use of LFDFT in this context. Thus, the treatment of LMCT, which is not yet possible with LFDFT, will constitute the next step methodological development.

The calculated spectra of [Ce(COT)$_2$]$^-$ present more complex features. The two strong absorption bands are due to the large spin-orbit coupling of the 3*d* electrons, and the fine structures results from the multipet levels of the $3d^9 4f^2$ configuration (see Figure 3b,d). The first absorption band is characterized by two peaks with small pre-edge shoulders, which is also observed in the experiment [68]. The second absorption band is characterized by three peaks with small post-edge shoulder, that can also be seen in the experimental data [68].

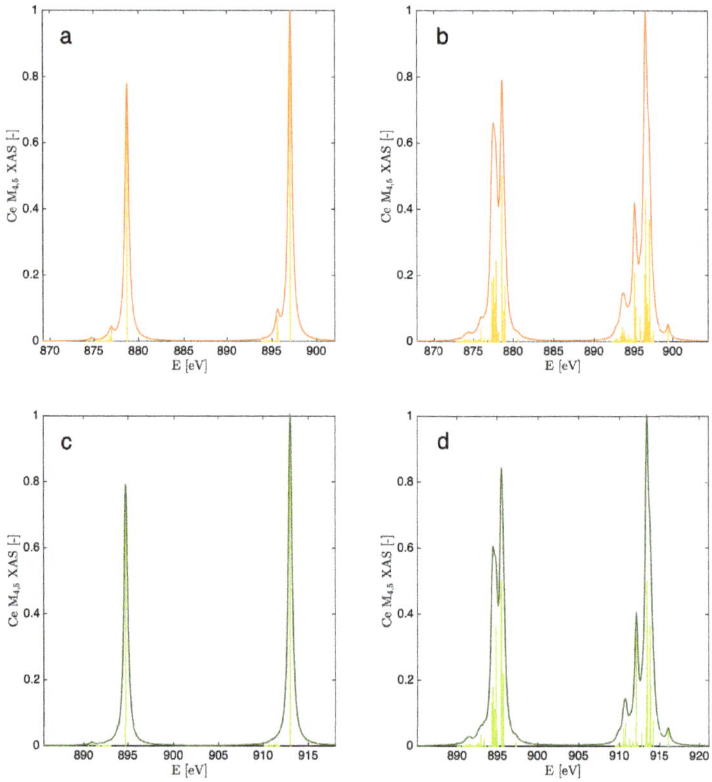

Figure 3. Calculated spectral profiles of the Ce $M_{4,5}$-edge XAS of [Ce(COT)$_2$] (left) and [Ce(COT)$_2$]$^-$ (right), obtained from the LFDFT calculations using the PBE [37] (**a**,**b**) and PBE0 [39,40] (**c**,**d**) DFT functional. The light orange and green color bars represent the oscillator strengths of the electric-dipole $4f^n \longleftarrow 3d^9 4f^{n+1}$, with $n = 0$, and 1. The Darker orange and green curves represent the broadening of the oscillator strengths with Lorentzian function with half-width at half maximum parameter of 0.25 eV.

4. Conclusions

The present work is aimed at describing the Ligand-Field Density-Functional Theory (LFDFT) method with practical examples that are chosen from current coordination chemistry topics. Open-shell *f* electrons still constitute a great challenge for computational chemists owing to strong electron-correlation effects within valence orbitals. Density Functional Theory (DFT) is nowadays very powerful for dealing with molecular and solid-state systems, and the ligand-field concept brings a suitable approach to treat multi-electronic interaction. It is shown here that LFDFT can be used to reasonably perform accurate calculations of coordination compounds with lanthanide elements. The energy levels of Eu^{3+} are calculated with reasonable uncertainties, showing also the influence of the choice of the DFT functional on the multiplet energies. The Ce $M_{4,5}$-edge XAS spectra of Ce^{3+} and Ce^{4+} are simulated with good agreement with the experimental data.

With this paper, we also want to state future developments in the LFDFT code. These developments will include: (1) the simulation of the $4f \longrightarrow 4f$ absorption and emission spectra, which on top of the energy levels will bring more complete understanding of the luminescence process; (2) the consideration of charge transfer model to take into account ligand orbitals in the active space of the LFDFT calculation; and (3) the development of a ligand-field concept for coordination compounds with two or multiple metallic centers. That is, in the perspective of elaborating more complete and user-friendly theoretical models for complex electronic structure problems.

Funding: This research received no external funding.

Institutional Review Board Statement: Not applicable.

Informed Consent Statement: Not applicable.

Data Availability Statement: Data is contained within the article.

Acknowledgments: The author gratefully acknowledges high performance computing (HPC) support, time and infrastructure from the Johannes Gutenberg University of Mainz (Mogon), and the HPC-Europa3 program. The author also recognizes support from the Software for Chemistry and Materials (www.scm.com (accessed on 31 March 2022)), namely for providing the ADF developper license file. The author also wishes to thank Claude Daul (University of Fribourg in Switzerland) for his constant support and for useful discussion. Finally, the author wishes to thank the editor for the invitation to this Special Issue.

Conflicts of Interest: The authors declare no conflict of interest.

References

1. Shibasaki, M.; Yoshikawa, N. Lanthanide Complexes in Multifunctional Asymmetric Catalysis. *Chem. Rev.* **2002**, *102*, 2187–2210. [CrossRef] [PubMed]
2. Trummer, D.; Searles, K.; Algasov, A.; Guda, S.A.; Soldatov, A.V.; Ramanantoanina, H.; Safonova, O.V.; Guda, A.A.; Copéret, C. Deciphering the Phillips Catalyst by Orbital Analysis and Supervised Machine Learning from Cr Pre-edge XANES of Molecular Libraries. *J. Am. Chem. Soc.* **2021**, *143*, 7326–7341. [CrossRef] [PubMed]
3. Bünzli, J.C.G. Benefiting from the Unique Properties of Lanthanide Ions. *Accounts Chem. Res.* **2006**, *39*, 53–61. [CrossRef] [PubMed]
4. Poe, T.N.; Molinari, S.; Beltran-Leiva, M.J.; Celis-Barros, C.; Ramanantoanina, H.; Albrecht-Schönzart, T.E. Influence of Outer-Sphere Anions on the Photoluminescence from Samarium(II) Crown Complexes. *Inorg. Chem.* **2021**, *60*, 15196–15207. [CrossRef] [PubMed]
5. Sessoli, R.; Powell, A.K. Strategies towards single molecule magnets based on lanthanide ions. *Coord. Chem. Rev.* **2009**, *253*, 2328–2341. [CrossRef]
6. Dreiser, J. Molecular lanthanide single-ion magnets: From bulk to submonolayers. *J. Phys. Condens. Matter* **2015**, *27*, 183203. [CrossRef]
7. Ramanantoanina, H.; Studniarek, M.; Daffé, N.; Dreiser, J. Non-empirical calculation of X-ray magnetic circular dichroism in lanthanide compounds. *Chem. Commun.* **2019**, *55*, 2988–2991. [CrossRef] [PubMed]
8. Daul, C. Density functional theory applied to the excited states of coordination compounds. *Int. J. Quantum Chem.* **1994**, *52*, 867–877. [CrossRef]
9. Chermette, H. Density functional theory: A powerful tool for theoretical studies in coordination chemistry. *Coord. Chem. Rev.* **1998**, *178–180*, 699–721. [CrossRef]
10. Schwarz, K. DFT calculations of solids with LAPW and WIEN2k. *J. Solid State Chem.* **2003**, *176*, 319–328. [CrossRef]
11. Grimme, S.; Waletzke, M. A combination of Kohn–Sham density functional theory and multi-reference configuration interaction methods. *J. Chem. Phys.* **1999**, *111*, 5645–5655. [CrossRef]
12. Szalay, P.G.; Müller, T.; Gidofalvi, G.; Lischka, H.; Shepard, R. Multiconfiguration Self-Consistent Field and Multireference Configuration Interaction Methods and Applications. *Chem. Rev.* **2012**, *112*, 108–181. [CrossRef] [PubMed]

13. Ghosh, S.; Verma, P.; Cramer, C.J.; Gagliardi, L.; Truhlar, D.G. Combining Wave Function Methods with Density Functional Theory for Excited States. *Chem. Rev.* **2018**, *118*, 7249–7292. [CrossRef] [PubMed]
14. Van den Heuvel, W.; Calvello, S.; Soncini, A. Configuration-averaged 4f orbitals in ab initio calculations of low-lying crystal field levels in lanthanide(iii) complexes. *Phys. Chem. Chem. Phys.* **2016**, *18*, 15807–15814. [CrossRef] [PubMed]
15. Pathak, S.; Lang, L.; Neese, F. A dynamic correlation dressed complete active space method: Theory, implementation, and preliminary applications. *J. Chem. Phys.* **2017**, *147*, 234109. [CrossRef]
16. Singh, S.K.; Atanasov, M.; Neese, F. Challenges in Multireference Perturbation Theory for the Calculations of the g-Tensor of First-Row Transition-Metal Complexes. *J. Chem. Theory Comput.* **2018**, *14*, 4662–4677. [CrossRef]
17. Burke, K. Perspective on density functional theory. *J. Chem. Phys.* **2012**, *136*, 150901. [CrossRef]
18. Cohen, A.J.; Mori-Sánchez, P.; Yang, W. Challenges for Density Functional Theory. *Chem. Rev.* **2012**, *112*, 289–320. [CrossRef]
19. Verma, P.; Truhlar, D.G. Status and Challenges of Density Functional Theory. *Trends Chem.* **2020**, *2*, 302–318. [CrossRef]
20. Schwarz, K.; Blaha, P. Solid state calculations using WIEN2k. *Comput. Mater. Sci.* **2003**, *28*, 259–273. [CrossRef]
21. Casida, M.; Huix-Rotllant, M. Progress in Time-Dependent Density-Functional Theory. *Annu. Rev. Phys. Chem.* **2012**, *63*, 287–323. [CrossRef] [PubMed]
22. Adamo, C.; Jacquemin, D. The calculations of excited-state properties with Time-Dependent Density Functional Theory. *Chem. Soc. Rev.* **2013**, *42*, 845–856. [CrossRef] [PubMed]
23. Hannachi, D.; Haroun, M.F.; Khireddine, A.; Chermette, H. Optical and nonlinear optical properties of Ln(Tp)2, where Ln = La,. . . ,Lu and Tp = tris(pyrazolyl)borate: A DFT + TD-DFT study. *New J. Chem.* **2019**, *43*, 14377–14389. [CrossRef]
24. Gutierrez, F.; Rabbe, C.; Poteau, R.; Daudey, J.P. Theoretical Study of Ln(III) Complexes with Polyaza-Aromatic Ligands: Geometries of $[LnL(H_2O)n]^{3+}$ Complexes and Successes and Failures of TD-DFT. *J. Phys. Chem. A* **2005**, *109*, 4325–4330. [CrossRef] [PubMed]
25. Momeni, M.R.; Brown, A. Why Do TD-DFT Excitation Energies of BODIPY/Aza-BODIPY Families Largely Deviate from Experiment? Answers from Electron Correlated and Multireference Methods. *J. Chem. Theory Comput.* **2015**, *11*, 2619–2632. [CrossRef]
26. Ramanantoanina, H.; Daul, C. A non-empirical calculation of 2p core-electron excitation in compounds with 3d transition metal ions using ligand-field and density functional theory (LFDFT). *Phys. Chem. Chem. Phys.* **2017**, *19*, 20919–20929. [CrossRef]
27. Atanasov, M.; Daul, C.; Rauzy, C. New insights into the effects of covalency on the ligand field parameters: A DFT study. *Chem. Phys. Lett.* **2003**, *367*, 737–746. [CrossRef]
28. Ramanantoanina, H.; Sahnoun, M.; Barbiero, A.; Ferbinteanu, M.; Cimpoesu, F. Development and applications of the LFDFT: The non-empirical account of ligand field and the simulation of the f–d transitions by density functional theory. *Phys. Chem. Chem. Phys.* **2015**, *17*, 18547–18557. [CrossRef]
29. Vlahović, F.; Perić, M.; Gruden-Pavlović, M.; Zlatar, M. Assessment of TD-DFT and LF-DFT for study of d– d transitions in first row transition metal hexaaqua complexes. *J. Chem. Phys.* **2015**, *142*, 214111. [CrossRef]
30. Ramanantoanina, H. A DFT-based theoretical model for the calculation of spectral profiles of lanthanide M4,5-edge X-ray absorption. *J. Chem. Phys.* **2018**, *149*, 054104. [CrossRef]
31. Vanquickenborne, L.; Ceulemans, A. Ligand-field models and the photochemistry of coordination compounds. *Coord. Chem. Rev.* **1983**, *48*, 157–202. [CrossRef]
32. Gerloch, M.; Harding, J.H.; Woolley, R.G. The context and application of ligand field theory. In *Inorganic Chemistry*; Springer: Berlin/Heidelberg, Germany, 1981; pp. 1–46.
33. Woolley, R.G. Ligand-field analysis of transition-metal complexes. *Int. Rev. Phys. Chem.* **1987**, *6*, 93–141. [CrossRef]
34. Hoffmann, R.; Alvarez, S.; Mealli, C.; Falceto, A.; Cahill, T.J.; Zeng, T.; Manca, G. From Widely Accepted Concepts in Coordination Chemistry to Inverted Ligand Fields. *Chem. Rev.* **2016**, *116*, 8173–8192. [CrossRef] [PubMed]
35. Baerends, E.; Ziegler, T.; Atkins, A.; Autschbach, J.; Baseggio, O.; Bashford, D.; Bérces, A.; Bickelhaupt, F.; Bo, C.; Boerrigter, P.; et al. *ADF 2021.1, SCM*; Theoretical Chemistry; Vrije Universiteit: Amsterdam, The Netherlands, 2021.
36. te Velde, G.; Bickelhaupt, F.M.; Baerends, E.J.; Guerra, C.F.; van Gisbergen, S.J.A.; Snijders, J.G.; Ziegler, T. Chemistry with ADF. *J. Comput. Chem.* **2001**, *22*, 931–967. [CrossRef]
37. Perdew, J.P.; Burke, K.; Ernzerhof, M. Generalized Gradient Approximation Made Simple. *Phys. Rev. Lett.* **1996**, *77*, 3865–3868. [CrossRef]
38. Stephens, P.J.; Devlin, F.J.; Chabalowski, C.F.; Frisch, M.J. Ab Initio Calculation of Vibrational Absorption and Circular Dichroism Spectra Using Density Functional Force Fields. *J. Phys. Chem.* **1994**, *98*, 11623–11627. [CrossRef]
39. Grimme, S. Accurate description of van der Waals complexes by density functional theory including empirical corrections. *J. Comput. Chem.* **2004**, *25*, 1463–1473. [CrossRef]
40. Ernzerhof, M.; Scuseria, G.E. Assessment of the Perdew–Burke–Ernzerhof exchange-correlation functional. *J. Chem. Phys.* **1999**, *110*, 5029–5036. [CrossRef]
41. Kang, J.K.; Musgrave, C.B. Prediction of transition state barriers and enthalpies of reaction by a new hybrid density-functional approximation. *J. Chem. Phys.* **2001**, *115*, 11040–11051. [CrossRef]
42. Lenthe, E.V.; Baerends, E.J. Optimized Slater-type basis sets for the elements 1–118. *J. Comput. Chem.* **2003**, *24*, 1142–1156. [CrossRef]

43. Lenthe, E.V.; Baerends, E.J.; Snijders, J.G. Relativistic regular two-component Hamiltonians. *J. Chem. Phys.* **1993**, *99*, 4597–4610. [CrossRef]
44. van Lenthe, E.; Baerends, E.J.; Snijders, J.G. Relativistic total energy using regular approximations. *J. Chem. Phys.* **1994**, *101*, 9783–9792. [CrossRef]
45. van Lenthe, E.; Ehlers, A.; Baerends, E.J. Geometry optimizations in the zero order regular approximation for relativistic effects. *J. Chem. Phys.* **1999**, *110*, 8943–8953. [CrossRef]
46. Singh, U.P.; Goel, N.; Singh, G.; Srivastava, P. Syntheses, structural and thermal studies of Eu(III) and Gd(III) complexes with 2, 6-dinitrophenol and 1, 10-phenanthroline/2, 2′-bipyridine ligands. *Inorganica Chim. Acta* **2012**, *387*, 294–307. [CrossRef]
47. Walter, M.D.; Booth, C.H.; Lukens, W.W.; Andersen, R.A. Cerocene Revisited: The Electronic Structure of and Interconversion Between Ce2(C8H8)3 and Ce(C8H8)2. *Organometallics* **2009**, *28*, 698–707. [CrossRef]
48. Georgieva, I.; Trendafilova, N.; Zahariev, T.; Danchova, N.; Gutzov, S. Theoretical insight in highly luminescent properties of Eu(III) complex with phenanthroline. *J. Lumin.* **2018**, *202*, 192–205. [CrossRef]
49. Bünzli, J.C.G.; Piguet, C. Taking advantage of luminescent lanthanide ions. *Chem. Soc. Rev.* **2005**, *34*, 1048. [CrossRef]
50. Thejo Kalyani, N.; Dhoble, S. Organic light emitting diodes: Energy saving lighting technology—A review. *Renew. Sustain. Energy Rev.* **2012**, *16*, 2696–2723. [CrossRef]
51. Binnemans, K. Interpretation of europium(III) spectra. *Coord. Chem. Rev.* **2015**, *295*, 1–45. [CrossRef]
52. Pan, Z.; Jia, G.; Duan, C.K.; Wong, W.Y.; Wong, W.T.; Tanner, P.A. Crystal Structure, Spectroscopy and Crystal Field Analysis of Substituted 1, 10-Phenanthroline–Europium Complexes. *Eur. J. Inorg. Chem.* **2011**, *2011*, 637–646. [CrossRef]
53. Zhang, Y.; Yang, W. A challenge for density functionals: Self-interaction error increases for systems with a noninteger number of electrons. *J. Chem. Phys.* **1998**, *109*, 2604–2608. [CrossRef]
54. de Silva, P.; Corminboeuf, C. Local hybrid functionals with orbital-free mixing functions and balanced elimination of self-interaction error. *J. Chem. Phys.* **2015**, *142*, 074112. [CrossRef] [PubMed]
55. Schwarz, K. Challenges for Theory and Computation. *Computation* **2017**, *5*, 49. [CrossRef]
56. Hüfner. *Optical Spectra of Transparent Rare Earth Compounds*; Academic Press: Cambridge, MA, USA, 1978.
57. Newman, D.J.; Ng, B. (Eds.) *Crystal Field Handbook*; Cambridge University Press: Cambridge, UK, 2000. [CrossRef]
58. Aravena, D.; Atanasov, M.; Neese, F. Periodic Trends in Lanthanide Compounds through the Eyes of Multireference ab Initio Theory. *Inorg. Chem.* **2016**, *55*, 4457–4469. [CrossRef]
59. Tchougréeff, A.L.; Dronskowski, R. Nephelauxetic effect revisited. *Int. J. Quantum Chem.* **2009**, *109*, 2606–2621. [CrossRef]
60. Tanner, P.A.; Yeung, Y.Y.; Ning, L. What Factors Affect the 5D0 Energy of Eu^{3+}? An Investigation of Nephelauxetic Effects. *J. Phys. Chem. A* **2013**, *117*, 2771–2781. [CrossRef]
61. Dorenbos, P. Lanthanide 4f-electron binding energies and the nephelauxetic effect in wide band gap compounds. *J. Lumin.* **2013**, *136*, 122–129. [CrossRef]
62. Cowan, R.D. *The Theory of Atomic Structure and Spectra*; University of California Press: Oakland, CA, USA, 1981.
63. de Groot, F. Multiplet effects in X-ray spectroscopy. *Coord. Chem. Rev.* **2005**, *249*, 31–63. [CrossRef]
64. de Groot, F.; Kotani, A. *Core Level Spectroscopy of Solids*; CRC Press: Boca Raton, FL, USA, 2008. [CrossRef]
65. Thole, B.T.; van der Laan, G.; Fuggle, J.C.; Sawatzky, G.A.; Karnatak, R.C.; Esteva, J.M. 3d X-ray-absorption lines and the $3d^9 4f^{n+1}$ multiplets of the lanthanides. *Phys. Rev. B* **1985**, *32*, 5107–5118. [CrossRef]
66. Edelstein, N.M.; Allen, P.G.; Bucher, J.J.; Shuh, D.K.; Sofield, C.D.; Kaltsoyannis, N.; Maunder, G.H.; Russo, M.R.; Sella, A. The Oxidation State of Ce in the Sandwich Molecule Cerocene. *J. Am. Chem. Soc.* **1996**, *118*, 13115–13116. [CrossRef]
67. Mooßen, O.; Dolg, M. Two interpretations of the cerocene electronic ground state. *Chem. Phys. Lett.* **2014**, *594*, 47–50. [CrossRef]
68. Smiles, D.E.; Batista, E.R.; Booth, C.H.; Clark, D.L.; Keith, J.M.; Kozimor, S.A.; Martin, R.L.; Minasian, S.G.; Shuh, D.K.; Stieber, S.C.E.; et al. The duality of electron localization and covalency in lanthanide and actinide metallocenes. *Chem. Sci.* **2020**, *11*, 2796–2809. [CrossRef] [PubMed]
69. Kerridge, A.; Coates, R.; Kaltsoyannis, N. Is Cerocene Really a Ce(III) Compound? All-Electron Spin–Orbit Coupled CASPT2 Calculations on M(η8-C8H8)2 (M = Th, Pa, Ce). *J. Phys. Chem. A* **2009**, *113*, 2896–2905. [CrossRef] [PubMed]

computation

MDPI

Article

Should We Gain Confidence from the Similarity of Results between Methods?

Pascal Pernot [1,*,†] and Andreas Savin [2,*,†]

1 Institut de Chimie Physique, UMR8000, CNRS, Université Paris-Saclay, 91405 Orsay, France
2 Laboratoire de Chimie Théorique, CNRS and UPMC Université Paris 06, Sorbonne Universités, 75252 Paris, France
* Correspondence: Pascal.Pernot@universite-paris-saclay.fr (P.P.); Andreas.Savin@lct.jussieu.fr (A.S.)
† These authors contributed equally to this work.

Abstract: Confirming the result of a calculation by a calculation with a different method is often seen as a validity check. However, when the methods considered are all subject to the same (systematic) errors, this practice fails. Using a statistical approach, we define measures for *reliability* and *similarity*, and we explore the extent to which the similarity of results can help improve our judgment of the validity of data. This method is illustrated on synthetic data and applied to two benchmark datasets extracted from the literature: band gaps of solids estimated by various density functional approximations, and effective atomization energies estimated by *ab initio* and machine-learning methods. Depending on the levels of bias and correlation of the datasets, we found that similarity may provide a null-to-marginal improvement in reliability and was mostly effective in eliminating large errors.

Keywords: statistics; methods comparison; benchmarking; band gaps; atomization energy

Citation: Pernot, P.; Savin, A. Should We Gain Confidence from the Similarity of Results between Methods? *Computation* **2022**, *10*, 27. https://doi.org/10.3390/computation10020027

Academic Editor: Henry Chermette

Received: 17 January 2022
Accepted: 7 February 2022
Published: 11 February 2022

1. Introduction

When all computational methods yield similar results, one often assumes that these cannot be wrong. However, logically, one cannot prove this: an argument is not necessarily right because the majority thinks so. One might, therefore, ask whether obtaining similar results with different methods gives a higher chance of achieving reliable results (one has to keep in mind that the better accuracy of a method when compared to another is a statistical assessment but is not necessarily valid for all systems [1,2].

In this paper, we propose and test a statistical approach to address this question in the context of computational approximations. The concepts of *reliability* and *similarity* are defined and measured by probabilities estimated from benchmark error sets. The interplay between reliability and similarity is estimated by conditional probabilities. *Reliability*, as defined here, is closely related to measures we used in previous studies, based on the empirical cumulative density function (ECDF) of error sets [3]. As for *similarity*, there is a link with correlation between error sets as illustrated in refs. [1,2]. Unlike correlation, similarity is affected by bias between methods, i.e., correlation does not imply similarity.

The following section (Section 2) presents the method. The Applications section (Section 3) illustrates the method on a toy dataset of normal distributions and on two real-world datasets. In order to be able to draw conclusions, we chose literature benchmark datasets with sufficient points to enable reliable numerical results, and a variety of methods encompassing various scenarios of bias and correlation. The main observations are summarized in the conclusion. The aim of this paper is to exemplify a statistical approach to similarity and not to draw general conclusions nor to recommend any of the studied methods.

2. Methodology

2.1. Frame

For a given computational method, M, and a given system, S, let the value calculated for a chosen property be denoted by $X(M, S)$. A benchmark provides reference values, $R(S)$. The error for the method M and the system S is given by

$$E(M, S) = X(M, S) - R(S) \quad (1)$$

2.2. Reliability and Similarity of Computational Results

A benchmark data set is expected to provide a large set of data. We can use statistical measures on this set to make assessments on the reliability of the computational method. Let us first define what we mean by the results of a calculation being *reliable* or being *similar*.

The computational method M is considered reliable for the system S if

$$|E(M, S)| = |X(M, S) - R(S)| < \epsilon_r \quad (2)$$

where the reliability threshold, ϵ_r, is chosen by the user of the method, depending on his needs. We consider here that two methods, M_1 and M_2 provide similar results for system S when

$$|X(M_1, S) - X(M_2, S)| = |E(M_1, S) - E(M_2, S)| < \epsilon_s \quad (3)$$

where the similarity threshold, ϵ_s, is also defined by the user. When we consider a set of methods, we say that the results of these methods are similar when all pairs of methods of the set yield similar results. If not specified otherwise, we will use, in this paper, $\epsilon_s = \epsilon_r = \epsilon$.

Figure 1 schematically presents the problem. The set of systems for which method M_1 is reliable is represented by a red disk; for method M_2, this is a blue disk. The systems for which the two methods are similar are contained in the gray disk. The overlapping region of the red (or blue) disk with the gray disk indicates the set of systems that are reliable with the method M_1 (or M_2), and, at the same time, close to the result provided by the other method.

Figure 1. A schematic representation of the properties of the systems. The region within the square represents the set of all benchmark systems. The red disk represents the set of systems for which method M_1 is reliable. The blue disk represents the set of systems for which method M_2 is reliable. The gray disk represents the set of systems for which methods M_1 and M_2 give similar results.

Let us define the following notations characterizing those sets, where the indices r and s refer to the reliability and similarity, respectively:

- N, the number of systems in the data set (corresponding to the white square in Figure 1).
- $N_s(M_1, M_2, \ldots; \epsilon_s)$, or $N_s(\epsilon_s)$ for brevity, the number of systems that yield similar results (within ϵ_s) using methods $M_1, M_2, \ldots$. (corresponding to the gray disk in Figure 1).
- $N_r(M, \epsilon_r)$, the number of systems for which method M is reliable (corresponding to the red or blue disk in Figure 1).
- $N_r(M, \epsilon_r \cap \epsilon_s)$, the number of systems for which method M is reliable and similar to the other methods (corresponding to the overlap region of the three disks in Figure 1).

2.3. Probabilities

If the data set is sufficiently large, we can estimate probabilities as frequencies from these numbers:

- The probability to obtain a reliable result with method M,

$$P_r(M, \epsilon_r) = \frac{N_r(M, \epsilon_r)}{N} \tag{4}$$

- The probability to obtain similar results for the set of considered methods,

$$P_s(M_1, M_2, \ldots; \epsilon_s) = P_s(\epsilon_s) = \frac{N_s(\epsilon_s)}{N} \tag{5}$$

  For a finite sample, the smallest value of ϵ_s for which $P_s(\epsilon_s) = 1$ is called the Hausdorff distance [4].
- The (conditional) probability to obtain reliable results with method M, given that this method is similar to the other methods in the set,

$$P_{r|s}(M, \epsilon_r, \epsilon_s) = \frac{N_r(M, \epsilon_r \cap \epsilon_s)}{N_s(\epsilon_s)} \tag{6}$$

- The (conditional) probability that a result with method M is similar to that of the other methods, given that it is reliable,

$$P_{s|r}(M, \epsilon_s, \epsilon_r) = \frac{N_r(M, \epsilon_r \cap \epsilon_s)}{N_r(M, \epsilon_r)} \tag{7}$$

with the limit values

$$P_{r|s}(M, \epsilon_r = \infty, \epsilon_s) = P_s(\epsilon_s) \tag{8}$$
$$P_{r|s}(M, \epsilon_r, \epsilon_s = \infty) = P_r(M, \epsilon_r) \tag{9}$$
$$P_{s|r}(M, \epsilon_s = \infty, \epsilon_r) = 1 \tag{10}$$
$$P_{s|r}(M, \epsilon_s, \epsilon_r = \infty) = P_s(\epsilon_s) \tag{11}$$

Furthermore, even for $\epsilon_s = \epsilon_r = \epsilon$, in general, $P_s(\epsilon) \neq P_r(M, \epsilon)$ and $P_{s|r}(M, \epsilon) \neq P_{r|s}(M, \epsilon)$, where the notations were shortened to imply the equality of both thresholds.

The main objective of this paper is to investigate whether choosing methods with similar results is a good criterion of reliability, i.e., to what extent $P_{r|s}(\epsilon_r, \epsilon_s) > P_r(\epsilon_r)$. Even if this aim is achieved, this does not go without a drawback: the systems for which similarity is not reached are eliminated from the study with a probability $1 - P_s(\epsilon_s)$. $P_{s|r}(M, \epsilon_s, \epsilon_r)$ gives us an indication about the quality of our selection criteria by similarity.

An important limitation of this approach is the sample size. Even for large data sets, it may happen that the number of similar results, $N_s(\epsilon_s)$ is small, e.g., because at least one of the methods yields results systematically different from that of the other methods or because ϵ_s was chosen too small. In such a case, the uncertainty of the empirical estimates becomes large.

2.4. Statistical Measures

Often, the distribution of errors is summarized by numbers, such as the mean error, the mean absolute error, and the standard deviation. Although these numbers convey some information, they sometimes hide the misconception that the distribution of errors is normal. In the cases we analyze below, as in most cases we studied previously [5], the distributions of errors are not normal. This justifies the use of probabilistic estimators, such as those presented in our previous work [1–3], or the ones introduced here.

A direct link can be made between the statistics based on the empirical cumulative distribution function (ECDF) of the absolute errors, presented in ref. [3], and some of those introduced above:

- The reliability probability $P_r(M, \epsilon_r)$ is equivalent to the ECDF of the absolute errors, noted $C(\epsilon)$ in our previous work.
- The qth percentile $Q_q(M)$ of the absolute errors is the value of ϵ_r, such as $P_r(M, \epsilon_r) = q/100$.

The conditional probabilities $P_{r|s}(M, \epsilon)$ and $P_{s|r}(M, \epsilon)$ will, thus, be represented as *conditional* ECDFs as a function of ϵ, generalizing our former probabilistic statistics.

3. Applications

3.1. Guidelines

In order to obtain a better understanding of the situations arising from data extracted from the chemical literature, let us first consider pairs of points generated randomly according to normal distributions: each point is assimilated to a "system", where the values on the abscissa are interpreted as "errors" for M_1 while those on the ordinate as "errors" for M_2. The results are presented in Figure 2. The panels on the left show the randomly produced "errors" (green dots).

The red stripe shows the *reliability region* for M_1, where $|E(M_1, S| < \epsilon_r$ (cf. Equation (2)), and the blue stripe shows the same for M_2. The gray stripe shows the region where the results produced by M_1 and M_2 are within $\pm\epsilon_s$ (Equation (3)). Some points are marked by numbers. The polygon with corners corresponding to the points (2, 4, 6, and 8) delimits the region where M_1 is both close to M_2 and is reliable. The polygon with corners corresponding to the points (1, 3, 5, and 7) delimits the region where M_2 is both close to M_1 and is reliable. The plots were drawn by choosing $\epsilon_r = \epsilon_s = 1$.

The ratio of the number of points in the red or blue stripe to the total number of points gives P_r. The ratio of the number of points in the gray stripe to the total number of points gives P_s. The ratio the number of points in the polygons (1, 3, 5, and 7) or (2, 4, 6, and 8) to the number of points in the gray stripe gives $P_{r|s}$. The ratio the number of points in the polygons to those in the red or blue stripe give $P_{s|r}$. The panels on the right show the dependence of the probabilities on $\epsilon_r = \epsilon_s$. The results for M_1 are in blue, those for M_2 in red. $P_r(M_i)$ are drawn as thin curves, $P_{r|s}$ as thick curves, and those for $P_{r|s}$ as dashed curves.

The top row is produced for errors centered at the origin (the mean errors are equal to zero for both methods; the variance is different for the two methods). In the second row, the mean errors are different and non-zero. In the third row, a correlation is introduced between the errors produced by the two methods. In the last row, the effect of correlation is enhanced. In the first three rows, the parameters are inspired from those obtained for the PBE/HSE06 pair (see Section 3.2), in the last row different parameters are used, namely those of PBE0/HSE06.

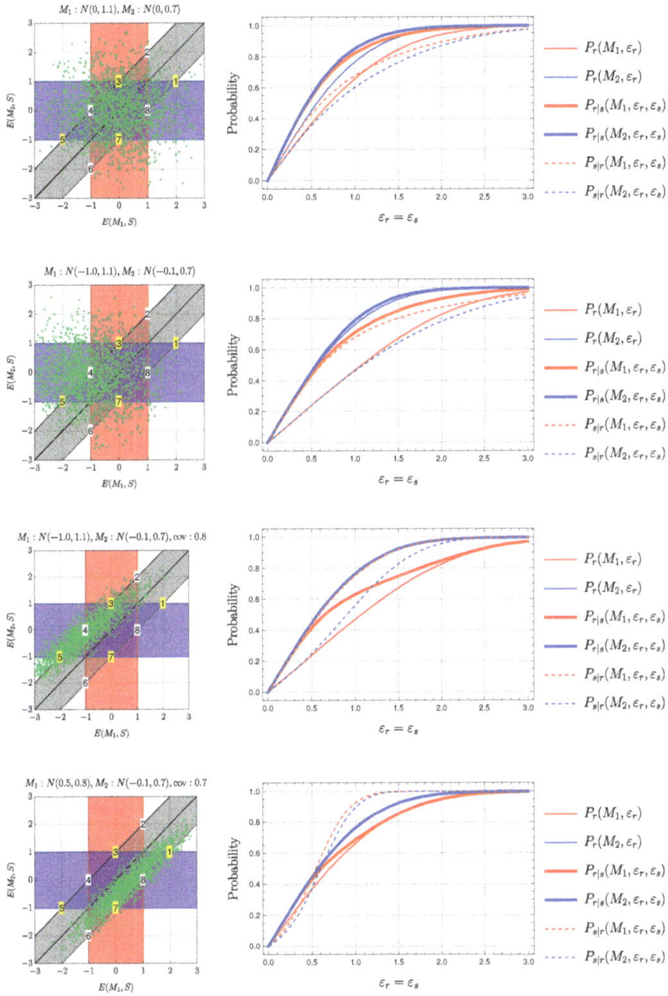

Figure 2. Examples of reliability and similarity configurations (**left**) and the corresponding probability curves (**right**) for two error sets sampled from normal distributions. See the text for description.

Let us start by discussing the first row. We see that, from the choice made for ϵ_r and ϵ_s, an important number of points is in the region where $|E(M_i, S)| < \epsilon_r$, ($i = 1$, or 2) and $|E(M_1, S) - E(M_2, S)| < \epsilon_s$. However, there are points that are within the reliable range for both methods (in the region where the red and blue stripes overlap) are not within the region of similarity (gray stripe).

This could be corrected by increasing ϵ_s to $\sqrt{2}\epsilon_r$, but there is a price to pay for it: for each of the methods, the number of points selected increases by including systems for which the method does not yield reliable results. Furthermore, we notice that there are points that are reliable with one method but not similar to the other method. There are also points that are similar but unreliable (inside the gray stripe but outside either the red or the blue stripe). Finally, there are points that where the methods are both unreliable and dissimilar (on white background). Let us now look at the evolution of probabilities with $\epsilon_r = \epsilon_s$ (top right panel). We see that M_1 is globally of worse quality than M_2, as $P_r(M_1) < P_r(M_2)$ (thin curves).

When first selecting the results by similarity and then checking the reliability, we see that the conditional probabilities $P_{r|s}$ are close for M_1 and M_2 and better than the P_r curves. Checking similarity has eliminated part of the good results (that were reliable with either M_1 or M_2) but provides a higher probability to obtain a good result. Note that, while $P_{r|s}$ is slightly better for M_2 than for M_1, the inverse is true for $P_{s|r}$, a consequence of the division by P_r.

Let us now shift the point cloud by analyzing it for the case when the mean errors are non-zero. If the shift for at least one of the methods is important, none of the "systems" produces similar results for the two methods (the point cloud is shifted outside the gray stripe). The figure shows an intermediate case, where the shift is not so important and plays a role mainly for M_1.

The similarity (gray stripe) essentially retains the results that are good for M_2 (within the blue stripe) because the number of points that are both similar and reliable for M_1 is reduced. As a result (second row, right panel), the similarity hardly improves the probability to obtain a good result for the better method (M_2) but eliminates a number of systems for which M_2 would provide reliable results. However, there is still an improvement for the method of lower quality (M_1).

Another effect reducing the improvement is the existence of positive correlation between the "errors". This is exemplified in the last two rows, where the position of the points are concentrated around a line. In the limit of perfect correlation, these points lie all on a line. If the mean errors make the lines lie in the similarity region, $P_{r|s} = P_r$: no gain is obtained through similarity. If the line lies outside the similarity region (outside the gray stripe), even worse, no point is selected by similarity: if we rely on similarity only, we cannot use any of the calculations.

Note that the correlation between data produces an increase in $P_{s|r}$: if a method is producing a reliable result, by correlation, it is likely that the other method produces also a reliable result, except when one of the methods is strongly biased compared to the other.

3.2. *The BOR2019 Dataset*

We consider a set of band gaps obtained for 471 systems with a selected set of density functional approximations (DFAs): LDA [6,7], PBE [8], PBEsol [9], SCAN [10], PBE0 [11,12], and HSE06 [13,14]. All the data were taken from Borlido et al. [15], and most summary statistics referred to below were reported in a previous study [1,2] (case BOR2019).

3.2.1. Performance of Individual Methods

The errors in the band gaps are quite large for this set of methods. The mean absolute errors lie between 0.5 eV (HSE06) and 1.2 eV (LDA), while Q_{95} varies between 1.7 eV (HSE06) and 3.2 eV (LDA) (Figure 3). The probability to have more reliable than unreliable results occurs at the median absolute error, which defines a minimal value $\epsilon_r = 0.33$ eV for HSE06—the best method in this set.

Figure 3 shows the dependence of $P_r(M, \epsilon)$ on ϵ. One can can safely qualify HSE06 as the best (most reliable) among the methods as $P_r(\text{HSE06}, \epsilon)$ is never smaller than any of the other $P_r(M, \epsilon)$ curves (within the sampling uncertainty). While PBE0 becomes competitive with HSE06 for $\epsilon > 1.7$ eV, it behaves rather like SCAN for the values of $\epsilon \approx 1$ eV and like the group of the three methods that perform worst (LDA, PBE, and PBEsol) for small ϵ values.

It is important to have in mind that, even if PBE0 and SCAN are identically reliable at the $\epsilon = 1$ threshold ($P_r(\text{PBE0}, \epsilon = 1) \simeq P_r(\text{SCAN}, \epsilon = 1)$), this is not necessarily true for the same subset of systems.

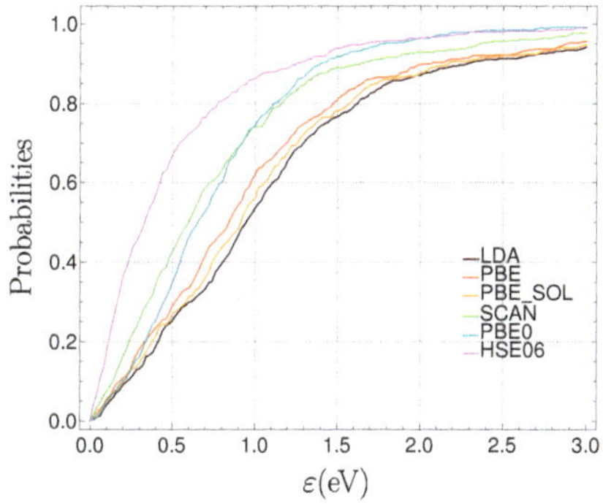

Figure 3. Empirical cumulative distribution functions for the absolute errors of the six DFAs considered in this paper. They correspond also to $P_r(M, \epsilon)$, the fraction of systems for which the DFA produces errors smaller than ϵ. The uncertainty bands are obtained by bootstrapping the ECDF and estimating 95% confidence intervals.

3.2.2. Similarity and Reliability

Figure 4 shows the absolute errors made by two methods (HSE06 and PBE or PBE0) and the distance between the results obtained with the two methods. We choose, for example, a threshold for the similarity of the two methods, $\epsilon_s = 1$ eV. We take the same value for the threshold defining a method reliable, $\epsilon_r = 1$ eV.

If we assume that the similarity of the results is a good criterion to select the reliable results, the points should lie either in the bottom left rectangle ($|X(M_1, S) - X(M_2, S)| < \epsilon_s, |E(M, S)| < \epsilon_r$), meaning that the selected results are reliable, or in the top right rectangle ($|X(M_1, S) - X(M_2, S)| > \epsilon_s, |E(M, S)| > \epsilon_r$), meaning that dissimilarity eliminates the bad results. However, we see many points in the top left rectangle ($|X(M_1, S) - X(M_2, S)| < \epsilon_s, |E(M, S)| > \epsilon_r$), showing that it is possible that similar results should be rejected.

This naturally shows up when the methods are highly correlated, as it is the case for HSE06 and PBE0. Furthermore, we notice the presence of points in the bottom right rectangle ($|X(M_1, S) - X(M_2, S)| > \epsilon_s, |E(M, S)| < \epsilon_r$), especially for the HSE06/PBE pair, indicating that the similarity criterion has eliminated good results obtained with one of the methods.

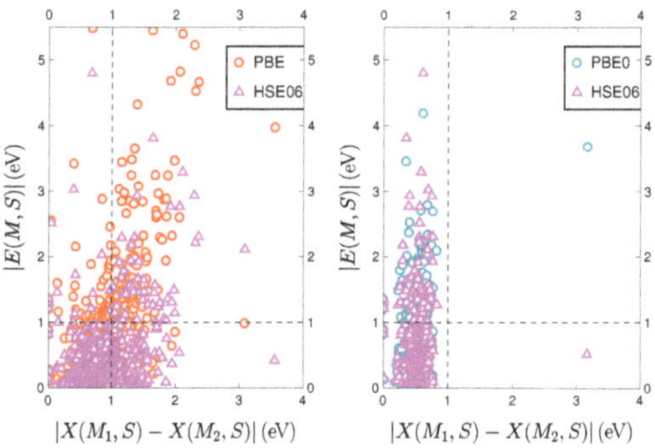

Figure 4. Similarity between HSE06 and PBE (left panel) and PBE0 (right panel) compared to the reliability of the three methods (PBE: red circles, PBE0: blue circles, and HSE06: purple triangles). The points correspond to the absolute errors made by the two methods, $|E(M, S)|$, Equation (2) (on the ordinate) and the distance between the results obtained by the two methods, $|X(M_1, S) - X(M_2, S)|$, Equation (3) (on the abscissa). The dashed lines exemplify choices for the thresholds for similarity, ϵ_s, and reliability, ϵ_r.

3.2.3. Impact of Similarity on Reliability

Let us now look at the probabilities as functions of ϵ (we take $\epsilon_s = \epsilon_r = \epsilon$), Figure 5. As a reference, we plot $P_r(M, \epsilon)$ (thin curves, identical to the ECDF curves in Figure 3), the estimation of reliability when no similarity check is made. The thick curves correspond to $P_{r|s}(M, \epsilon)$, the probability to obtain with method M errors smaller than ϵ if the results of method M are within $\pm\epsilon$ of the other method(s). The dashed curves indicate $P_{s|r}(M, \epsilon)$, the probability of a reliable result obtained with method M to be in the subset selected by similarity.

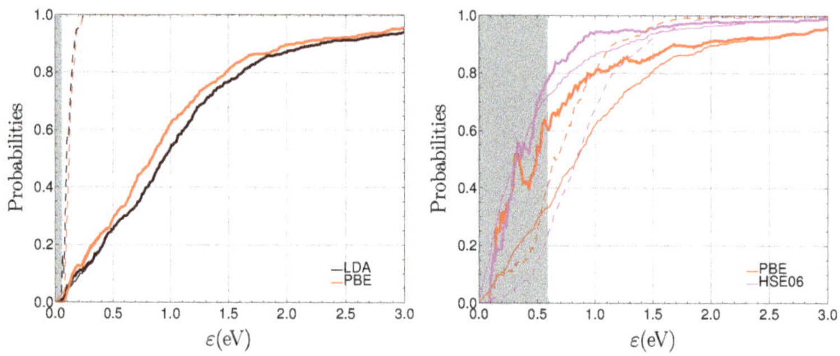

Figure 5. Probabilities for the pairs LDA/PBE (**left panel**) and PBE/HSE06 (**right panel**): $P_r(M, \epsilon)$ (thin curves), $P_{r|s}(M, \epsilon)$ (thick curves), and $P_{s|r}(M, \epsilon)$ (dashed curves). The gray rectangle covers the region where the selected sample size is less than 100.

The data set contains originally 471 systems. However, by making selections, e.g., of systems where the DFAs yield similar results, the size of the sample is reduced, and the use of statistical estimates is hampered. We estimate that, below 100 selected systems, the statistics become unreliable. The region for which the size of the sample is smaller than 100 is marked by a gray rectangle in Figure 5.

We notice that, for LDA and PBE, which provide close reliability curves, practically no distinction can be made between the thin and thick curves: similarity has no impact on reliability. We also see that $P_{s|r}(M, \epsilon) \approx 1$ for almost the whole range of ϵ: if one method gives a reliable result for a system, the other one is very likely to give a reliable result too. Thus, the size of the sample of similar results is reaching a size comparable to that of the complete sample already for a small value of ϵ (the gray rectangle is very thin).

The situation changes when we compare PBE to HSE06. The region where the size of the sample of similar results is below 100 reaches a large value of $\epsilon \approx 0.6$ eV. For $\epsilon > 0.6$ eV, we notice an improvement for each of the methods: $P_{r|s}(M, \epsilon) > P_r(M, \epsilon)$. However, we notice that the improvement of the worse of the two methods is not compensating the difference of quality between the two methods.

Even as ϵ increases beyond 0.6 eV, $P_{s|r}(M, \epsilon)$ is at first relatively small: if one method gives a reliable result, the probability that the other provides a reliable result too, is relatively small. Without surprise, the risk of the better of the two methods (HSE06) to eliminate systems by selection is higher than that of the worse of the two methods (PBE), cf. dashed curves in Figure 5. In this case, one should take the result provided by the better of the two methods, not, e.g., the average of the results of the two methods.

The improvement has to be paid: for some of the systems, the methods provide results that are not similar, and are not taken into consideration - we have no answer to give for these systems. Figure 6 shows an example by choosing PBE, finding how many systems from the data set are similar (within ϵ) to those obtained with another method. The graph confirms that an important number of systems are lost, unless one declares similarity by choosing a large value for ϵ.

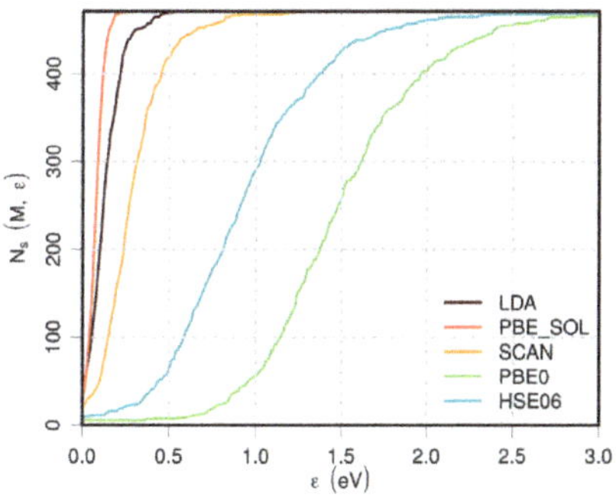

Figure 6. Number of systems that yield band gaps close to those obtained with PBE, for different methods, as a function of ϵ.

Let us attempt to condensate the results by looking at the values of ϵ for which the probabilities of having an absolute error smaller than ϵ is 0.95, $Q_{95}(M)$, cf. Table 1. This provides only an exploration of the behavior at large ϵ. Nevertheless, it leads to the conclusions above: LDA and PBE do not gain by using similarity: $Q_{95}(\text{LDA}) = 3.1$ eV and $Q_{95}(\text{PBE}) = 2.9$ eV, even after similarity is imposed. However, $Q_{95}(\text{HSE06})$ decreases from 1.7 to 1.3 eV when the similarity with PBE is taken into account.

Table 1. $Q_{95}(M_1)$, in eV, for the method indicated by the row (M_1), when similar to the method described by the column (M_2), for $\epsilon_s = \epsilon_r$.

$M_1 \backslash M_2$	LDA	PBE	PBEsol	SCAN	PBE0	HSE06
LDA	3.1	3.1	3.1	3.1	2.1	3.1
PBE	2.9	2.9	2.9	2.9	2.1	2.9
PBEsol	3.0	3.0	3.0	3.0	2.2	3.0
SCAN	2.4	2.4	2.4	2.4	2.1	2.4
PBE0	1.4	1.4	1.4	1.5	1.8	1.8
HSE06	1.1	1.3	1.1	1.7	1.7	1.7

We can expect the errors of different DFAs to be highly correlated [2]. (For example, recall that making the approximation valid for the uniform electron gas is a basic ingredient in almost all DFAs.) In other words, this could mean that if one method is right, all are right, and if one method is wrong, all are wrong: little improvement can be expected from agreement between methods.

Another measure of similarity is Spearman's rank correlation coefficient (Table 2). This varies between 0.76 (LDA and PBE0), and 0.99/1.00 (within the group of lower performance: LDA, PBE, and PBEsol). For PBE and HSE06, it takes an intermediate value (0.83). The correlation coefficients gives a hint for grouping the methods; however, it is more difficult to extract from them the information given in Figure 5 than it is from $Q_{95}(M)$.

Table 2. Rank correlation matrix between error sets.

$M_1 \backslash M_2$	LDA	PBE	PBEsol	SCAN	PBE0	HSE06
LDA	1	0.99	1.00	0.95	0.76	0.81
PBE	-	1	1.00	0.97	0.78	0.83
PBEsol	-	-	1	0.96	0.77	0.82
SCAN	-	-	-	1	0.83	0.87
PBE0	-	-	-	-	1	0.98
HSE06	-	-	-	-	-	1

Another problem of using the correlation index is its invariance with respect to a monotonous transformation of the calculated values (a linear transformation for the Pearson correlation). If one of the methods is biased and another not, these methods are not likely to give similar results, despite a high correlation index. Of course, this dissimilarity can be reduced by correcting the bias, typically, by subtracting the estimated mean error from the values obtained.

Figure 7 shows the probability that the results of two methods are similar (within ϵ). The similarity of LDA and PBE can be recognized immediately, as well as the dissimilarity between LDA and PBE0 or HSE06. One can also notice the improvement after centering the errors (i.e., correcting the bias by subtracting the mean signed error for each of the methods). At the same time, the difference between methods (PBE0 and HSE06) is reduced.

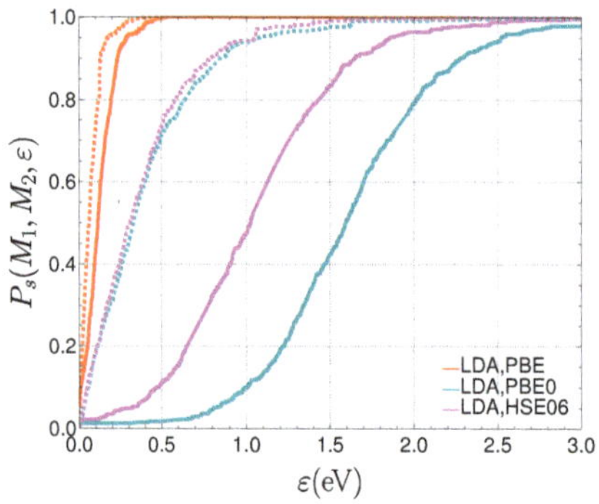

Figure 7. Probabilities $P_s(M_1, M_2; \epsilon)$ that a pair of methods (M_1, M_2) yields similar results (within ϵ) for (LDA and PBE), (LDA and PBE0), and (LDA and HSE06). The dashed curves are obtained after centering the errors.

One may want to summarize the information present in $P_s(M_1, M_2; \epsilon)$ by its mean, $\mu_s(M_1, M_2)$, and standard deviation, $\sigma_s(M_1, M_2)$:

$$\mu_s(M_1, M_2) = \int_0^\infty \epsilon[1 - P_s(M_1, M_2; \epsilon)]d\epsilon \tag{12}$$

$$\sigma_s^2(M_1, M_2) = \int_0^\infty \epsilon^2[1 - P_s(M_1, M_2; \epsilon)]d\epsilon - \mu_s^2 \tag{13}$$

The numerical results are given in Table 3. The similarity of LDA, PBE, and PBEsol is well visible from these numbers.

Table 3. The mean and the standard deviation of the probability distribution of having two DFAs giving similar results, $\mu_s(M_1, M_2)(\sigma_s(M_1, M_2))$, Equations (12) and (13).

$M_1 \backslash M_2$	LDA	PBE	PBEsol	SCAN	PBE0
PBE	0.1(0.1)	-	-	-	-
PBEsol	0.1(0.1)	0.1(0.0)	-	-	-
SCAN	0.4(0.3)	0.3(0.2)	0.4(0.2)	-	-
PBE0	1.6(0.6)	1.5(0.5)	1.6(0.5)	1.2(0.4)	-
HSE06	1.1(0.5)	0.9(0.5)	1.0(0.5)	0.6(0.3)	0.6(0.2)

Let us now increase the number of methods that we are considering. Taking into account the closeness of the results of LDA, PBE, and PBEsol, we do not expect anything considering the similarity of these three methods. However, one might ask whether comparing PBE, HSE06, and SCAN, or PBE, HSE06, and PBE0 provides any improvement. In the first case, Q_{95}(HSE06) stays at 1.3 eV; in the second, it slightly increases to 1.4 eV.

Increasing the number of methods for similarity checks does not provide necessarily an improvement on reliability (as one increases the number of "bad" methods to compare with). All six methods provides, at best, $Q_{95}(M) \approx 1.3$ eV, while the best value in Table 1 is of 1.1 eV. This can be also seen in Figure 8, the analogue of Figure 5, showing the probabilities obtained when similarity among all six methods is taken into account. This also shows the increase of the region of poor sampling.

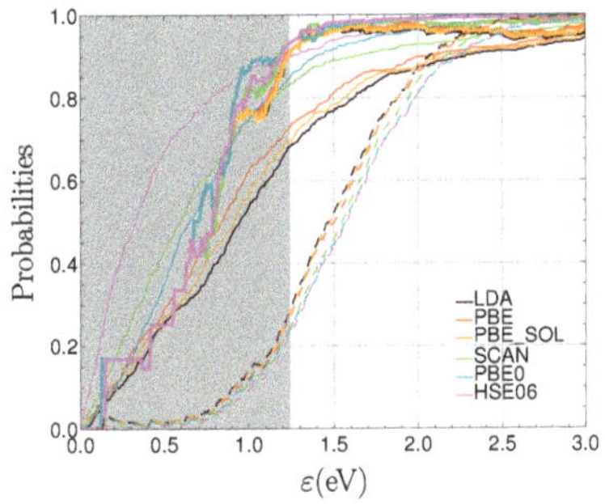

Figure 8. Probabilities : $P_r(M,\epsilon)$ for M in the set of 6 methods (thin curves), $P_{r|s}(M,\epsilon)$ (thick curves), and $P_{s|r}(M,\epsilon)$ (dashed curves). The gray rectangle covers the region where the selected sample size is less than 100.

3.2.4. Eliminating Strange Results?

The distribution of errors in density functional approximations is often not normal [5]. This can be seen in Figure 9. It seems that similarity confirms (in part) the prejudice that a strange behavior of one method is not repeated by another, different method. After restricting the data set to similar values, the distribution of errors is more compact. This explains the lowering of the $Q_{95}(M)$. Recall, however, that wrong results obtained with both methods are not excluded.

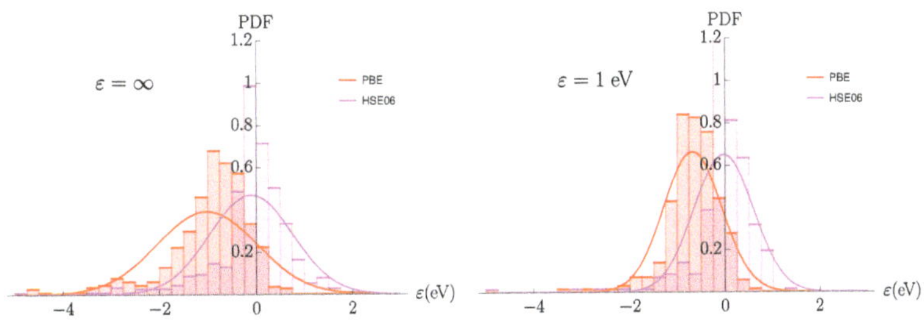

Figure 9. Histograms showing the distribution of errors before and after introducing similarity (**left**, and **right** panel for $\epsilon = \infty$ and $\epsilon = 1$ eV, respectively), for PBE (red) and HSE06 (blue). The normal distributions using the mean and standard deviation of these error distributions are shown as curves.

3.3. *The ZAS2019 Dataset*

The effective atomization energies (EAE) for the QM7b dataset [16], for molecules up to seven heavy atoms (C, N, O, S, and Cl) are issued from the study by Zaspel et al. [17]. We consider here values for the cc-pVDZ basis set, and the prediction error for 6211 systems for the SCF, MP2, and machine-learning (SLATM-L2) methods with respect to CCSD(T) values as analyzed by Pernot et al. [18].

In contrast to the case of the DFAs presented in the previous section (Table 2), the errors in this dataset present negligible rank correlation coefficients (smaller than 0.1 in absolute value). Similarity will, thus, be dominated by the bias in the errors and their dispersion. The $P_r(M,\epsilon)$ and $P_{r|s}(M,\epsilon)$ curves are shown in Figure 10. When comparing HF to MP2, one sees that both methods benefit from similarity as soon as $\epsilon > 0.2$ kcal/mol.

Naturally, HF benefits much more from the similarity selection than MP2: its Q_{95} decreases from 6.1 to 4.2 kcal/mol, while Q_{95} for MP2 decreases slightly from 3.4 to 2.9 kcal/mol. A similar behavior is observed in the comparison of HF to SLATM-L2 with a larger onset of improvement for SLATM-L2 ($\epsilon \sim 0.5$ kcal/mol). For HF, Q_{95} decreases from 6.1 to 3.8 kcal/mol and for SLATM-L2 from 4.7 to 2.5 kcal/mol. The comparison of MP2 to SLATM-L2 provides an intermediate case, where both methods present more balanced improvements: for MP2, Q_{95} decreases from 3.4 to 2.4 kcal/mol and for SLATM-L2 from 4.7 to 1.9 kcal/mol.

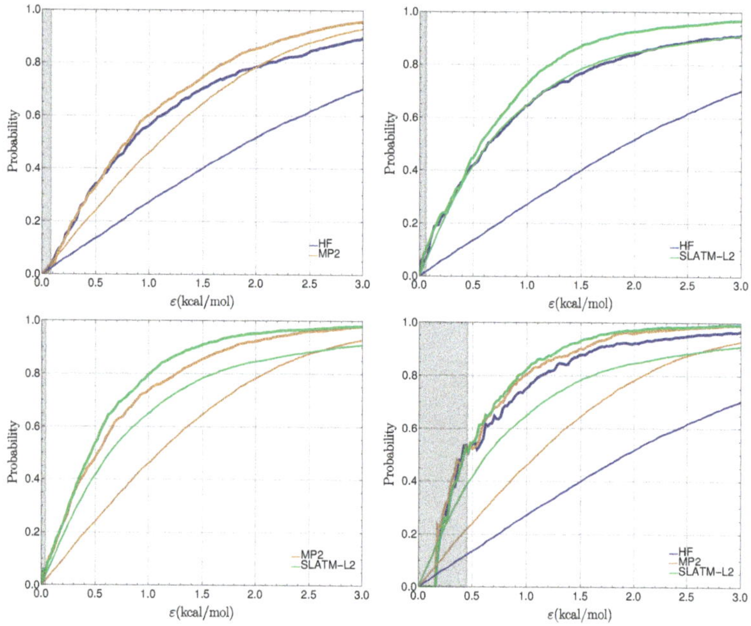

Figure 10. $P_r(M, \epsilon)$ (thin curves) and $P_{r|s}(M, \epsilon)$ (thick curves) for the pairs and triple in the set (HF, MP2, and SLATM-L2).

Adding HF to the MP2/SLATM-L2 pair produces a marginal gain for the latter methods, whereas HF presents a strong gain in reliability: the final Q_{95} values are 2.5 (HF), 1.8 (MP2) and 1.7 kcal/mol (SLATM-L2). However, this comes at the price of a large number of system rejections: for $\epsilon \sim 2.0$ kcal/mol, only 1/4th of the 6211 systems are selected by their similarity. For comparison, this number is about 2/3rd for the MP2/SLATM-L2 comparison.

In this context of uncorrelated error sets with different accuracy levels, one sees that similarity selection has a notable positive impact on the reliability of predictions by any of the methods, even the most accurate ones. It is striking that MP2 or SLATM-L2 might benefit from comparison with HF, but, as already discussed for band gaps (Figure 9) , this proceeds mainly by elimination of systems with large errors.

4. Conclusions

We asked whether picking only results that are similar to different methods would improve the accuracy of their predictions (in spite of possibly eliminating a significant part of the calculations done). The use of probabilities to treat reliability and similarity was illustrated on two benchmark data sets, one of band gap calculations with different density functional approximations, the other of effective atomization energies with two *ab initio* methods and one machine-learning method.

For the properties and methods studied, the thresholds for reliability and similarity were chosen quite generously. For the band gap data set, we found that similarity of the density functional results had only a marginal impact on improving the prediction accuracy. This is consistent with previous findings that the differences between density functional approximations are less important when considering the error distributions [1,2], or taking into account experimental uncertainty [19].

For the effective atomization energies data set, in which the error sets are uncorrelated, notable improvements of reliability after similarity selection were observed for all methods, even the most accurate ones. Roughly, we observed two categories of results:

1. methods that always give close results, for which similarity is irrelevant; and
2. methods for which an improvement can be achieved, especially by eliminating certain systems that behave strangely with one or the other methods—similarity is mainly effective for eliminating large errors.

Note that the size of the data sets might have an impact on the uncertainty of all the statistics. For the smaller datasets, this uncertainty might be comparable with the observed differences between statistics. Bootstrapping approaches, such as the ones used in our previous works [1,3], could be used to this effect. This was not the focus of the present study, and uncertainty management will be considered in forthcoming research.

Author Contributions: Conceptualization, A.S. and P.P.; methodology, A.S. and P.P.; software, A.S. and P.P.; validation, A.S. and P.P.; formal analysis, A.S. and P.P.; investigation, A.S. and P.P.; resources, A.S. and P.P.; data curation, A.S. and P.P.; writing—original draft preparation, A.S. and P.P.; writing—review and editing, A.S. and P.P. All authors have read and agreed to the published version of the manuscript.

Funding: This research received no external funding.

Institutional Review Board Statement: Not applicable.

Informed Consent Statement: Not applicable.

Data Availability Statement: Data is contained within the article.

Acknowledgments: This paper is dedicated to Karlheinz Schwarz, who not only was a pioneer in the development of density functional theory, but also of computer programs using it for large systems. This led to a large number of applications, enabling us to perform statistical analyses as presented in this paper.

Conflicts of Interest: The authors declare no conflict of interest.

References

1. Pernot, P.; Savin, A. Probabilistic performance estimators for computational chemistry methods: Systematic Improvement Probability and Ranking Probability Matrix. I. Theory. *J. Chem. Phys.* **2020**, *152*, 164108. [CrossRef] [PubMed]
2. Pernot, P.; Savin, A. Probabilistic performance estimators for computational chemistry methods: Systematic Improvement Probability and Ranking Probability Matrix. II. Applications. *J. Chem. Phys.* **2020**, *152*, 164109. [CrossRef] [PubMed]
3. Pernot, P.; Savin, A. Probabilistic performance estimators for computational chemistry methods: The empirical cumulative distribution function of absolute errors. *J. Chem. Phys.* **2018**, *148*, 241707; Erratun in *J. Chem. Phys.* **2019**, *150*, 219906. [CrossRef] [PubMed]
4. Hausdorff, F. *Set Theory*; Chelsea: London, UK, 1978.
5. Pernot, P.; Savin, A. Using the Gini coefficient to characterize the shape of computational chemistry error distributions. *Theor. Chem. Acc.* **2021**, *140*, 24. [CrossRef]
6. Dirac, P.A.M. Note on Exchange Phenomena in the Thomas Atom. *Math. Proc. Camb. Philos. Soc.* **1930**, *26*, 376–385. [CrossRef]
7. Vosko, S.H.; Wilk, L.; Nusair, M. Accurate spin-dependent electron liquid correlation energies for local spin density calculations: A critical analysis. *Can. J. Phys.* **1981**, *58*, 1200–1211. [CrossRef]
8. Perdew, J.P.; Burke, K.; Ernzerhof, M. Generalized Gradient Approximation Made Simple. *Phys. Rev. Lett.* **1996**, *77*, 3865–3868. [CrossRef] [PubMed]
9. Perdew, J.P.; Ruzsinszky, A.; Csonka, G.I.; Vydrov, O.A.; Scuseria, G.E.; Constantin, L.A.; Zhou, X.; Burke, K. Restoring the Density-Gradient Expansion for Exchange in Solids and Surfaces. *Phys. Rev. Lett.* **2008**, *100*, 136406; Erratun in *Phys. Rev. Lett.* **2009**, *102*, 039902. [CrossRef] [PubMed]

10. Zhang, Y.; Kitchaev, D.A.; Yang, J.; Chen, T.; Dacek, S.T.; Sarmiento-Perez, R.A.; Marques, M.A.L.; Peng, H.; Ceder, G.; Perdew, J.P.; et al. Efficient first-principles prediction of solid stability: Towards chemical accuracy. *Npj Comput. Mater.* **2018**, *4*, 9. [CrossRef]
11. Perdew, J.P.; Ernzerhof, M.; Burke, K. Rationale for mixing exact exchange with density functional approximations. *J. Chem. Phys.* **1996**, *105*, 9982–9985. [CrossRef]
12. Adamo, C.; Barone, V. Toward reliable density functional methods without adjustable parameters: The PBE0 model. *J. Chem. Phys.* **1999**, *110*, 6158–6170. [CrossRef]
13. Heyd, J.; Scuseria, G.E.; Ernzerhof, M. Hybrid functionals based on a screened Coulomb potential. *J. Chem. Phys.* **2003**, *118*, 8207–8215; Erratun in *J. Chem. Phys.* **2006**, *124*, 219906. [CrossRef]
14. Krukau, A.V.; Vydrov, O.A.; Izmaylov, A.F.; Scuseria, G.E. Influence of the exchange screening parameter on the performance of screened hybrid functionals. *J. Chem. Phys.* **2006**, *125*, 224106. [CrossRef] [PubMed]
15. Borlido, P.; Aull, T.; Huran, A.W.; Tran, F.; Marques, M.A.L.; Botti, S. Large-scale benchmark of exchange-correlation functionals for the determination of electronic band gaps of solids. *J. Chem. Theor. Comput.* **2019**, *15*, 5069–5079. [CrossRef] [PubMed]
16. Montavon, G.; Rupp, M.; Gobre, V.; Vazquez-Mayagoitia, A.; Hansen, K.; Tkatchenko, A.; Müller, K.R.; Anatole von Lilienfeld, O. Machine learning of molecular electronic properties in chemical compound space. *New J. Phys.* **2013**, *15*, 095003. [CrossRef]
17. Zaspel, P.; Huang, B.; Harbrecht, H.; von Lilienfeld, O.A. Boosting Quantum Machine Learning Models with a Multilevel Combination Technique: Pople Diagrams Revisited. *J. Chem. Theory Comput.* **2019**, *15*, 1546–1559. [CrossRef] [PubMed]
18. Pernot, P.; Huang, B.; Savin, A. Impact of non-normal error distributions on the benchmarking and ranking of Quantum Machine Learning models. *Mach. Learn. Sci. Technol.* **2020**, *1*, 035011. [CrossRef]
19. Savin, A.; Pernot, P. Acknowledging User Requirements for Accuracy in Computational Chemistry Benchmarks. *Z. Anorg. Allg. Chem.* **2020**, *646*, 1042–1045. [CrossRef]

computation MDPI

Article

Accuracy and Precision in Electronic Structure Computation: Wien2k and FPLO

Manuel Richter [1,*], Seo-Jin Kim [2], Klaus Koepernik [1], Helge Rosner [2] and Arnulf Möbius [3]

[1] Institute for Theoretical Solid State Physics, Leibniz IFW Dresden, Helmholtzstr. 20, D-01069 Dresden, Germany; k.koepernik@ifw-dresden.de
[2] MPI-CPfS Dresden, Nöthnitzer Str. 40, D-01187 Dresden, Germany; seojin.kim@cpfs.mpg.de (S.-J.K.); rosner@cpfs.mpg.de (H.R.)
[3] Institut für Physik, Technische Universität Chemnitz, D-09107 Chemnitz, Germany; arnulf.moebius@t-online.de
* Correspondence: m.richter@ifw-dresden.de

Abstract: Electronic structure calculations in the framework of density functional theory are based on complex numerical codes which are used in a multitude of applications. Frequently, existing experimental information is used as a gauge for the reliability of such codes. However, their results depend both on the chosen exchange-correlation energy functional and on the specific numerical implementation of the Kohn-Sham equations. The only way to disentangle these two items is a direct comparison of two or more electronic structure codes. Here, we address the achievable numerical accuracy and numerical precision in the total energy computation of the two all-electron density-functional codes Wien2k and FPLO. Both codes are based on almost independent numerical implementations and largely differ in the representation of the Bloch wave function. Thus, it is a highly encouraging result that the total energy data obtained with both codes agree within less than 10^{-6}. We here relate the term numerical accuracy to the value of the total energy E, while the term numerical precision is related to the numerical noise of E as observed in total energy derivatives. We find that Wien2k achieves a slightly higher accuracy than FPLO at the price of a larger numerical effort. Further, we demonstrate that the FPLO code shows somewhat higher precision, i.e., less numerical noise in E than Wien2k, which is useful for the evaluation of physical properties based on derivatives of E.

Keywords: density functional theory; DFT codes; electronic structure calculation; numerical accuracy and precision

Citation: Richter, M.; Kim, S.-J.; Koepernik, K.; Rosner, H.; Möbius, A. Accuracy and Precision in Electronic Structure Computation: Wien2k and FPLO *Computation* **2022**, *10*, 28. https://doi.org/10.3390/computation10020028

Academic Editor: Henry Chermette

Received: 6 January 2022
Accepted: 7 February 2022
Published: 11 February 2022

1. Introduction

Computations of electronic structure provide an important input to research in physics, chemistry, and materials science, and density functional theory (DFT) is behind a major part of such computations. While DFT is a formally exact theory, its practical use relies on the implementation of approximative exchange-correlation functionals. A second but not less important ingredient to DFT applications consists in the development and verification of numerical electronic structure codes. Here, we consider the latter subject.

The task to solve the Kohn-Sham equations of DFT is well defined, but its solution is far from being trivial. The mathematical complexity of this system of non-linear integro-differential equations is mirrored by the complexity of its solvers, numerical codes with typically several 10^5 source lines. Codes of such a size and complexity bear two unavoidable problems: approximations and coding errors. Fortunately, a large number of codes has been developed in past decades by independent teams. These codes partly rely on completely different approximations for, e.g., the representation of the wave function. Thus, many of them can be considered as virtually independent implementations with little risk to suffer from the same source of inaccuracy. Hence, deviations from the numerically exact solution can be considered as code-specific and uncorrelated, at least for codes with different wave function representations. As noteworthy exceptions, we mention standard algebra routines that are frequently taken from common software packages and methods for k-space integrations. The latter are usually implemented independently in each code but may lead to method-specific convergence behavior with the number of k-points.

In 2016, a comparison was published regarding the volume-dependent total energy, $E(V)$, calculated by means of 15 different DFT codes for 71 elements of the periodic table [1]. The goal of that work was to evaluate the achievable precision of Kohn-Sham solvers under well-defined common choices of exchange-correlation functional and treatment of relativistic effects. Thereby, *precision* was defined as the scatter among computed equation-of-state data, in distinction to *accuracy* which would describe the deviation of the mean DFT result from experiment [1].

In the present work which is again solely focused on the comparison of computations, we will use the terms accuracy and precision in a slightly different sense. Motivated by the idea that the total energy is a variational quantity, lower total energy will be identified with higher numerical accuracy in the common meaning of a smaller systematic error [2]. This choice comes with the known caveat that numerical problems like basis set over-completeness can reduce the resulting total energy in an unpredictable manner [3] and have to be excluded by appropriate data analysis. Furthermore, our definition only makes sense if all calculations make use of one and the same exchange-correlation approximation. With this condition fulfilled, the numerical accuracy of a modern DFT code is mainly determined by the completeness of the basis set for the representation of the Bloch states.

Regarding precision as a description of random errors [2], we will associate this term with numerical noise along the $E(V)$ curve. This choice is motivated by the following consideration. DFT computations are usually quasi-deterministic. Thus, it is not possible to perform a statistical analysis for a single point on $E(V)$. However, different points on $E(V)$ are known to show quasi-random deviations from averaging approximants to $E(V)$, such as polynomial fits. The default numerical settings of advanced codes are nowadays tight enough to keep these quasi-random contributions to $E(V)$ well below its systematic deviations from a hypothetical, exact solution of the Kohn-Sham problem. In other words, the numerical precision of $E(V)$ is much higher than its numerical accuracy. However, this may not be the case for derivatives of the total energy which are of interest on their own. For the sake of brevity, we will drop the qualifier "numerical" for the considered accuracy and precision in most places below.

Seven of the codes taken into account in the abovementioned comparison are so-called all-electron codes which do not rely on approximative pseudopotentials but include the complete nuclear potential and the electronic core wave functions in the calculation. Considering the scatter of results only among these seven codes used with their numerically best settings, the data produced by Wien2k [4] turned out to be close to the barycenter [1]. This finding confirms the status of Wien2k as the "gold standard" of DFT codes: a well-deserved result of several decades of dedicated work by the developer team of Wien2k, formed and headed by Karlheinz Schwarz.

The predictive computation of physical properties by means of DFT codes has been widely established in past decades. If applied to a large set of materials, such a property screening is faced with resource considerations. Disregarding so-called $O(N)$ methods – they are not among the most commonly used DFT codes and are often geared towards certain subsets of compounds like insulators –, the required CPU time and main storage grow with the third and second power, respectively, of the Hamilton matrix rank. Hence, a code with a small matrix, i.e., a small basis set, is usually superior to a code with a large basis set in terms of computation time and main storage demand. However, small basis sets frequently limit the achievable accuracy. Thus, it is important to gauge the quality of codes with small basis sets against a confirmed quality standard such as Wien2k both in terms of accuracy and precision. In the current study, we compare Wien2k and the full-potential local-orbital (FPLO) code [5] with each other. They are based on completely different concepts to represent the Bloch wave function and the FPLO code is equipped with a much smaller basis set than Wien2k. However, both these codes belong to the all-electron category of electronic structure codes. This allows a direct comparison of their total energies.

To establish a procedure for comparing the quality of two codes concerning their predictive power beyond total energies, we advance from the previous equation-of-state comparison [1] to the consideration of total energy derivatives which give access to more detailed thermodynamic information. For example, the first derivative of the total energy with respect to the magnetization equals the external field necessary to stabilize this magnetization, and a zero second derivative indicates a metamagnetic transition. Yet more subtle is the detection of an electronic topological transition (ETT), also called Lifshitz transition, on the sole basis of total-energy data. For hexagonal Osmium, a previous experimental study claimed visibility of an ETT as a kink in the ratio of lattice parameters c/a vs. pressure p [6]. Theoretical equation-of-state data should in principle allow to identify such a peculiarity. In later work, however, the existence of this particular ETT was doubted, see Ref. [7] and references therein.

Here, we demonstrate for a particular case that a conventional ETT can be detected in the third derivative of $E(a)$ as obtained from current DFT codes, a denoting the lattice parameter. The reader may object that ETTs can simply be found by careful inspection of band structure and density of states. This was indeed our starting point in seeking to distinguish the traces of ETTs from numerical noise. We achieved a substantial reduction of the latter by modifying the default numerical settings for both considered codes. This allowed us to obtain even d^3E/da^3 relations with high resolution. The obtained precision enabled us to perform convincing Birch-Murnaghan fits using the minimum possible number of data points.

Our exercise focuses on the comparatively simple case of face centered cubic (fcc) Aluminium. This metal was already the target of very early Fermi surface investigations by electronic structure theory [8]. Later, Fermi surface changes under different types of pressure were predicted [9] and an ETT was observed by nuclear magnetic resonance experiments on Al under hydrostatic pressure [10].

Details of the computational methods are provided in the following Section 2. Results and the related discussion are presented in Section 3, and Section 4 summarizes this work.

2. Methods

The Wien2k calculations were performed in the version 19.1 of this code [4]. The parameter Rk_{max} was varied in the range from 4 to 12, and the muffin-tin sphere radius R_{mt} was chosen as 2.39 Bohr radii for Aluminium. As convergence conditions, 10^{-6} for density and 10^{-8} Rydberg for the total energy were used in almost all calculations. The only exception was one data point for $a = 0.3646$ nm, $Rk_{\text{max}} = 8$, for which only a density convergence of 10^{-5} could be achieved.

FPLO was used in the version FPLO-18.00-52 [11]. Self-consistency of the iterations was considered to be achieved upon stability of both density and total energy. The default values, 10^{-6} for density and 10^{-8} Hartree for energy, were taken as related thresholds if not indicated otherwise. A specific modification of the numerical mesh for orbital-potential matrix elements is explained and discussed in Section 3.3.

All calculations, performed with either of the codes, were conducted in a scalar relativistic mode according to Koelling and Harmon [12] and using the generalized gradient approximation (GGA) in the parameterization by Perdew, Burke, and Ernzerhof (PBE) [13]. Integrations in reciprocal space were done with a linear tetrahedron method including Blöchl corrections [14]. It should be noted that broadening techniques are less suited for the detection of ETTs.

The k-point numbers of the specific calculations are defined in Section 3 using the following notation: A mesh with $n \times n \times n$ intervals in the full Brillouin zone, the irreducible part of which was used in a self-consistent calculation, is denoted as sc$\text{K}n$; a mesh with $m \times m \times m$ intervals in the full Brillouin zone, the irreducible part of which was used in a single-step calculation with an input density from a calculation sc$\text{K}n$, is denoted as $\text{K}m$-sc$\text{K}n$.

Calculations were performed for fcc Aluminium (space group 225) with lattice parameters a between 0.36 nm and 0.41 nm in steps of 0.0005 nm. In certain cases, the step size was chosen as 0.0001 nm.

In our evaluation of the $E(a)$ data, numerical differentiation plays an important role. This approach, however, requires a trade-off between the conflicting demands for the best possible resolution and the lowest possible noise. Here, we applied two approaches. The first one, called "direct evaluation" henceforth, uses generalizations of the midpoint formula of differentiation. For a given set of data points with equidistant arguments,

$$d^2E/da^2|_{a_i} = (E(a_{i+1}) - 2E(a_i) + E(a_{i-1}))/(a_i - a_{i-1})^2$$

and

$$d^3E/da^3|_{(a_i+a_{i+1})/2} = (E(a_{i+2}) - 3E(a_{i+1}) + 3E(a_i) - E(a_{i-1}))/(a_i - a_{i-1})^3$$

were evaluated for $(a_i - a_{i-1}) = 0.0005$ nm. This value is a carefully chosen compromise between a high noise level for small step sizes and a suppression of physical singularities for large step sizes. A considerable further noise reduction can be achieved by starting from third-order polynomial fits within a moving window. This second approach, called "moving window differentiation" below, was carried out on equidistant data sets with $(a_i - a_{i-1}) = 0.0001$ nm and 33 data points in each fit.

3. Results and Discussion

3.1. *Effect of the k-Mesh on the Density of States*

The density of states (DOS), g, is a primary quantity obtained in any electronic structure calculation on extended systems. It has widely been used for the interpretation of numerous experimental results like photon- or electron-spectroscopy data, thermodynamic properties, or electron-nucleus interaction data. Here, we focus on the DOS at the Fermi level $g(e_F)$. It is related to the low-temperature specific heat and also to the Knight shift. Last but not least, singularities in the dependence of $g(e_F)$ on the volume (or lattice parameter, or pressure) signal Lifshitz transitions. The aim of this subsection is to sensitize the reader to the importance of choosing a dense k-mesh in order to achieve a precise DOS. Henceforth, we use $g(e_F)$ as abbreviation for $g(e_F(a), a)$.

Figure 1 shows $g(e_F)$ of fcc Aluminium vs. lattice parameter a for different k-meshes. Consider the upper panel of the figure first. The default k-mesh scK12 (red line) provides a smooth behavior of $g(e_F)$ in the whole range of the lattice parameter. It does, however, not meet the expectation of a root-like dependence for a nearly-free electron situation. For the next denser mesh scK24, a local maximum of $g(e_F)$ appears close to $a = 0.394$ nm. This maximum could be related to an ETT, but the yet denser mesh scK50 produces a much more structured curve with several local maxima or kinks. The majority of these singularities can be assigned to numerical minigaps arising from wrong band connections. Such minigaps are more numerous but smaller in the scK100 results, and they are almost invisible in the scK200 curve.

The latter two data sets each exhibit two kinks that indeed originate from ETTs. They are marked by the vertical lines and named ETT1 and ETT2. Furthermore, data obtained with a very dense k-mesh K400-scK200 are presented in the lower panel of the figure and in its inset. The related curve is piece-wise quasi-analytic. Here, another but less prominent Lifshitz transition, ETT0, is recognizable close to $a = 0.392$ nm, see the inset of Figure 1.

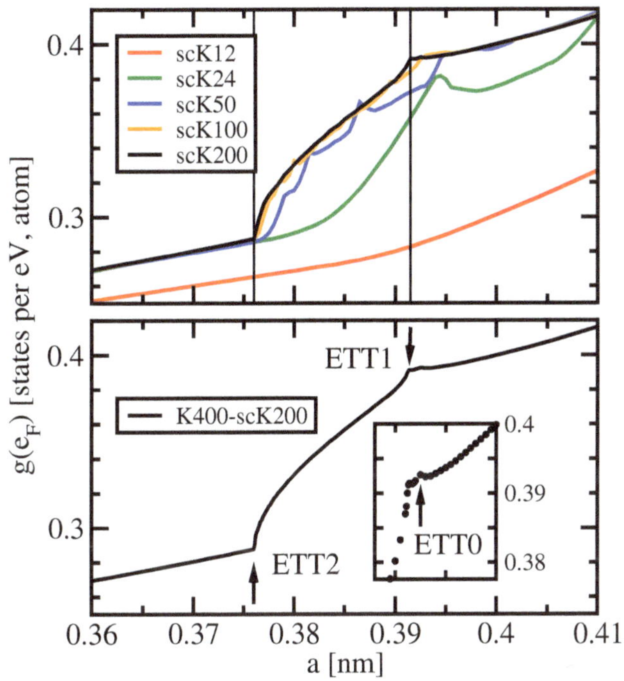

Figure 1. Total density of states at the Fermi level vs. lattice parameter, obtained with the FPLO code using different k-meshes as specified in the legends. The positions of two prominent ETTs are indicated by vertical lines in the upper panel and by arrows in the lower panel. Another, less prominent ETT is indicated by an arrow in the inset, which has an a-axis identical to that of the main graph. The upper panel shows data points which were obtained at a distance of 0.0005 nm. They are connected by straight lines. The lower panel presents data obtained at the same distance completed by data at a distance of 0.0001 nm in the vicinity of the ETTs. The data points are connected by lines in the lower main panel and depicted as circles in the inset. The K400-scK200 data were computed with tightened convergence criteria of 10^{-7} (density) and 10^{-9} Hartree (total energy) in the scK200-step.

All three ETTs detected above can be identified in the schematic Fermi surface plots shown in Figure 2. Therein, the well-known third-zone monster of fcc Aluminium is shown in the left-hand part within the complete Brillouin zone and as the lowest detail in the right-hand part, both for the equilibrium situation.

Consider the right-hand part of Figure 2. Upon reduction of a, the β-orbit merges with its neighbor and the monster become multiply connected (arrow in the second lowest detail). This ETT0 is barely visible in the DOS, compare the inset of Figure 1. A slight further reduction of a disconnects the Fermi surface and lets the α-orbits vanish, as shown in the third detail. This transition was named ETT1 above. Finally, the ellipsoid around symmetry point U vanishes in another Lifshitz transition, named ETT2. The latter two transitions are well-resolved in the DOS, provided it is computed with appropriate precision, see Figure 1.

To avoid the occurrence of the abovementioned artificial singularities of the DOS, resulting from wrong band connections, broadening techniques are frequently used for k-integration and even for the calculation of the DOS. Such approaches, however, do not allow for the identification of van Hove singularities as indicators of ETTs. Hence, the use of a linear tetrahedron method with a very dense k-mesh is the only safe way to identify an ETT in the DOS. An alternative way is to calculate Fermi surfaces using an equally dense mesh. We stress that a band structure at symmetry lines is usually not sufficient for this aim since changes of the Fermi surface topology frequently happen at general k-points, compare Figure 2.

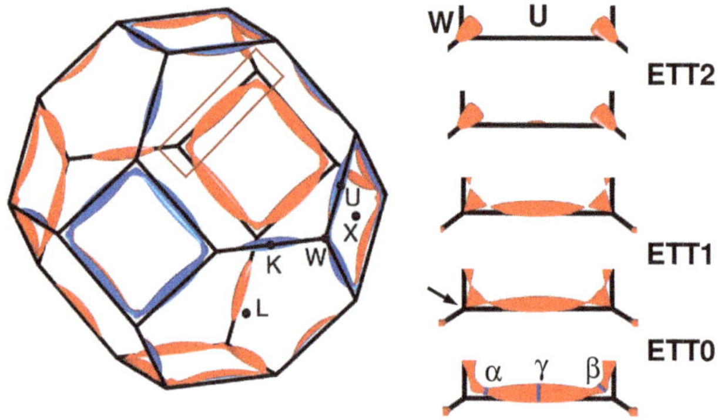

Figure 2. Schematic Fermi surface of Aluminium as obtained with the FPLO code. **Left part**: Third-zone equilibrium-volume Fermi surface in the fcc Brillouin zone with symmetry points. Blue (red) color indicates the occupied (unoccupied) side of the Fermi surface. **Right part**: variation of a detail of the Fermi surface close to one of the square edges as indicated by a brown rectangle in the left part. The lattice parameter shrinks from bottom to top. The lowest detail shows the equilibrium-volume situation with three extremal orbits (α, β, γ) indicated in the common notation [9] . Our denominations ETT0, ETT1, and ETT2 refer to the Lifshitz transitions detected in Figure 1.

The limited precision of DOS data evaluated by using a coarse k-mesh also results in large errors in the predicted values of Sommerfeld parameters, Pauli susceptibilities, and Knight shifts. To illustrate this remark, Table 1 compiles the relative differences between the well-converged value of $g(e_F)$ obtained from K400-scK200 and $g(e_F)$ data obtained from less dense meshes for three specific lattice parameters. It shows that the default mesh, which is in many cases sufficient for fairly accurate self-consistent calculations, provides an error of up to 30% in $g(e_F)$ at the considered lattice parameters. Even the rather dense mesh scK100 yields an error of up to 2%.

Table 1. Relative differences $1 - g(e_F)_{\text{other}}/g(e_F)_{400}$ between $g(e_F)_{400}$, computed from K400-scK200, and $g(e_F)_{\text{other}}$, computed from other k-meshes, for lattice parameters close to the equilibrium structure and close to ETT1 and ETT2. All data were obtained with the FPLO code.

Mesh	scK12	scK24	scK50	scK100	scK200
$1 - g(e_F)_{\text{other}}/g(e_F)_{400}$ at $a = 0.4040$ nm	0.234	0.051	−0.0008	0.0002	−0.0002
$1 - g(e_F)_{\text{other}}/g(e_F)_{400}$ at $a = 0.3915$ nm	0.278	0.087	0.049	0.019	−0.0004
$1 - g(e_F)_{\text{other}}/g(e_F)_{400}$ at $a = 0.3760$ nm	0.078	0.007	0.007	0.0008	0.0004

The data presented in Table 2 corroborate the above note that k-meshes of usual density are sufficient for the computation of integral properties like the equilibrium lattice parameter a_0. For each of the considered k-meshes, a_0 was obtained as the minimum of a parabola through the three total energy values computed for $a = 0.40415, 0.40420, 0.40425$ nm. These lattice parameter values are close to the GGA equilibrium which we identify with $a_{0,\text{scK400}} = 0.4042144$ nm, denoting a_0 computed with scK400. Already for scK36, the relative error $1 - a_{0,\text{scKxx}}/a_{0,\text{scK400}}$ is below 2×10^{-5}. Note, this precision is somewhat better than the stability of the Aluminium sample lattice parameter in an experiment without explicit and accurate temperature control. For this situation, temperature stability of 1 K and thermal expansion of 2.3×10^{-5} K^{-1} are assumed. After increasing the density of k-points by a factor of 8, that is for scK72, the deviation is already reduced to 2.5×10^{-6}, far better than the achievable precision of any single X-ray experiment.

Table 2. Relative differences $1 - a_{0,\text{scKxx}}/a_{0,\text{scK400}}$ between equilibrium lattice parameter $a_{0,\text{scK400}}$, computed from scK400, and $a_{0,\text{scKxx}}$, computed from other k-meshes. All data were obtained with the FPLO code.

Mesh	scK12	scK24	scK36	scK48	scK72	scK100	scK200
$10^5 \cdot (1 - a_{0,\text{scKxx}}/a_{0,\text{scK400}})$	2.1	−8.7	1.9	0.7	−0.25	0.02	0.05

3.2. Preliminary Considerations—Wien2k

Before a serious numerical investigation by means of any DFT code for a periodic system can be started, two important decisions regarding the numerical settings have to be taken. The first one, choosing the method and mesh density for k-space integration, was discussed in the previous subsection. The second one is the choice of the basis set for the Bloch wave function representation. In the Wien2k code, the extend of the basis set is encoded in the parameter Rk_{max}, the product of the smallest muffin-tin sphere radius and the largest k-vector of the plane wave basis set used in the interstitial space. In addition, the number of local orbitals depends on Rk_{max}.

In order to check the impact of Rk_{max} on the accuracy, we performed calculations for a number of integer values around the default value $Rk_{\text{max}} = 7$, using a dense k-mesh K100-scK50. Figure 3 shows the total energy vs. Rk_{max} of fcc Aluminium for $a = 0.404$ nm, which is close to the GGA equilibrium. The default value of Rk_{max} is indicated by a red circle. We note that the total energy diminishes monotonically with growing Rk_{max} between $4 \leq Rk_{\text{max}} \leq 12$. Hence, the accuracy improves with growing Rk_{max} at least till $Rk_{\text{max}} = 12$. The energy seems to be converged within about 0.1 meV for this largest tested basis set. The default setting $Rk_{\text{max}} = 7$ provides a total energy about 7 meV above the lowest value.

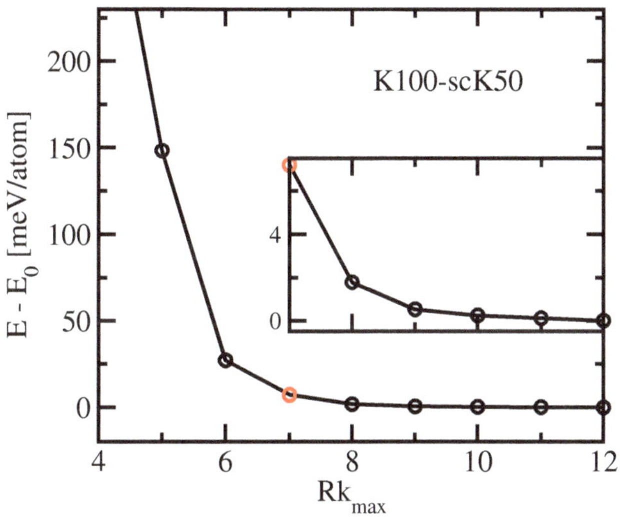

Figure 3. Total energy vs. Rk_{max} of fcc Aluminium for $a = 0.404$ nm obtained with the Wien2k code. All data were computed with K100-scK50. The lowest obtained total energy value, $E_0 = -6{,}607{,}524.417$ meV/atom, was chosen as reference. The inset shows part of the data using a zoomed energy axis. Its Rk_{max}-axis agrees with that of the main plot. Red circles indicate the result for the default value of Rk_{max} and black circles all other calculated points.

Next, we turn to the influence of Rk_{max} on the precision by evaluating total energy derivatives with respect to the lattice parameter. Figure 4 shows d^2E/da^2 (left panel) and d^3E/da^3 (right panel) vs. a, as obtained using Wien2k for a lattice parameter step size of 0.0005 nm. Note that, in order to avoid strong overlap, the individual data sets are offset against each other as detailed in the figure caption. For both the second and third derivatives, the numerical noise is smallest for $Rk_{max} = 8$ or 9, slightly above the default value $Rk_{max} = 7$. For both $Rk_{max} < 8$ and $Rk_{max} > 9$, the precision decreases. We stress that, in the case of third derivatives, the noise level is larger than the absolute mean value of the derivative for all values of Rk_{max}. This fact will necessitate data smoothing in the search for ETTs, see Section 3.5.

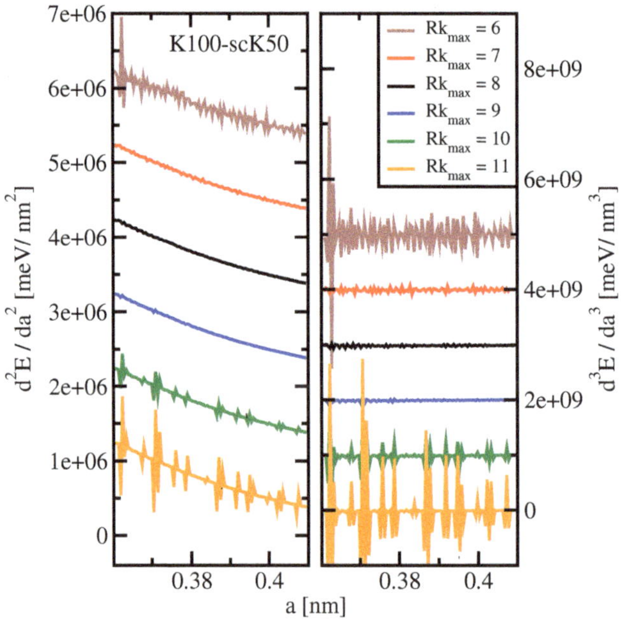

Figure 4. Derivatives of the total energy of fcc Aluminium with respect to the lattice parameter vs. lattice parameter. The total energies were obtained with the Wien2k code for different values of Rk_{max} as detailed in the legend. **Left panel**: second derivative, d^2E/da^2; the data are offset against the original data by $(11 - Rk_{max}) \cdot 10^6$ meV/nm^2. **Right panel**: third derivative, d^3E/da^3; the data are offset against the original data by $(11 - Rk_{max}) \cdot 10^9$ meV/nm^3. The derivatives are obtained by direct evaluation, see Section 2, for a lattice parameter step size of 0.0005 nm. All data were computed with K100-scK50.

One possible explanation for, above $Rk_{max} = 9$, the noise unexpectedly increasing with increasing number of plane waves could be numerical problems arising from near completeness. In such a situation, the newly added basis vectors contain more and more redundant directions in Hilbert space. Shifting (a tiny) wave function weight from one to another of these directions will not alter the charge density in any appreciable manner. However, which directions are finally chosen to contribute to the wave function is increasingly determined by unavoidable numerical noise from various places of the code in a quasi-random fashion, the less important the added directions are. Such noise will reflect in derivatives of the total energy. We have no hypothesis that could explain the observed increase of noise if a smaller basis than the default one is chosen.

3.3. *Preliminary Considerations—FPLO*

The precision of the total energy computation, viz. the noise of $E(V)$, can be evaluated in two ways. In the previous subsection, we used numerical differentiation for this aim. Now, we consider the comparison to a global fit of the $E(V)$ data set: the subtraction of a Birch-Murnaghan fit, $E_{BM-fit}(V)$, from the original data removes a large and smooth contribution.

Figure 5 compares the results of this procedure when applied to two FPLO data sets obtained with different treatment of orbital-potential matrix elements. In both cases, the rather dense k-mesh K200-scK50 and default settings concerning all other parameters were used. The black and red curves present the lattice parameter dependences of the difference $E(a) - E_{\mathrm{BM-fit}}(a)$ obtained with default and modified integration meshes for the orbital-potential matrix elements, respectively. Along the whole range of considered lattice parameters, these energy differences amount to less than 10^{-7} parts of the total energy.

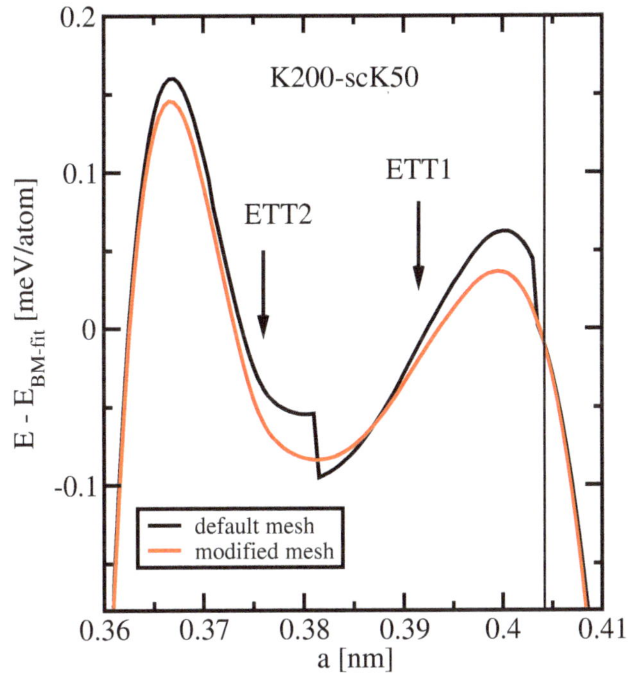

Figure 5. Difference between the total energy calculated with the FPLO code and the total energy obtained with a related third-order Birch-Murnaghan fit vs. lattice parameter. All data were obtained with K200-scK50. The black line shows data obtained with default numerical settings and the red line shows data obtained with a modified mesh for the computation of orbital-potential matrix elements as explained in the text. The arrows indicate the positions of two ETTs, identified in Section 3.1. The vertical line marks the equilibrium lattice parameter derived from the Birch-Murnaghan fit to data obtained with the modified integration mesh.

The general behavior of both $E(a) - E_{\mathrm{BM-fit}}(a)$ curves is typical of a fourth-order polynomial, as expected for the residual of a third-order fit. However, the black curve exhibits two well-visible discontinuities at $a = 0.381 \ldots 0.3815$ nm and $a = 0.403 \ldots 0.4035$ nm, as well as another, barely visible one at $a = 0.3705 \ldots 0.371$ nm. The energy residual $E(a) - E_{\mathrm{BM-fit}}(a)$ jumps at these lattice parameter values by less than 0.05 meV, which is about two orders of magnitude smaller than the energy variations usually encountered. Hence, these discontinuities are not visible in most presentations. They are, moreover, not important in most applications. In the present study, however, they are essential.

A possible idea would be to relate the discontinuities in $E(a) - E_{\mathrm{BM-fit}}(a)$ to Lifshitz transitions. However, the two ETTs identified in Section 3.1 occur at other lattice parameters, were no singularity in $E(a) - E_{\mathrm{BM-fit}}(a)$ is visible in Figure 5. The true origin of these discontinuities lies in a time-saving approximation in the evaluation of the orbital-potential matrix elements. This numerical integration is performed on a three-dimensional grid, which combines a radial mesh with angular meshes at each of the radial points. In FPLO, computing time is saved by using angular meshes with lower density for smaller radii. Since, the transition from smaller to larger meshes with varying radius happens at reasonably chosen fixed distances from the nuclei, the number of grid points changes with varying lattice parameter, resulting in the observed discontinuities.

The remedy is to choose the most dense angular mesh at all radii, which increases the CPU time only by about 10%. This modification has been applied to compute the data shown in the red curve of Figure 5. There, no visible deviation from a smooth behavior is observed and the noise level is well below 0.01 meV. For this reason, we used the modified mesh for the evaluation of the orbital-potential matrix elements for all calculations presented in this work, except for the default data in Figure 5.

Finally, we remark that such small precision-related effects in the total energy as visualized in Figure 5 can only be detected by considering the difference between total energy and a Birch-Murnaghan or another appropriate polynomial fit. The parameter values of the Birch-Murnaghan fits used here are given in Table 3. They are equivalent to each other, i.e., the precision of these parameters is not reduced by applying the default integration mesh instead of the modified one as long as the fit is performed with a large enough number of data points.

Table 3. Parameter values obtained in the third-order Birch-Murnaghan fits used for Figure 5. V_0 denotes the equilibrium atomic volume, B_0 and B_0' are the bulk modulus and its pressure derivative.

Quantity	V_0 **[nm^3]**	B_0 **[GPa]**	B_0'
Default mesh values	0.0165056	79.05	4.29
Modified mesh values	0.0165050	79.06	4.29

3.4. Comparison of Total Energies

We now compare the accuracies reached by the Wien2k and FPLO codes. This will be done by considering the equation of state for two different basis sets of each code. Figure 6 shows results obtained with default basis sets (Wien2k: $Rk_{\text{max}} = 7$; FPLO: local states $2s, 2p, 3s, 3p, 3d, 4s, 4p$) and with basis sets for enhanced accuracy (Wien2k: $Rk_{\text{max}} = 9$; FPLO: local states $2s, 2p, 3s, 3p, 3d, 4s, 4p, 4d, 4f, 5s, 5p$). For the Wien2k code, the enhanced basis set was defined according to the precision considerations in Section 3.2; for the FPLO code, the enhanced basis set as defined in the appendix to Ref. [1] was used. We remark that the Rk_{max} settings for Wien2k in the mentioned reference were 6.5 for default and 10.0 for enhanced accuracy calculations [1].

We learn from Figure 6 that the characterization of Wien2k as a highly accurate code is once more confirmed. For the default settings, the ground-state energy obtained by FPLO is about 10 meV per atom above the corresponding value obtained by the Wien2k code; considering the enhanced settings, this difference is reduced to about 4 meV per atom. At lattice spacings much smaller than the equilibrium spacings, the order is, however, reversed for the enhanced settings: the FPLO energy is found to be about 4 meV below the Wien2k value (see upper inset). A larger value of Rk_{max}, which could be chosen at the price of reduced precision, would not change this picture, since it would reduce the total energy by merely 0.5 meV compared with the value for $Rk_{\text{max}} = 9$. We remark that the different sequences of the total energy values obtained by Wien2k and FPLO for equilibrium spacing, on the one hand, and for much smaller values of a, on the other hand, are not very surprising: DFT codes are usually optimized to achieve high accuracy close to the equilibrium state. For the case of high pressure, numerical settings may have to be adapted specifically.

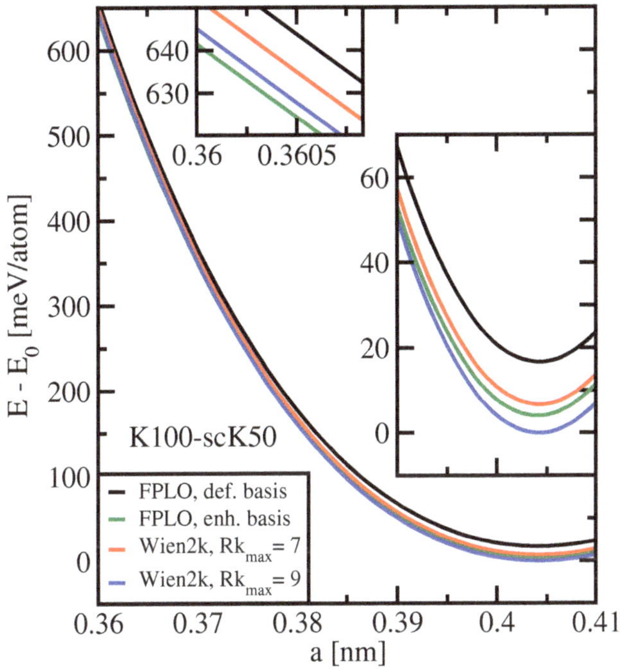

Figure 6. Comparison of total energies of fcc Aluminium vs. lattice parameter calculated with two different codes and two different basis sets for each code as detailed in the legend. All data were obtained with K100-scK50. The lowest total energy, $E_0 = -6{,}607{,}523.903$ meV/atom, was chosen as reference. The insets show zoomed regions of the main plot. The a-axis of the lower inset agrees with that of the main plot.

Table 4 presents a comparison of the parameter values obtained by means of Birch-Murnaghan fits to all four data sets. These fits yield virtually the same results: $a = 0.4040 \ldots 0.4042$ nm, bulk modulus $B_0 = 79$ GPa, and its pressure derivative $B_0' = 4.3$.

Table 4. Comparison of third-order Birch-Murnaghan fits applied to the data shown in Figure 6. Rows denoted with Rk_{max} values refer to Wien2k, rows denoted with a basis type refer to FPLO. V_0 denotes the equilibrium atomic volume, B_0 and B_0' are the bulk modulus and its pressure derivative, respectively.

Quantity	V_0 [nm^3]	B_0 [GPa]	B_0'
$Rk_{\mathrm{max}} = 7$	0.0165067	79.14	4.29
$Rk_{\mathrm{max}} = 9$	0.0165049	78.95	4.31
default basis	0.0165050	79.06	4.29
enhanced basis	0.0164844	78.85	4.30

It is very remarkable that two completely independent codes yield total energies with an absolute difference of less than 5 meV and a relative difference of less than 10^{-6} along the whole range of lattice parameters. These findings confirms the reliability of *both* codes because the probability of accidental error compensation is extremely low. Note finally that Aluminium is usually considered to be a nearly-free-electron metal. Thus, it is expected to be best described by plane-wave related methods. The above results prove that, even in this case, a local orbital based code achieves almost the celebrated accuracy of an augmented plane wave code.

We would like to conclude this evaluation of accuracies with a comparison of the related numerical efforts. One anticipates that a better accuracy requires a larger effort. This expectation is indeed confirmed by the typical computation times presented in Table 5. We observed that, as a rough estimate, the Wien2k calculations consume about one order of magnitude more CPU resources than the FPLO calculations. This difference originates mainly from the following: Wien2k uses about twice as many basis functions as contained in the enhanced basis of FPLO.

Table 5. Typical computation times on Intel Xeon CPUs with 2.6 GHz clock frequency to calculate the data shown in Figure 6. Rows denoted with Rk_{max} values refer to Wien2k, rows denoted with a basis type refer to FPLO. The runtime of each of the cores for one iteration cycle is called t_{core}. It is given for serial mode (1 core) and, in the case of Wien2k, for parallel mode with 16 cores.

	t_{core} [s] (1 core)	t_{core} [s] (16 cores)
$Rk_{\mathrm{max}} = 7$, scK50	21	9
$Rk_{\mathrm{max}} = 7$, K100	134	20
$Rk_{\mathrm{max}} = 9$, scK50	31	10
$Rk_{\mathrm{max}} = 9$, K100	190	23
default basis, scK50	2	-
default basis, K100	13	-
enhanced basis, scK50	8	-
enhanced basis, K100	48	-

3.5. *Search for ETTs by Means of Total Energy Derivatives*

In Section 3.1, carefully inspecting the lattice parameter dependence of the density of states at the Fermi level, we detected three Lifshitz transitions. Two of them, termed ETT1 and ETT2, are clearly visible in $g(e_{\mathrm{F}})$, see Figure 1, provided the k-mesh is K100 or better; another one, ETT0, is hardly visible. Close to a three-dimensional ETT related to the bottom of a band at e_{c}, the total energy contains a contribution proportional to $(e_{\mathrm{F}} - e_{\mathrm{c}})^{2.5}$ [15]. In other situations, the sign can be changed. Assume linear dependence of $e_{\mathrm{F}} - e_{\mathrm{c}}$ on a close to the transition. Consequently, the ETT should be signaled by a divergence of the third derivative d^3E/da^3.

We now ask whether the respective optimum total energy precision of Wien2k and FPLO is sufficient for the identification of such divergences. The upper panel of Figure 7 shows d^3E/da^3 vs. a as obtained by means of both codes with K100-scK50 and subsequent direct numerical differentiation (d). In both cases, the step size of a was set to 0.0005 nm. Despite the use of rather dense k-meshes and although the settings of both codes were optimized with respect to low numerical noise, there is no sign of a divergence visible in the upper panel data close to any of the expected critical lattice parameters, which are indicated by vertical lines.

The fluctuations of the Wien2k data (black curve) are considerably stronger than those of the FPLO data (red), which points to a better precision of the latter. We remark that FPLO data generated with enhanced basis set as described in Section 3.4 (not included in Figure 7) exhibit up to four times stronger noise of d^3E/da^3 than data obtained using the default basis (red curve).

The lower panel of Figure 7 shows data obtained for the setting K200-scK100 using yet denser k-meshes than above (blue, Wien2k and orange, FPLO), and for K400-scK200 (green, FPLO). The two former data sets were computed by using a moving window differentiation (m) with 33 equidistant data points (distance 0.0001 nm). In this way, a clear singularity is seen at the expected position of ETT2, which is yet more pronounced in the FPLO K400-scK200 data set which was obtained by direct evaluation of the derivative (distance 0.0005 nm). On the one hand, the moving window differentiation reduces the numerical noise considerably; on the other hand, it broadens the singularity which is more pronounced in the latter data. Note that the individual data are offset against each other as detailed in the figure caption.

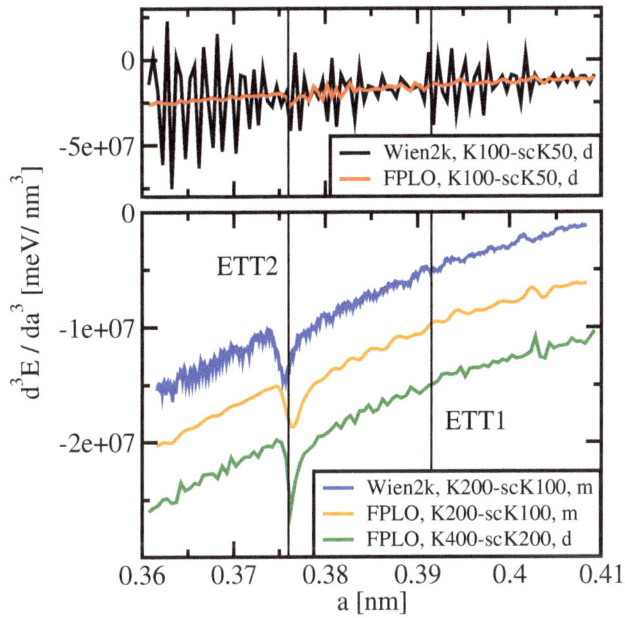

Figure 7. Third derivative of the total energy with respect to lattice parameter a in relation to electronic topological transitions. **Upper panel**: Comparison of Wien2k data obtained with $Rk_{\max} = 8$ (same data as shown in the right panel of Figure 4, black curve) with FPLO data (red). Both calculations were performed using the K100-scK50 setting with total energy convergence of 10^{-8} Rydberg. The derivatives were obtained by direct evaluation, indicated in the legends by "d". **Lower panel**: Comparison of Wien2k data obtained with $Rk_{\max} = 8$ (blue, with 10^7 meV/nm^3 offset) with FPLO data (orange, with 0.5×10^7 meV/nm^3 offset). Both data sets were obtained using the K200-scK100 setting with total energy convergence of 10^{-8} Rydberg. The derivatives were calculated using a moving window with 33 data points, indicated by "m", see Section 2. Another FPLO data set, K400-scK200 with total energy convergence of 10^{-9} Hartree and direct evaluation of the derivative, is shown in green. The vertical lines indicate the positions of ETT1 and ETT2 as obtained in Section 3.1.

None of the presented data sets shows any peculiarity exceeding numerical fluctuations at the position of ETT1 or of ETT0. We remind that the root-like singularities in $g(e_F)$ caused by ETT0 and ETT1 are less strong than that one caused by ETT2, see Figure 1. This difference may result from the differing natures of the respective topological changes: ETT0 and ETT1 are the opening and the closing of connections between Fermi surface parts, whereas ETT2 is the vanishing of an isolated part of the Fermi surface, see Figure 2.

To conclude this section, we have shown that an electronic topological transition can in principle be detected in a higher derivative of the total energy. To this end an extreme precision is required, such that the total energy curve appears very smooth as compared to usual code applications. Our calculations have shown that ETT2 is indeed visible as a singularity in $d^3E/da^3(a)$, whereas we could not detect the other ETT in this way. The other kinks observed are not related to topological transitions. In consequence, the total energy is not suited to detect such transitions. This implies that similar analyses of experimental data, which have a naturally lower precision than a highly tuned computer code, will be challenging. These strict statements, however, do no longer hold if the ETT is of lower than three-dimensional nature, in which case the transition can become first order and more easily detectable [16]. The differentiation procedures used here are very sensitive and might be helpful both in detecting first or second order phase transitions or in disproving their existence, such as in the case of β-Zirconium. There, contrary to earlier claims, no evidence for isostructural phase transitions was recently found [17].

3.6. Birch-Murnaghan Fits Using Minimum Number of Data Points

We have shown in Section 3.1 that the calculation of the equilibrium lattice constant is well converged already for moderately large k-meshes. Now, we would like to find out whether it is possible to utilize precise total energy data for evaluating the equilibrium volume, the bulk modulus, and its pressure-derivative with the minimum number of data points needed. For this aim, third-order Birch-Murnaghan fits were performed with only four total energy values obtained using FPLO with K100-scK50 setting. The equidistant fit points are distributed within a fit region of 0.0015 . . . 0.012 nm around the equilibrium lattice parameter.

Figure 8 shows results for the three experimentally accessible fit parameters vs. the width of the fit region. For comparison, the results of two Birch-Murnaghan fits to 25 and 13 data points, respectively, are shown as well. As expected from the results presented in Table 2, the equilibrium volume is almost independent of the width of the fit region. Furthermore, the variation of the bulk modulus, 0.1 GPa within the considered range, is negligible. Only the value of B_0' seems to be unstable if the fit region is chosen smaller than 0.004 nm. This observation is consistent with the fact that the stability of numerical derivatives worsens with their order increasing.

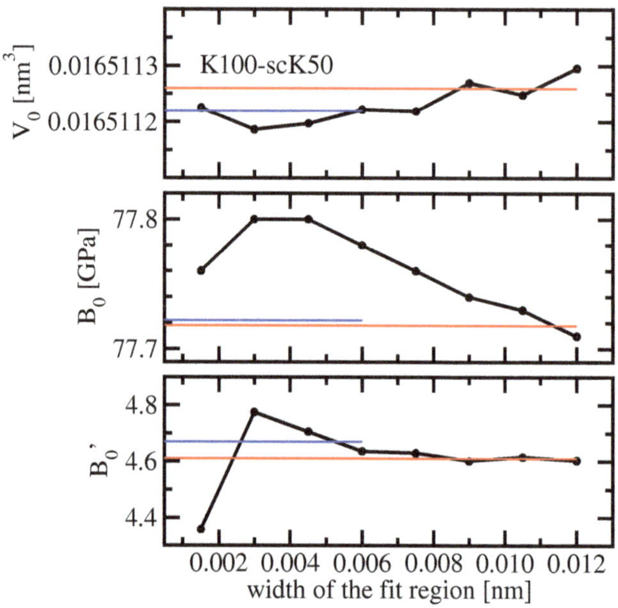

Figure 8. Results of third-order Birch-Murnaghan fits to few data points. The equilibrium volume V_0 (upper panel), the equilibrium bulk modulus B_0 (middle panel), and its pressure-derivative B_0' (lower panel) were obtained from FPLO total energy data for K100-scK50 setting. The circles connected with black lines show results obtained from four $E(a)$ data points, centered at the equilibrium lattice parameter and equally distributed over the fit region. The red (blue) lines show fit results obtained from 25 (13) $E(a)$ data points equally distributed within 0.012 (0.006) nm around the equilibrium lattice parameter.

The possibility to use a minimum number of precise data points to numerically evaluate quantities derived from the total energy can be useful e.g. for screening studies. However, this chance comes at the price of a sufficiently dense k-mesh. As long as specific effects of band structure or DOS, such as ETTs, Weyl points or magnetic instabilities, are not important, it can be useful to apply broadening techniques with a less dense k-mesh. They offer the advantages to reduce the related effort and to smoothen the self-consistent iteration.

4. Summary

Electronic structure calculation packages such as the famous Wien2k code, developed and maintained by Karlheinz Schwarz and his group for decades, are meanwhile indispensable tools for the interpretation of experimental data as well as for the prediction of materials properties. In the latter case, there is no standard to gauge the result but by other codes.

Doing so, we have confirmed the classification of Wien2k as the "gold standard" concerning total energy accuracy. This high numerical accuracy is achieved, as in the case of similar codes [1], by using a large and elaborated basis set for the representation of the Bloch wave function. The FPLO code, taken in this study for comparison, uses a smaller basis set. In consequence, it can only reach slightly lower total energy accuracy than Wien2k, but it has the advantage of considerably smaller numerical effort.

Either of the two compared codes allows for the detection of one of the known electronic topological transitions of Aluminium under pressure by appropriate evaluation of $d^3E/da^3(a)$. The other two discussed ETT, which are of different topological nature than the former one, are only visible in the density of states and in the Fermi surface. A surprising finding is the high total energy precision of FPLO, demonstrated by comparing d^3E/da^3 with related Wien2k data.

The total energy precision can be improved in both codes by refining the k-mesh. While this behavior is expected, further strong improvements of the numerical precision are obtained in the case of Wien2k by optimizing the value of $Rk_{\max}$ and in the case of FPLO by modifying a real-space integration mesh. Thus, it also became possible to perform meaningful third-order Birch-Murnaghan fits to the minimum number of four data points of the obtained very precise $E(V)$.

Concerning both codes, a counterintuitive finding has to be mentioned: tuning the respective basis sets toward optimum numerical accuracy results in a reduced numerical precision. This feature presumably originates from incipient linear dependence of the basis functions.

Finally, we note that our approach to quantify the numerical precision by means of total energy derivatives is not only applicable to the two codes considered in the present study, but can easily be extended to other methods.

Author Contributions: Conceptualization and methodology, M.R., A.M. and H.R.; software, K.K.; investigation, all authors; writing—original draft preparation, M.R. and K.K.; writing—review and editing, all authors. All authors have read and agreed to the published version of the manuscript.

Funding: This research received no external funding.

Data Availability Statement: The data presented in this study are available on request from the corresponding author.

Acknowledgments: We are are grateful to Ulrike Nitzsche for her efforts in establishing and maintaining the IFW high-performance computational environment.

Conflicts of Interest: The authors declare no conflict of interest.

References

1. Lejaeghere, K.; Bihlmayer, G.; Björkman, T.; Blaha, P.; Blügel, S.; Blum, V.; Caliste, D.; Castelli, I.E.; Clark, S.J.; Dal Corso, A.; et al. Reproducibility in density functional theory calculations of solids. *Science* **2016**, *351*, aad3000. [CrossRef] [PubMed]
2. Accuracy and Precision. Available online: https://en.wikipedia.org/wiki/Accuracy_and_precision (accessed on 5 January 2022).
3. Eschrig, H. *Optimized LCAO Method and the Electronic Structure of Extended Systems*; Akademie-Verlag: Berlin, Germany, 1988.
4. Schwarz, K.; Blaha, P. Solid state calculations using WIEN2k. *Comput. Mater. Sci.* **2003**, *28*, 259–273. [CrossRef]

5. Koepernik, K.; Eschrig, H. Full-potential nonorthogonal local-orbital minimum-basis band-structure scheme. *Phys. Rev. B* **1999**, *59*, 1743. [CrossRef]
6. Occelli, F.; Farber, D.L.; Badro, J.; Aracne, C.M.; Teter, D.M.; Hanfland, M.; Canny, B.; Couzinet, B. Experimental Evidence for a High-Pressure Isostructural Phase Transition in Osmium. *Phys. Rev. Lett.* **2004**, *93*, 095502. [CrossRef] [PubMed]
7. Koudela, D.; Richter, M.; Möbius, A.; Koepernik, K.; Eschrig, H. Lifshitz transitions and elastic properties of Osmium under pressure. *Phys. Rev. B* **2006**, *74*, 214103. [CrossRef]
8. Harrison, W.A. Fermi Surface in Aluminum. *Phys. Rev.* **1959**, *116*, 555. [CrossRef]
9. Joss, W.; Monnier, R. Fermi surface of aluminium under homogeneous strain. *J. Phys. F Met. Phys.* **1980**, *10*, 9. [CrossRef]
10. Meissner, T.; Goh, S.K.; Haase, J.; Richter, M.; Koepernik, K.; Eschrig, H. Nuclear magnetic resonance at up to 10.1 GPa pressure detects an electronic topological transition in aluminum metal. *J. Phys. Condens. Matter* **2014**, *26*, 015501. [CrossRef] [PubMed]
11. Available online: https://www.fplo.de/ (accessed on 5 January 2022).
12. Koelling, D.D.; Harmon, B.N. A technique for relativistic spin-polarised calculations. *J. Phys. C Solid State Phys.* **1977**, *10*, 3107. [CrossRef]
13. Perdew, J.P.; Burke, K.; Ernzerhof, M. Generalized Gradient Approximation Made Simple. *Phys. Rev. Lett.* **1996**, *77*, 3865. [CrossRef] [PubMed]
14. Blöchl, P.; Jepsen, O.; Andersen, O.K. Improved tetrahedron method for Brillouin-zone integrations. *Phys. Rev. B* **1994**, *49*, 16223. [CrossRef] [PubMed]
15. Blanter, Y.M.; Kaganov, M.I.; Pantsulaya, A.V.; Varlamov, A.A. The theory of electronic topological transitions. *Phys. Rep.* **1994**, *245*, 159. [CrossRef]
16. Rosner, H.; Koudela, D.; Schwarz, U.; Handstein, A.; Hanfland, M.; Opahle, I.; Koepernik, K.; Kuz'min, M.; Müller, K.H.; Mydosh, J.; et al. Magneto-elastic lattice collapse in YCo_5. *Nat. Phys.* **2006**, *2*, 469–472. [CrossRef]
17. O'Bannon, E.; Söderlind, P.; Sneed, D.; Lipp, M.; Cynn, H.; Smith, J.; Park, C.; Jenei, Z. High pressure stability of β-Zr: No evidence for isostructural phase transitions. *High Press. Res.* **2021**, *41*, 247–266. [CrossRef]

computation

MDPI

Article

Density Functional Theory Study of Metal and Metal-Oxide Nucleation and Growth on the Anatase TiO_2(101) Surface

Leila Kalantari *, Fabien Tran and Peter Blaha

Institute of Materials Chemistry, Vienna University of Technology, Getreidemarkt 9/165-TC, A-1060 Vienna, Austria; tran@theochem.tuwien.ac.at (F.T.); pblaha@theochem.tuwien.ac.at (P.B.)

* Correspondence: leila.kalantari@tuwien.ac.at

Abstract: Experimental studies have shown the possible production of hydrogen through photocatalytic water splitting using metal oxide (MO_y) nanoparticles attached to an anatase TiO_2 surface. In this work, we performed density functional theory (DFT) calculations to provide a detailed description of the stability and geometry of M_xO_y clusters M = Cu, Ni, Co, Fe and Mn, x = 1–5, and y = 0–5 on the anatase TiO_2(101) surface. It is found that unsaturated 2-fold-coordinated O-sites may serve as nucleation centers for the growth of metal clusters. The formation energy of Ni-containing clusters on the anatase surface is larger than for other M clusters. In addition, the Ni_n adsorption energy increases with cluster size n, which makes the formation of bigger Ni clusters plausible as confirmed by transition electron microscopy images. Another particularity for Ni-containing clusters is that the adsorption energy per atom gets larger when the O-content is reduced, while for other M atoms it remains almost constant or, as for Mn, even decreases. This trend is in line with experimental results. Also provided is a discussion of the oxidation states of M_5O_y clusters based on their magnetic moments and Bader charges and their possible reduction with oxygen depletion.

Keywords: DFT; anatase TiO_2(101) surface; adsorption energy; Bader charge

Citation: Kalantari, L.; Tran, F.; Blaha, P. Density Functional Theory Study of Metal and Metal-Oxide Nucleation and Growth on the Anatase TiO_2(101) Surface. *Computation* **2021**, *9*, 125. https://doi.org/10.3390/computation9110125

Academic Editor: Henry Chermette

Received: 12 October 2021
Accepted: 5 November 2021
Published: 19 November 2021

1. Introduction

The over-exploitation of the fossil energies leads to a significant increase of CO_2 in the atmosphere, resulting in severe climate problems. It is therefore absolutely necessary to replace them by alternative energy sources. When using sun light as energy source, however, materials with specialized properties are necessary, and TiO_2 is such a versatile material that has numerous applications in catalysis, photocatalysis, and solar energy [1]. TiO_2 crystallizes in three major different structures: rutile (tetragonal), anatase (tetragonal) and brookite (rhombohedral). Other structures exist as well, as for example cotunnite, that has been synthesized at high pressures and is one of the hardest polycrystalline materials known. However, only rutile and anatase play an important role in the applications of TiO_2. A well-suited model compound for photocatalytic water splitting is anatase. It has a slightly larger band gap than rutile ($\sim$3.0 eV for rutile and $\sim$3.2 eV for anatase) [2–6] and shows better performance [7], since it also has a longer electron-hole pair life time, which makes anatase more suitable for photocatalytic applications [8]. To improve the water splitting photocatalyst activity, the use of co-catalysts is necessary and the most widely used are Pt and Pd for photocatalytic reduction, and IrO_2 and RuO_2 for the oxidation of water. However, these are expensive materials based on rare noble metals. To achieve large-scale industrial applications of photocatalytic water splitting, the development of new co-catalysts based on cheap and widely available elements remains an important issue. With regard to novel abundant co-catalysts for photocatalytic water splitting, research efforts focus on 3*d* transition metals (M) oxides, which are known for their excellent catalytic properties and applications in industry, research and nature.

Recently, small clusters of M oxides were produced by deposition of M-acetylacetonate precursors on TiO_2 and subsequent calcination in air. In particular, the Ni (and Cu) nanoparticles showed pronounced activities for water splitting, but little is known about the atomic structure of these nanoparticles [9]. This inspired the present work, where we used density functional theory (DFT) [10,11] to study possible adsorption sites of a 3*d* transition metal atom (Cu, Ni, Co, Fe, and Mn), which can be the nucleation sites for bigger M clusters (2 to 5 atoms) or M oxides clusters on TiO_2(101) anatase, the lowest-energy surface of anatase [12]. There are a couple of previous theoretical works on the M adsorption on anatase, studying single M atom adsorption and diffusion [13], or the adsorption of small Ni [14] or Cu and Cu-oxide [15,16] clusters. A nice review related to this topic is given by Zhou and Dong [17]. However, note that most of these calculations used a generalized gradient approximation (GGA) in DFT for the correlated 3*d* electrons, which may not be accurate enough.

2. Methods

Spin-polarized DFT calculations of the adsorption of M_xO_y clusters (M = Cu, Ni, Co, Fe or Mn, $x = 1, 2, 3, 5$, and $y = 0, 1, 2, 3, 4, 5$) on the TiO_2(101) surface were performed using the full-potential augmented plane wave plus local orbitals method [18,19] as implemented in the WIEN2k code [20,21]. We employed the PBEsol exchange-correlation functional [22], which is a GGA and yields lattice parameters of bulk anatase TiO_2, $a = b = 3.77$ Å and $c = 9.54$ Å that are in good agreement with experimental data [23]. In order to treat the correlated 3*d* electrons of the M atoms a Hubbard U correction was used [24]. We used $U_{\text{eff}} = U - J$ of 5 eV for Cu, Ni, and Co and 4 eV for Mn and Fe. These values are deliberately chosen smaller than what is typically used in strongly correlated oxides, since in our case also less ionic (metallic) clusters were investigated and the correlation may not be always so strong. We note that small changes of U_{eff} would not affect our basic results and conclusions.

The (101) surface was modeled by a symmetric slab with a thickness of three layers of TiO_2 (see Figure 1, where only one layer is shown for better visibility) and a vacuum region of 16 Å between the slabs. In order to reduce the interactions between the adsorbed clusters in neighboring cells a 3×1 supercell was used for all surfaces with adsorbed M atoms. Moreover, such a supercell is also helpful to release the interfacial strain energy and find more stable configurations. A $2 \times 2 \times 1$ Γ-centered mesh of **k**-points is used during relaxation (a $5 \times 4 \times 1$ mesh for the final results) and a basis-set size corresponding to $R_{\text{MT}}^{\text{min}} K_{\text{max}} = 7$, where $R_{\text{MT}}^{\text{min}}$ is the smallest atomic sphere radius and K_{max} the largest reciprocal lattice vector, was used. All surface models were relaxed until all residual forces were below 1 mRy/bohr.

Consistent R_{MT} and $R_{\text{MT}}^{\text{min}} K_{\text{max}}$ values have been used for calculating the adsorption energy E_{ads} of clusters on anatase surface, which is defined as

$$\begin{aligned} E_{\text{ads}} &= \frac{1}{x+y}[E(\text{TiO}_2(101)) + xE(\text{M}) + y\frac{1}{2}E(\text{O}_2) \\ &\quad -E(\text{M}_x\text{O}_y/\text{TiO}_2(101))], \end{aligned} \tag{1}$$

where $E(\text{M})$ and $E(\text{O}_2)$ are the total energies of the free M atom and an O_2 molecule, and $E(\text{TiO}_2(101))$ is for the bare TiO_2(101) surface. Bader's quantum theory of atoms in molecules [25] was used to calculate the atomic magnetic moments and charges.

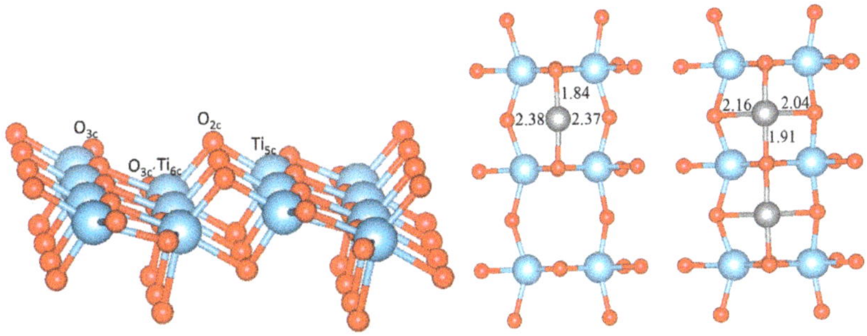

Figure 1. Side and top views of the top-most layer of the $TiO_2(101)$ surface (in the 3×1 supercell) with a single adsorbed Ni atom (middle) and two Ni atoms (right structure). Red, blue and gray spheres correspond to O, Ti and Ni atoms, respectively. Selected Ni-O bond distances (in Å) are also given. O_{3c} and $O_{3c'}$ are threefold coordinated oxygen with two Ti_{6c} and one Ti_{5c}, and two Ti_{5c} and one Ti_{6c} nearest neighbors, respectively.

3. Results and Discussion

3.1. Perfect Anatase $TiO_2(101)$ Surface

As shown in Figure 1, the anatase $TiO_2(101)$ surface has a stepped structure. Threefold coordinated O atoms (O_{3c} and $O_{3c'}$) and sixfold coordinated Ti atoms (Ti_{6c}) are fully saturated and have bulk-like coordination, whereas the twofold O_{2c} and fivefold Ti_{5c} atoms are under-coordinated. The O_{2c} atoms are located at the ridges of the saw-tooth-like structure, and after optimization they relax inwards by ~0.16 Å with respect to bulk TiO_2. The three-fold coordinated O_{3c} and $O_{3c'}$ atoms relax outwards by ~0.07 Å, while the Ti_{5c} and Ti_{6c} atoms relax inwards by ~0.15 Å and ~0.16 Å respectively, so that the surface exhibits a slightly buckled structure.

3.2. Pure Metal Clusters on the Anatase $TiO_2(101)$ Surface

Upon calcination of the M-acetylacetonate precursors, single M atoms may adsorb, diffuse and form larger clusters and finally oxidize on the surface. It is thus natural to study as a first step the possible adsorption sites of a single M atom or a small cluster. We have considered several possible initial configurations of the adsorbed metal atoms and clusters M_x with $x = 1, 2, 3$, and 5. The most stable structures, which are the same for all M (Cu, Ni, Co, Fe or Mn), are presented in Figures 1–3 in the case of Ni clusters with $x = 1, 2$ and 5. The nearest-neighbor distances are shown in Table 1, while the adsorption energy E_{ads} and magnetic moment are available in Table 2.

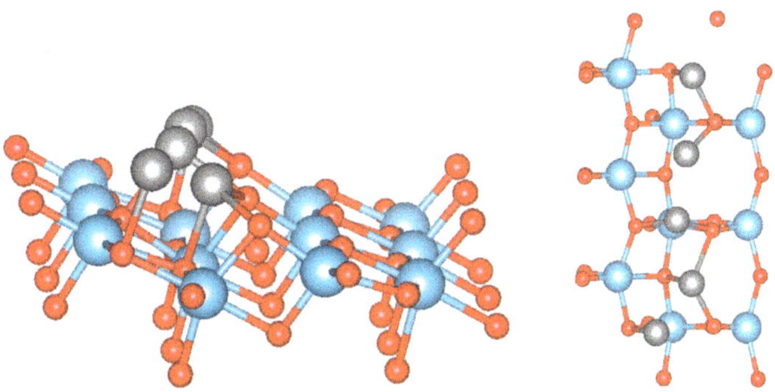

Figure 2. Side and top views of the top-most layer of the $TiO_2(101)$ surface with an adsorbed Ni_5 cluster in a chain-like structure. The color coding is as in Figure 1.

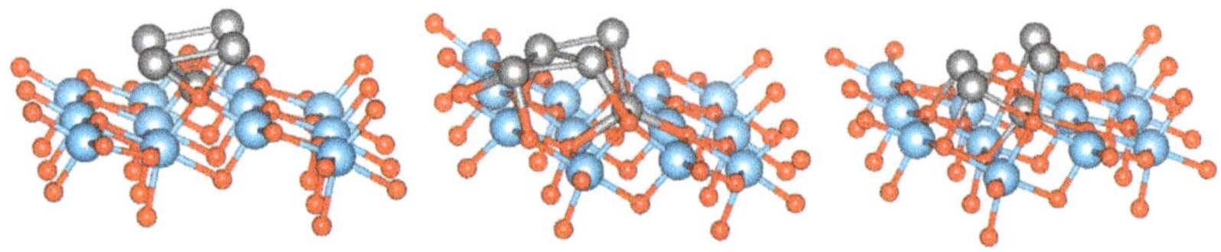

Figure 3. Side view of the starting (**left**) and relaxed structures of Ni_5 (**middle**) and Fe_5 clusters (**right**) adsorbed on the $TiO_2(101)$ surface in a crystalline structure. The color coding is as in Figure 1.

Table 1. Shortest M-O bonds (in Å) between a M atom (of the pure metal clusters adsorbed on the $TiO_2(101)$ surface) and O atoms (O_{2c} and O_{3c}) of TiO_2. $M_5(1)$ and $M_5(2)$ refer to chain-like and crystalline structures, respectively, and $Mn_5(2a)$ corresponds to a bcc like structure with antiferromagnetic configuration. The results for different M atoms are separated by semi-colons. "(2)" means two bonds of same length.

	MO Bulk	MO_{2c} at the Surface	M-O_{3c} at the Surface
TiO_2-Cu	(CuO)1.97/(Cu_2O)1.86	1.87(2)	2.33, 2.47
TiO_2-Cu_2		(1.88, 1.89); (1.88, 1.89)	(2.34, 2.52); (2.34, 2.52)
TiO_2-Cu_3		1.90(2); 1.90(2); 1.90(2)	(2.28, 2.46); (2.28, 2.46); (2.29, 2.46)
TiO_2-$Cu_5(1)$		1.97; 2.11; (2.09, 2.10); (2.04, 2.15); 2.83	(2.81, 2.71); (2.31, 2.90); 2.30; 2.42; -
TiO_2-$Cu_5(2)$		1.86; -; -; 2.94; 1.86	-; -; -; 2.00; 2.75
TiO_2-Ni	2.08	1.84(2)	2.37, 2.38
TiO_2-Ni_2		(1.91, 1.92); (1.91, 1.92)	(2.04, 2.16); (2.04, 2.16)
TiO_2-Ni_3		1.89(2); 1.89(2); 1.89(2)	(2.02, 2.31); (2.02, 2.31); (2.02, 2.31)
TiO_2-$Ni_5(1)$		2.02; 1.99; (2.04, 2.22); (2.13, 2.46); 2.86	2.04; 2.75; 2.10; 2.04; 2.05
TiO_2-$Ni_5(2)$		(1.96, 1.94); 2.13; 1.98; -; 2.07	(2.04, 2.16); -; -; 2.01; 2.07
TiO_2-Co	2.13	1.89(2)	2.03, 2.71
TiO_2-Co_2		1.97(2); (1.88, 1.89)	(2.03, 2.04); (2.03, 2.76)
TiO_2-Co_3		(1.99, 2.00); (2.03, 1.91); (1.90, 2.02)	(2.05, 2.09); (2.04, 2.12); (2.04, 2.29)
TiO_2-$Co_5(1)$		2.08; 2.30; (2.06, 2.09); (2.03, 2.06); 2.06	(2.15, 2.30); 2.04; 2.13; 2.15; 2.86
TiO_2-$Co_5(2)$		2.02; -; 1.91; 2.90; 2.07	2.09; -; -; 1.96; 2.19
TiO_2-Fe	2.16	1.89(2)	2.08, 2.23
TiO_2-Fe_2		(1.88, 1.91); (1.88, 1.91)	(2.10, 2.11); (2.10, 2.11)
TiO_2-Fe_3		1.88(2); 1.88(2); 1.88(2)	(2.11, 2.14); (2.11, 2.14); (2.11, 2.14)
TiO_2-$Fe_5(1)$		2.03; 2.09; (2.00, 2.09); (2.05, 2.15); 2.98	(2.21, 2.25); 2.09; 2.14; 2.06; 2.02
TiO_2-$Fe_5(2)$		(2.06, 2.05); 2.04; 2.04; 2.22; 2.25	(2.14, 2.04); -; -; 2.11; 2.08
TiO_2-Mn	2.22	1.93(2)	2.14, 2.30
TiO_2-Mn_2		1.94(2); 1.94(2)	(2.11, 2.28); (2.11, 2.28)
TiO_2-Mn_3		(1.91, 1.92); (1.92, 1.93); (1.92, 1.93)	(2.16, 2.30); (2.16, 2.27); (2.16, 2.27)
TiO_2-$Mn_5(1)$		2.17; 2.19; (2.03, 2.08); (2.10, 2.03); 3.27	2.12; 2.13; 2.16; 2.16; 2.07
TiO_2-$Mn_5(2)$		(2.02, 2.02); 2.08; 2.08; 2.17; 2.18	(2.03, 2.04); -; -; 2.09
TiO_2-$Mn_5(2a)$		(2.14, 2.15); 2.09; 2.09; 2.07; 2.07	2.15; -; -; 2.15; 2.16

A single adsorbed M atom prefers to adsorb at the bridge site between two unsaturated oxygens (O_{2c}) atoms, similar as found in previous studies [13,14]. The corresponding M-O bond length (it is the shortest for Ni, 1.84 Å) follows in general the trend found in the bulk M oxides (except for the Cu-O and Ni-O distances, which are interchanged), but in all cases is much shorter (and therefore indicates a stronger bonding) than in the bulk MO (e.g., 2.08 Å in NiO). The particular short bond distance for Ni goes hand in hand with the O_{2c}-Ni-O_{2c} bond angle, which reaches 177° and forms a nearly linear bond, while for the other metals this varies from 167° (Fe) to 150° (Co). On the other hand the two larger M-O_{3c} bond distances are also almost equal (but very large) for Ni, but for all other M form a short and long bond in an asymmetric position (see Table 1). All M atoms except Cu possess a magnetic moment (see Table 2), where as expected Mn has the largest moment, while it is the smallest for Ni. It should be noted that PBEsol without U would lead to a non-magnetic Ni atom and the moments for the other M would be smaller but still sizeable. The adsorption energy on the $TiO_2(101)$ surface (Table 2) is the largest for Ni (3.46 eV) and the smallest for Cu (2.24 eV). Comparing our results with literature the adsorption energy of Ni is 0.4 eV larger than in references [13,14] and the Ni position is symmetric with respect to the O_{3c} atoms. On the other hand, our calculated adsorption energy for Fe is 0.7 eV smaller than that in reference [13], while it agrees well for Cu and Co. These differences could arise due to the neglect in references [13,14] of a Hubbard U correction for the M atom or the use of pseudopotentials.

Table 2. Adsorption energy (in eV/atom) and atomic magnetic moment (in μ_B) for pure metal clusters adsorbed on the $TiO_2(101)$ surface. $M_5(1)$ and $M_5(2)$ refer to the chain-like and crystalline structures, respectively, and $Mn_5(2a)$ corresponds to the bcc structure with antiferromagnetic configuration. The cohesive energies for bulk MO are 2.75, 4.20, 3.72, 3.41, and 3.34 eV for Cu, Ni, Co, Fe, and Mn, respectively.

Solid	E_{ads}	Magnetic Moment
TiO_2-Cu	2.24	0
TiO_2-Cu_2	2.08	0
TiO_2-Cu_3	1.99	0
TiO_2-$Cu_5(1)$	1.90	0
TiO_2-$Cu_5(2)$	2.26	0
TiO_2-Ni	3.46	0.25
TiO_2-Ni_2	4.11	0.97(2)
TiO_2-Ni_3	3.94	0.98(3)
TiO_2-$Ni_5(1)$	3.68	1.66; 1.08; 1.31; 1.11; 1.35
TiO_2-$Ni_5(2)$	4.49	−0.97; 1.22; 0.74; −1.10; 0.87
TiO_2-Co	3.19	2.12
TiO_2-Co_2	3.04	2.12; 2.11
TiO_2-Co_3	3.00	2.15; 2.28; 2.28
TiO_2-$Co_5(1)$	2.73	2.03; 2.53; 2.52; 2.18; 2.44
TiO_2-$Co_5(2)$	2.87	2.34; 2.00; 2.66; 1.96; 2.04
TiO_2-Fe	2.71	3.61
TiO_2-Fe_2	2.57	3.51; 3.50
TiO_2-Fe_3	2.48	3.52(3)
TiO_2-$Fe_5(1)$	2.18	3.62; 3.79; 3.64; 3.48; 3.71
TiO_2-$Fe_5(2)$	2.03	3.00; 3.75; 3.72; 3.58; 3.57
TiO_2-Mn	2.40	4.66
TiO_2-Mn_2	2.19	4.66(2)
TiO_2-Mn_3	2.08	4.61; 4.61; 4.63
TiO_2-$Mn_5(1)$	1.78	4.90; 4.94; 4.66; 4.66; 4.83
TiO_2-$Mn_5(2)$	1.10	0.45; 5.00; 4.99; 4.57; 4.57
TiO_2-$Mn_5(2a)$	1.44	4.90; −5.04; 5.03; −4.73; −4.73

If 2 or 3 M atoms are adsorbed on the surface, the most stable structures are still obtained when the M atoms sit at the O_{2c} bridge sites (the structure on the right side in Figure 1 shows 2 M atoms adsorbed on the surface), very similar to a single atom adsorption. Interestingly, for all M atoms except Ni the adsorption energy per M atom decreases slightly (Figure 4 and Table 2). For Ni it increases by 0.65 eV. The reason seems to be that the Ni atom is smaller than the others and an almost linear chain fits perfectly for Ni but not in the other cases, where this additional strain reduces E_{ads} slightly. Most magnetic moments remain unchanged, except for Ni, where it increases to 0.97/0.98 μ_B. In our 3×1 supercell, 3 M atoms form an infinite linear chain of M-O_{2c} pairs.

For 5 adsorbed M atoms, we tried several different starting structures in the 3×1 supercell. The most promising structures are based on the chain-like structure with 2 additional atoms (labelled $M_5(1)$, see Figure 2) or a structure with a more "crystalline type" nucleated on a single M bridge atom (labelled $M_5(2)$, see Figure 3). The additional

2 atoms "destroy" the symmetric chain-like 3-atom structure above the O_{2c} atoms forming a quite irregular array of M atoms with bonds to O_{2c}, O_{3c} and $O_{3c'}$ atoms. Still this structure seems to be the most stable for Fe and Mn although E_{ads} is smaller than for less M atoms. The crystalline $M_5(2)$ structures behave very differently for the bcc metals Mn and Fe and the fcc metals Co, Ni and Cu. While for the former the cluster structure stays basically close to the starting one and form a "bcc" like cluster (Figure 3), for the latter the cluster rearranges and can be considered as a nucleus for an fcc like cluster. In fact, their E_{ads} is larger than for the corresponding chain-like $M_5(1)$ structures and for Ni and Cu even the highest within the series.

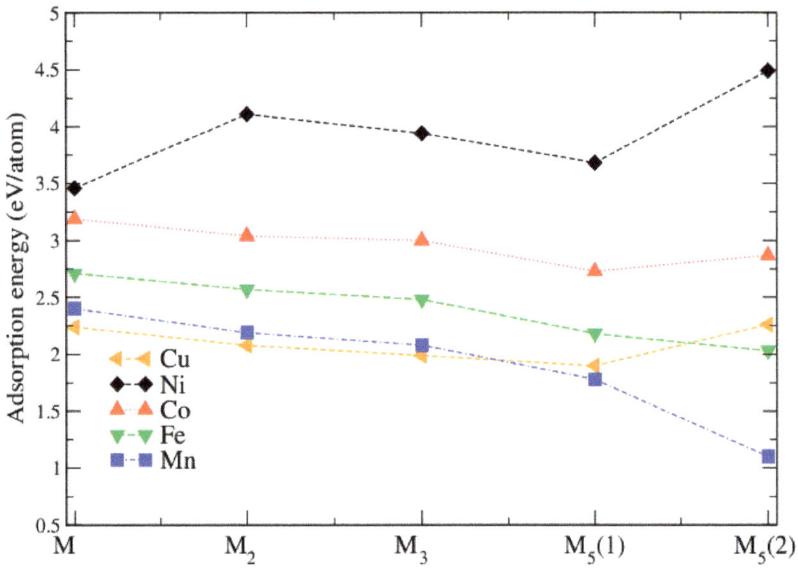

Figure 4. Adsorption energy (in eV/atom) for different metal clusters M_x. $M_5(1)$ and $M_5(2)$ correspond to chain-like and crystalline structures, respectively.

For Mn_5 in a crystalline bcc structure, we also started from different antiferromagnetic configurations (labelled $Mn_5(2a)$), but the adsorption energies were not much larger than in the ferromagnetic case.

The partial density of states (PDOS) for the $TiO_2(101)$ surface and for the energetically most favorable configuration of one metal atom on the surface are shown in Figure 5. For the $TiO_2(101)$ surface, both the valance band maximum (VBM) and the conduction band minimum (CBM) are composed of a mixture of O-2p and Ti-3d states, with the CBM having mostly 3d character while the VBM has more O-2p character, which indicates a mixed ionic and covalent bonding. After adsorption of a M atom on the anatase $TiO_2(101)$ surface, the Fermi level moves to the conduction band and the system becomes formally metallic. In the case of adsorbed Cu and Ni atoms, almost all the d states of Cu/Ni are concentrated in the band gap below the Fermi energy, and they slightly hybridize with the O-p and Ti-d states of the Ti and O atoms which are the closest to the Cu/Ni atom. With an adsorbed Co, Fe, or Mn atom the d states are not only in the band gap but also show a peak at the VBM for spin-up and inside the conduction band for spin-down states.

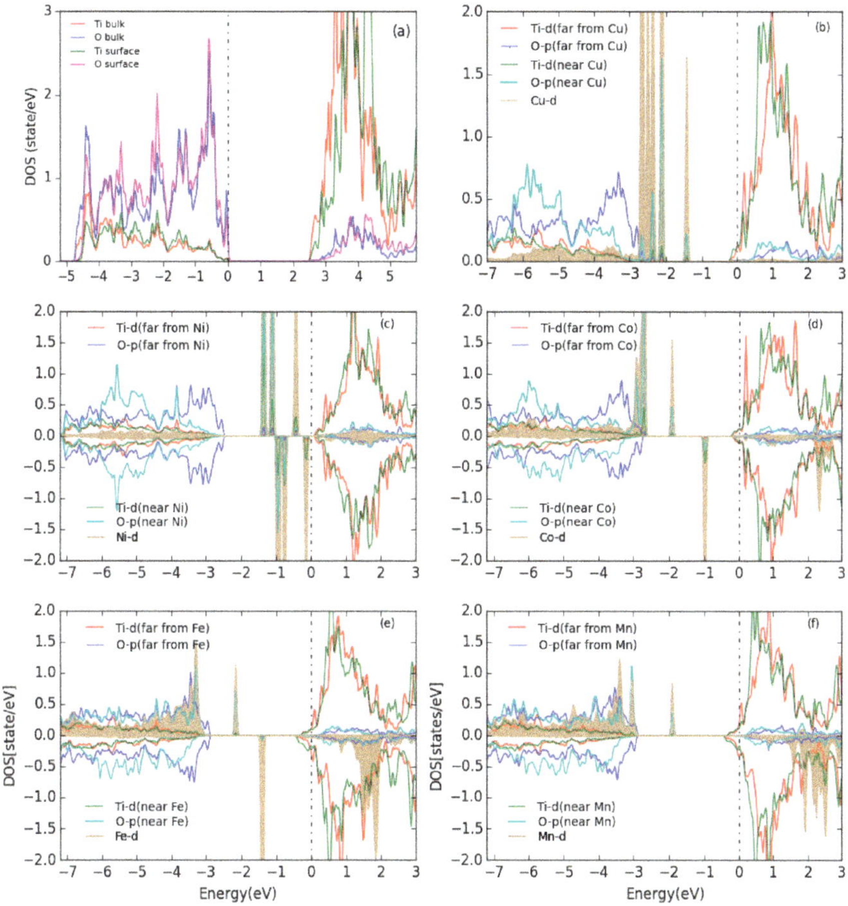

Figure 5. PDOS for (**a**) the bare $TiO_2(101)$ surface and (**b**–**f**) for one metal atom adsorbed on the $TiO_2(101)$ surface. The Fermi level is indicated by a vertical dashed line. All Ti and O PDOS are from surface atoms. The *d* partial DOS of the metals are scaled down by a factor of 3. The plots are aligned at the TiO_2 CBM and the Fermi energy is set at 0 eV.

3.3. M_xO_y on the Anatase $TiO_2(101)$ Surface

Turning now to the adsorption of metal-oxides clusters M_xO_y on the anatase $TiO_2(101)$ surface, several 2D and 3D (starting) configurations among the numerous possible ones were considered. Some of the most stable structures are shown in Figures 6 and 7, and the adsorption energies are shown in Table 3 for the Ni_xO_y clusters. It can be seen that E_{ads} is almost constant with the increase of cluster size indicating that bigger Ni clusters can easily be formed on the anatase $TiO_2(101)$ surface, which is in agreement with the experimental result [9]. The most favorable structure is $Ni_{10}O_9$ with a NaCl like structure (shown in Figure 7).

It was found experimentally [9] that the oxidation state of the M atoms changed after a hydrogen evolution reaction (HER) experiment. In order to study the M oxidation state theoretically, we started from the chain-like M_5 clusters (Figure 6) and added oxygen atoms systematically until a fully oxidized M cluster has been obtained. Figure 8 compares the adsorption energy of different M_5O_y M oxide clusters on the surface and also includes for comparison the cohesive energy of the bulk pure metals and bulk metal monoxides. It can be seen that, among all metal-oxide clusters, Ni_xO_y have the highest adsorption energy. All adsorption energies are lower than the cohesive energies of the corresponding bulk monoxide and the pure bulk metal. Thus they form non-crystalline clusters on the surface. It can be seen that Ni clusters are a bit more stable with less oxygen, while for the other metals the adsorption energy remains almost constant, or, in particular for Mn, is reduced with less oxygen. This is in agreement with experimental transition electron microscopy (TEM) results [9], where cluster formation was seen in particular for Ni. In addition, XPS on calcinated NiO_x-TiO_2 samples shows mainly Ni^{2+}, but after HER experiments a substantial increase of Ni^0 compared to Ni^{2+} appears. On the other hand, XPS on MnO_x-TiO_2 samples shows only Mn^{2+}, even after the HER experiments.

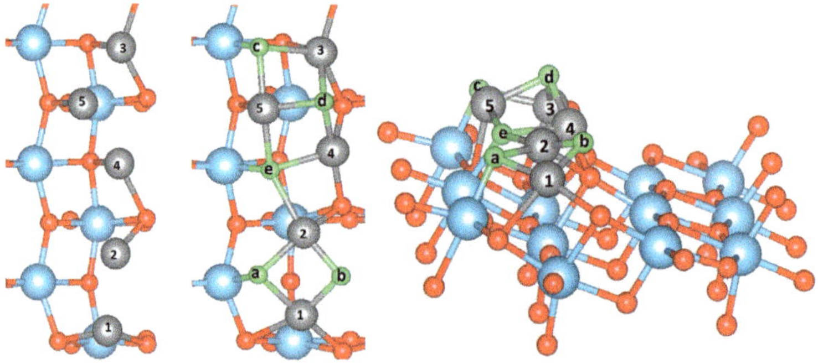

Figure 6. Top view of the Ni_5 cluster (**left**) and, top and side view of Ni_5O_5 cluster (**middle** and **right**) adsorbed on $TiO_2(101)$ surface in chain-like structure. Red, blue, gray and green spheres correspond to O, Ti, Ni and added O, respectively. From Ni_5 to Ni_5O_5, the O atoms are added in the order O_a, O_b, O_c, O_d and O_e.

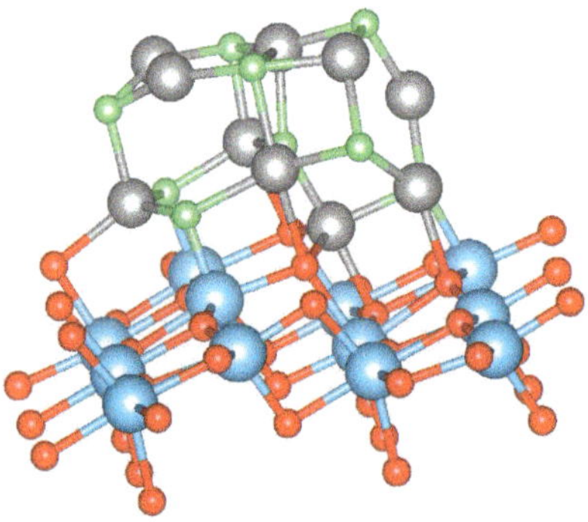

Figure 7. $Ni_{10}O_9$ cluster with a NaCl-like structure adsorbed on the $TiO_2(101)$ surface. The color coding is as in Figure 6.

Theoretically, the oxidation state is more difficult to estimate. One can use for instance Bader's method [26], which uses the electron density to calculate the gradient vector field and searches for surfaces of zero flux. The charge enclosed within this zero flux surface (basin) can be used to define the total electronic charge of an atom. It must be mentioned that one never gets the full nominal charges, but these Bader charges are always smaller than anticipated. In addition, in particular for open structures like surfaces, these charges may sometimes correlate more with the specific geometry and the distances and number of other atoms around the M atom. For transition-metal atoms, the magnetic moment can also be a good indication of the charge state in some cases. For instance, only Cu^{2+} ions possess a magnetic moment, while both Cu^+ and neutral Cu are non-magnetic. Similarly, one can quite well distinguish, e.g., Fe^{2+} and Fe^{3+} based on their smaller or larger magnetic moments (provided they remain in high-spin states). The magnetic moments and the Bader charges of each atom in M_5O_y clusters are shown in bar charts in Figures 9 and 10. The largest magnetic moments for each M atom increase as expected from 0.6 μ_B for Cu^{2+} to 1.6, 2.6, 3.7 and 4.9 μ_B in the series Ni, Co, Fe and Mn. This is due to the number of spin-down 3*d* electrons that gets reduced when dealing with high-spin states and fully occupied spin-up states.

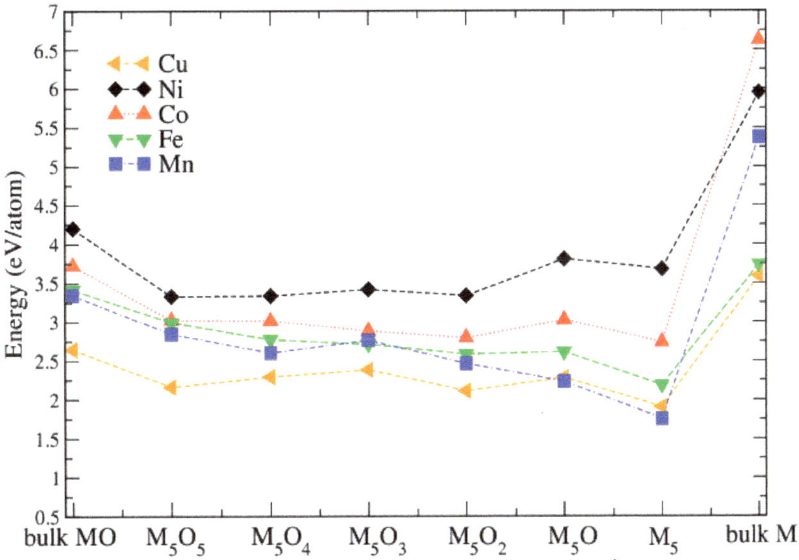

Figure 8. Adsorption energy (in eV/atom) of M_xO_y clusters on the $TiO_2(101)$ surface. Also shown is the cohesive energy of bulk MO (left-most data) and bulk M (right-most data).

Table 3. Adsorption energy (in eV/atom) of Ni_xO_y clusters on $TiO_2(101)$ surface. The largest values are in bold.

Cluster	E_{ads}
NiO(1)	**3.55**
NiO(2)	3.14
NiO(3)	3.19
NiO(4)	3.14
Ni_2O_2(1)	3.07
Ni_2O_2(2)	2.59
Ni_2O_2(3)	**3.58**
Ni_3O_3(1)	3.32
Ni_3O_3(2)	3.17
Ni_3O_3(3)	2.75
Ni_3O_3(4)	**3.51**
Ni_4O_4(1)	**3.42**
Ni_4O_4(2)	2.87
Ni_5O_5	**3.35**
$Ni_{10}O_9$	**3.63**

For the charge state of Cu clusters, both methods mentioned above give quite reliable results. Bader charges of about one and magnetic moments of 0.6 μ_B indicate Cu^{2+}, while Bader charges of 0.6 indicate Cu^{+} and even smaller Bader charges neutral Cu, both with vanishing magnetic moments. Apparently, in the fully oxidized cluster all Cu are in the 2+ charge state, while removal of an O_e atom (see Figure 6) leads to a drastic reduction of the charge state of the neighboring Cu_4 and Cu_5 atoms.

For Ni, however, the situation is more complicated. Even in the fully oxidized cluster some Ni moments are completely quenched because the Ni-O bond length (1.77 Å) is for these Ni atoms much smaller than for the magnetic ions (1.97 Å) and it is difficult to estimate the charge states based on the magnetic moment. On the other hand the Bader charges show a very systematic decrease with oxygen removal, and the experimentally observed increase of neutral Ni correlates reasonably with these Bader charges. Going to the extreme case of Mn, the magnetic moments stay almost constant around 5 μ_B indicative of Mn^{2+} in a high-spin state in agreement with experiment. The Bader charges are less convincing, but nevertheless they never indicate a neutral charge state of a Mn atom.

In order to get more information about the electronic structure, the PDOS of various Ni-O clusters is shown in Figure 11. For a Ni_5 cluster (Figure 11b) we see a wide range of Ni-d states within the gap of TiO_2, but also some spin-up states mixed into the O-2p valence band and some spin-down states in the empty conduction band, indicating the different bonding situation of the different Ni atoms. Upon oxidation a more pronounced formation of upper and lower Hubbard bands can be seen and the remaining Ni-d states in the gap sharpen until for the $Ni_{10}O_9$ cluster (Figure 11f) quite sharp Ni-d states of (distorted) e_g and t_{2g} character appear similar as in bulk NiO.

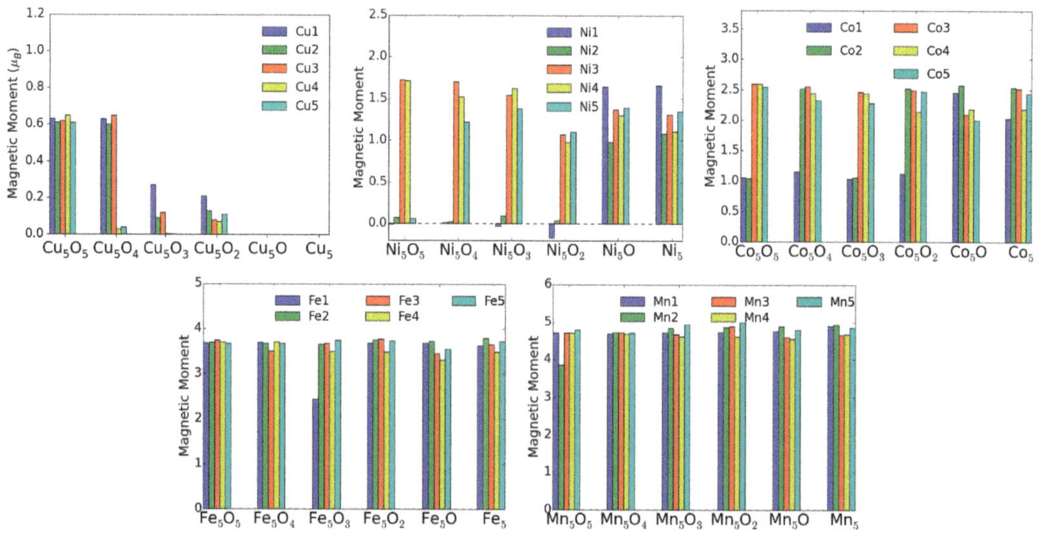

Figure 9. Atomic magnetic moments (in μ_B) of M atoms in M_xO_y clusters adsorbed on $TiO_2(101)$ surface.

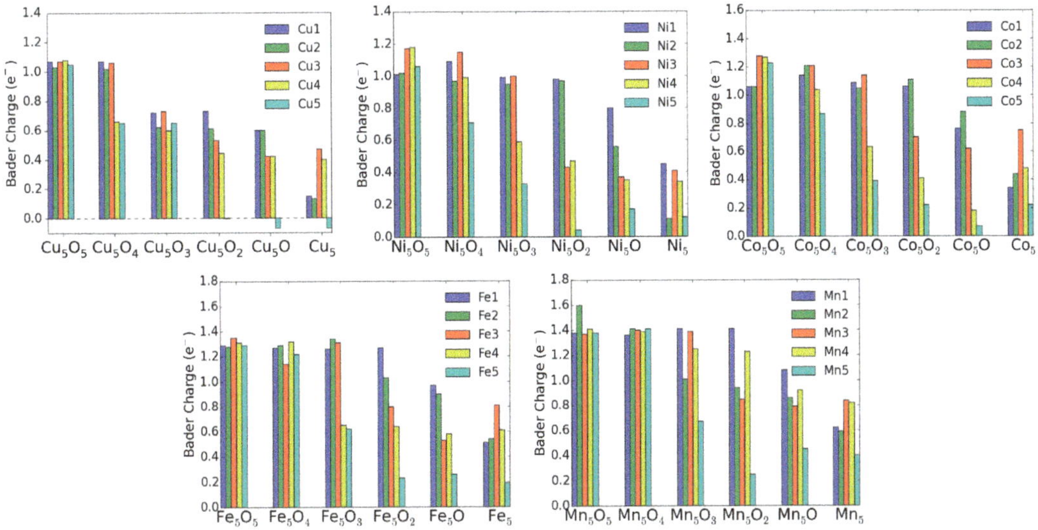

Figure 10. Bader charges of M atoms in M_xO_y clusters adsorbed on $TiO_2(101)$ surface.

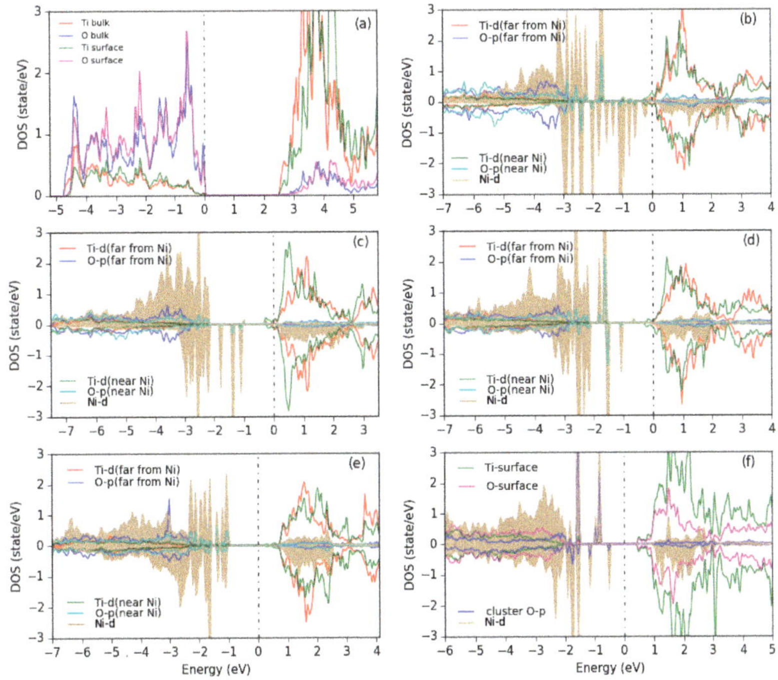

Figure 11. PDOS for (**a**) the bare anatase $TiO_2(101)$ surface, showing the surface atoms and the bulk-layer states, (**b**) Ni_5, (**c**) Ni_5O, (**d**) Ni_5O_3, (**e**) Ni_5O_5 clusters in chain-like structures and (**f**) a $Ni_{10}O_9$ cluster with NaCl-like structure adsorbed on the surface. All Ti and O PDOS are from surface atoms only. The Ni PDOS is shaded. The plots are aligned at the TiO_2 CBM and the Fermi energy is set at 0 eV.

4. Summary and Conclusions

Spin-polarized DFT calculations were carried out to investigate possible stable structures of five different transition metal clusters, namely Cu, Ni, Co, Fe and Mn adsorbed on the anatase TiO_2(101) surface. The adsorption site for single atoms is in all cases the bridge site between two unsaturated oxygen (O_{2c}) atoms. We found that among all tested metals Ni has the largest adsorption energy, and for all metals except Ni the adsorption energies decrease with increased number of adsorbed atoms, so that probably only Ni prefers to form bigger clusters on the TiO_2 surface in agreement with experimental TEM results [9]. Clusters with five M atoms can form either a fcc-like structure (Cu, Ni, Co) or still remain in a chain-like structure (Fe, Mn) as is common for smaller clusters.

We also studied several oxidized metals clusters M_xO_y, and in order to get more information on the charge state of the M atom a systematic study of M_5O_y (y = 0–5) clusters on the TiO_2(101) surface was made. We found that Ni forms more stable structures with reduced oxidation, while for the other metals the adsorption energy is almost constant or is reduced. An analysis of the Bader charges and magnetic moments allows to reveal the corresponding charges of the M atoms, and we found, in agreement with the experimental results [9], that Ni can be more easily reduced during a HER experiment, while Mn atoms will keep their 2+ oxidation state.

Author Contributions: L.K.: investigation, formal analysis, writing—original draft, software, visualization. F.T.: writing—review and editing. P.B.: supervision, writing—review and editing. All authors have read and agreed to the published version of the manuscript.

Funding: This research was funded by the TU-D doctoral college (TU Vienna).

Institutional Review Board Statement: Not applicable.

Informed Consent Statement: Not applicable.

Data Availability Statement: The data presented in this study are available in insert article.

Conflicts of Interest: The authors declare no conflict of interest.

References

1. Diebold, U. The surface science of titanium dioxide. *Surf. Sci. Rep.* **2003**, *48*, 53–229. [CrossRef]
2. Khan, M.M.; Pradhan, D.; Sohn, Y. *Nanocomposites for Visible Light-Induced Photocatalysis*; Springer: Cham, Switzerland, 2017; Volume 101.
3. Ameta, R.; Ameta, S.C. *Photocatalysis: Principles and Applications*; CRC Press: Boca Raton, FL, USA, 2016.
4. Ibhadon, A.O.; Fitzpatrick, P. Heterogeneous photocatalysis: Recent advances and applications. *Catalysts* **2013**, *3*, 189–218. [CrossRef]
5. Ni, M.; Leung, M.K.; Leung, D.Y.; Sumathy, K. A review and recent developments in photocatalytic water-splitting using TiO_2 for hydrogen production. *Renew. Sustain. Energy Rev.* **2007**, *11*, 401–425. [CrossRef]
6. Jafari, T.; Moharreri, E.; Amin, A.S.; Miao, R.; Song, W.; Suib, S.L. Photocatalytic water splitting—The untamed dream: A review of recent advances. *Molecules* **2016**, *21*, 900. [CrossRef] [PubMed]
7. Liu, L.; Zhao, H.; Andino, J.M.; Li, Y. Photocatalytic CO_2 reduction with H_2O on TiO_2 nanocrystals: Comparison of anatase, rutile, and brookite polymorphs and exploration of surface chemistry. *ACS Catal.* **2012**, *2*, 1817–1828. [CrossRef]
8. Xu, M.; Gao, Y.; Moreno, E.M.; Kunst, M.; Muhler, M.; Wang, Y.; Idriss, H.; Wöll, C. Photocatalytic activity of bulk TiO_2 anatase and rutile single crystals using infrared absorption spectroscopy. *Phys. Rev. Lett.* **2011**, *106*, 138302. [CrossRef] [PubMed]
9. Schubert, J.S.; Popovic, J.; Haselmann, G.M.; Nandan, S.P.; Wang, J.; Giesriegl, A.; Cherevan, A.S.; Eder, D. Immobilization of Co, Mn, Ni and Fe oxide co-catalysts on TiO_2 for photocatalytic water splitting reactions. *J. Mater. Chem. A* **2019**, *7*, 18568–18579. [CrossRef]
10. Hohenberg, P.; Kohn, W. Inhomogeneous electron gas. *Phys. Rev.* **1964**, *136*, B864. [CrossRef]
11. Kohn, W.; Sham, L.J. Self-consistent equations including exchange and correlation effects. *Phys. Rev.* **1965**, *140*, A1133. [CrossRef]
12. Hebenstreit, W.; Ruzycki, N.; Herman, G.S.; Gao, Y.; Diebold, U. Scanning tunneling microscopy investigation of the TiO_2 anatase (101) surface. *Phys. Rev. B* **2000**, *62*, R16334–R16336. [CrossRef]
13. Alghannam, A.; Muhich, C.L.; Musgrave, C.B. Adatom surface diffusion of catalytic metals on the anatase TiO_2(101) surface. *Phys. Chem. Chem. Phys.* **2017**, *19*, 4541–4552. [CrossRef]
14. Wang, Y.; Su, Y.; Zhu, M.; Kang, L. Ni cluster nucleation and growth on the anatase TiO_2(101) surface: A density functional theory study. *RSC Adv.* **2015**, *5*, 16582–16591. [CrossRef]

15. Iyemperumal, S.K.; Fenton, T.G.; Gillingham, S.L.; Carl, A.D.; Grimm, R.L.; Li, G.; Deskins, N.A. The stability and oxidation of supported atomic-size Cu catalysts in reactive environments. *J. Chem. Phys.* **2019**, *151*, 054702. [CrossRef]
16. Sharma, P.K.; Cortes, M.A.; Hamilton, J.; Han, Y. Surface modification of TiO_2 with copper clusters for band gap narrowing. *Catal. Today* **2019**, *321*, 9–17. [CrossRef]
17. Zhou, X.; Dong, H. A Theoretical perspective on charge separation and transfer in metal oxide photocatalysts for water splitting. *ChemCatChem* **2019**, *11*, 3688–3715. [CrossRef]
18. Singh, D.J.; Nordström, L. *Planewaves, Pseudopotentials and the LAPW Method*, 2nd ed.; Springer: Berlin, Germany, 2006.
19. Karsai, F.; Tran, F.; Blaha, P. On the importance of local orbitals using second energy derivatives for d and f electrons. *Comput. Phys. Commun.* **2017**, *220*, 230–238. [CrossRef]
20. Blaha, P.; Schwarz, K.; Madsen, G.K.H.; Kvasnicka, D.; Luitz, J.; Laskowski, R.; Tran, F.; Marks, L.D. *WIEN2k: An Augmented Plane Wave plus Local Orbitals Program for Calculating Crystal Properties*; Vienna University of Technology: Vienna, Austria, 2018.
21. Blaha, P.; Schwarz, K.; Tran, F.; Laskowski, R.; Madsen, G.K.H.; Marks, L.D. WIEN2k: An APW+ lo program for calculating the properties of solids. *J. Chem. Phys.* **2020**, *152*, 074101. [CrossRef] [PubMed]
22. Perdew, J.P.; Ruzsinszky, A.; Csonka, G.I.; Vydrov, O.A.; Scuseria, G.E.; Constantin, L.A.; Zhou, X.; Burke, K. Restoring the density-gradient expansion for exchange in solids and surfaces. *Phys. Rev. Lett.* **2008**, *100*, 136406. [CrossRef]
23. Schossberger, F. Über die Umwandlung des Titandioxyds. *Zeitschrift Kristallographie* **1942**, *104*, 358–374. [CrossRef]
24. Anisimov, V.I.; Zaanen, J.; Andersen, O.K. Band theory and Mott insulators: Hubbard U instead of Stoner I. *Phys. Rev. B* **1991**, *44*, 943–954. [CrossRef]
25. Bader, R.F.W. *Atoms in Molecules: A Quantum Theory*; Oxford University Press: Oxford, UK, 1990.
26. Bader, R.F.W. Atoms in molecules. *Acc. Chem. Res.* **1985**, *18*, 9–15. [CrossRef]

computation

Article

Hexatetra-Carbon: A Novel Two-Dimensional Semiconductor Allotrope of Carbon

Mosayeb Naseri [1,2], Jaafar Jalilian [3], Dennis R. Salahub [2,*], Maicon Pierre Lourenço [4] and Ghasem Rezaei [3]

1 Department of Physics, Kermanshah Branch, Islamic Azad University, Kermanshah 67189-97551, Iran; mosayeb.naseri@ucalgary.ca
2 CMS—Center for Molecular Simulation, Department of Chemistry, Department of Physics and Astronomy, IQST—Institute for Quantum Science and Technology, Quantum Alberta, University of Calgary, 2500 University Drive NW, Calgary, AB T2N 1N4, Canada
3 Department of Physics, College of Sciences, Yasouj University, Yasouj 75918-74934, Iran; jaafarjalilian@gmail.com (J.J.); grezaei@yu.ac.ir (G.R.)
4 Departamento de Química e Física, Centro de Ciências Exatas, Naturais e da Saúde (CCENS), Universidade Federal do Espírito Santo, Alegre, Vitória 29500-000, Brazil; maiconpl01@gmail.com
* Correspondence: dsalahub@ucalgary.ca

Abstract: Employing first-principles calculations based on density functional theory (DFT), we designed a novel two-dimensional (2D) elemental monolayer allotrope of carbon called hexatetra-carbon. In the hexatetra-carbon structure, each carbon atom bonds with its four neighboring atoms in a 2D double layer crystal structure, which is formed by a network of carbon hexagonal prisms. Based on our calculations, it is found that hexatetra-carbon exhibits a good structural stability as confirmed by its rather high calculated cohesive energy −6.86 eV/atom, and the absence of imaginary phonon modes in its phonon dispersion spectra. Moreover, compared with its hexagonal counterpart, i.e., graphene, which is a gapless material, our designed hexatetra-carbon is a semiconductor with an indirect band gap of 2.20 eV. Furthermore, with a deeper look at the hexatetra-carbon, one finds that this novel monolayer may be obtained from bilayer graphene under external mechanical strain conditions. As a semiconductor with a moderate band gap in the visible light range, once synthesized, hexatetra-carbon would show promising applications in new opto-electronics technologies.

Keywords: density functional theory; hexatetra-carbon; electrical properties

Citation: Naseri, M.; Jalilian, J.; Salahub, D.R.; Lourenço, M.P.; Rezaei, G. Hexatetra-Carbon: A Novel Two-Dimensional Semiconductor Allotrope of Carbon. *Computation* **2022**, *10*, 19. https://doi.org/10.3390/computation10020019

Academic Editor: Henry Chermette

Received: 27 December 2021
Accepted: 20 January 2022
Published: 25 January 2022

Publisher's Note: MDPI stays neutral with regard to jurisdictional claims in published maps and institutional affiliations.

1. Introduction

The discovery of the interesting behavior of graphene [1] has motivated further theoretical and experimental investigations in order to find possible new stable two-dimensional (2D) materials, especially mono-elemental 2D monolayer materials [2–7]. As carbon exhibits a large number of different allotropes, such as graphite, diamond, C_{60} fullerene [8], nanotube [9], carbon nano-cone [10], and nanochain [11], 2D carbon mono-elemental monolayer materials beyond graphene have attracted significant attention from both theoretical and experimental fields of study. However, the electronic structure of graphene limits its application in designing electronic nano-devices due to its semi-metallic gapless nature. Therefore, finding new 2D mono-elemental monolayer materials with a semiconductor behavior is technologically important.

In recent years, the possibility of free standing 2D carbon allotropes beyond graphene has been explored. In addition, a number of new carbon allotrope monolayers has been proposed [12–15]. Although these 2D materials, such as graphdiyne [12], penta-graphene [13], and phagraphene [14] are metastable compared with graphene, only a few have been successfully synthesized. Moreover, these new 2D carbon monolayers exhibit very interesting properties, such as anisotropic Dirac cones, inherent ferromagnetism, high catalytic activity, and potential superconductivity related to the high density of states at the Fermi level.

This demonstrates that structural properties and the crystal configuration of 2D carbon allotropes effectively influence the electronic, optical, and other chemo-physical properties of these materials [12–19]. For instance, graphdiyne [12] is a predicted flat one-atom-thick allotrope of carbon with a Dirac cone in its band structure, in which the Dirac points are located at the *K* point. On the other hand, penta-graphene is a new 2D carbon allotrope with semiconducting properties that stabilizes in a buckled structure composed entirely of carbon pentagons, and resembles the Cairo pentagonal tiling with an intrinsic indirect band gap of about 3.25 eV [13]. Moreover, phagraphene, which is a monolayer sheet of carbon with a structure composed of 5-6-7 carbon rings has distorted Dirac cones [14].

On the other hand, carbon nanostructures have tetragons, pentagons, and hexagons as their main basic building blocks. For instance, cubane (C_8H_8) is a synthetic hydrocarbon molecule formed by eight carbon atoms positioned at the corner of a cube. It is attached to its three neighboring carbon atoms and a hydrogen atom with tetragonal top and side views [20], 2D graphene, and 3D graphite, which are formed by carbon hexagons. In addition, penta-graphene consists entirely of pentagons of carbon atoms. Moreover, the C_{60} molecule is formed by 12 pentagons, which are separated by 20 hexagons with a soccer ball shape [8].

In this paper, using first-principles calculations based on DFT, we propose a novel 2D elemental monolayer allotrope of carbon, which is called hexatetra-carbon due to its hexagonal and tetragonal top and side views. In a crystal network of the newly proposed 2D carbon allotrope, each carbon atom binds with its four neighboring atoms in a 2D double layer crystal structure, which is formed by hexagonal prisms. The meta-stability of our newly designed monolayer is shown by the cohesive energy and phonon calculations. Furthermore, evaluating the electrical properties of hexatetra-carbon shows that it exhibits a semiconductor behavior with a moderate band gap.

The current paper is organized as follows: The computational method used in our study is presented in Section 2. In Section 3, the details of the structural properties are outlined and the stability of the proposed structure is discussed. The electrical characteristics of the designed hexatetra-carbon monolayer and its potential applications are described in Section 4. Finally, in Section 5, the paper is concluded.

2. Computational Methods

To obtain accurate structural and electrical properties of the proposed 2D material, the full-potential linearized-augmented plane wave (FP-LAPW) scheme was utilized [21]. This scheme is based on the DFT implemented in the WIEN2k computational package [22], in which the generalized gradient approximation (GGA) parameterized by Perdew–Burke–Ernzerhof (PBE) was used [23]. Moreover, since the GGA method underestimates band gaps, to obtain more reliable band gaps, the screened short range hybrid functional exchange correlation implemented in the WIEN2k code [24] was employed for band gap calculations. Moreover, the Monkhorst–Pack scheme [25] was used to sample the Brillouin zone by a 24 × 24 × 1 k-mesh, where we chose $RK_{max} = 7$, $Gmax = 14\ Ry^{1/2}$, and $l_{max} = 10$. To avoid interlayer interactions, a large vacuum distance of 20 Å along the non-periodic direction was utilized. With regards to dynamic stability, an evaluation of all the calculations was conducted with the Quantum Espresso (QE) package [26]. Furthermore, the Martin–Troullier norm-conserving pseudopotential [27] was used to treat the core electrons, while the valence electronic wave functions were expanded using an energy cut-off of 80 Ry. However, for the investigation of structural properties of bilayer graphene, we considered the van der Waals correction in our calculations.

3. Structural Properties and the Stability of Hexatetra-Carbon Monolayer

The design of the 2D hexatetra-carbon monolayer was initiated by examining the effect of vertical compressive strains on the structural, electronic, and optical properties of bilayer graphene in its AA-stacking configuration. Recently, several research works have reported the influence of strain on the different physical characteristics of bilayer graphene.

Moreover, it is well-known that a crystal structure of a bilayer graphene stabilizes at the interlayer distance of about h_0 = 3.4 Å layer separation, in which the long-range van der Waals (vdW) interaction plays an important role in the structural properties of the materials. In this case, when evaluating the effect of vertical compressive strains on the different properties of AA-stacked bilayer graphene, with the aim of preserving the stability of the initial AA-stacked bilayer graphene, researchers mostly consider vertical strains that influence a maximum variation of 20% in interlayer separation, i.e., $h = h_0 \pm 20\%\, h_0$ (in which the long-ranged van der Waals (vdW) interaction should be taken into account). In these conditions, not only the stability of the AA-stacked bilayer graphene is preserved, but also its crystal structure configuration is retained. However, one may be curious to know the effect of higher values of compressive vertical strains, in which the layer separation distance of h < 2.0 Å for the long-range van der Waals (vdW) interaction has no important effect. On this basis, we found a new unprecedented 2D allotrope of carbon called hexatetra-carbon during our structure searches. Figure 1 shows the optimized structure of our designed hexatetra-carbon monolayer from different views. A unit cell of hexatetra-carbon monolayer consists of four carbon atoms with optimized lattice constants of a = b = 2.65 Å. In a crystal network of the hexatetra-carbon, C atoms are distributed in two different exactly planar atomic planes with a vertical distance of about 1.55 Å. Specifically, each C atom of the hexatetra-carbon bonds to four neighboring C atoms to form a tetra-coordinated carbon structure, in which the C–C bond lengths are about 1.55 Å, longer than those in graphene (1.43 Å) and almost equal to the C1–C2 bond lengths in penta-graphene [13].

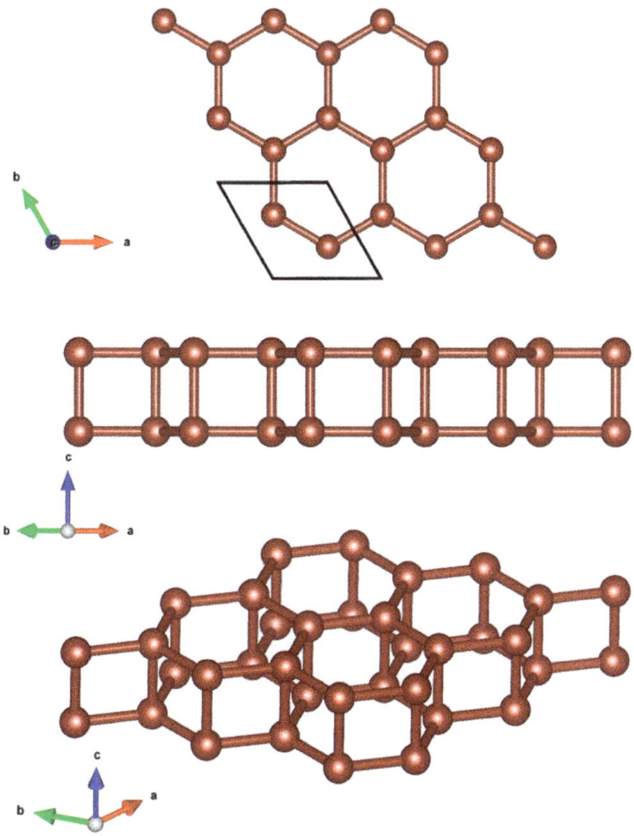

Figure 1. Crystal structure of 2D hexatetra-carbon from different views.

Furthermore, before an evaluation could be performed for the electrical properties and potential applications of the hexatetra-carbon monolayer, it is imperative to analyze the stability of the monolayer structure. This was carried out by first evaluating its energetic stability through a calculation of its cohesive energy, which is given by $E_{coh} = \frac{E_{hexatetra-carbon} - 4E_c}{4}$ (E_c, $E_{hexatetra-carbon}$ are the total energies of a single C atom, and a unit cell of the hexatetra-carbon monolayer, respectively). Based on our calculation, the 2D hexatetra-carbon monolayer has a cohesive energy of about −6.86 eV/atom. For comparison, we also calculated the cohesive energy of graphene, penta-graphene, and T-carbon [28], which is a 3D carbon allotrope obtained by replacing each atom in diamond with a carbon tetrahedron at the same theoretical level. The values were about −8.01, −7.02, −6.34 eV/atom, respectively, which confirm that the proposed 2D hexatetra-carbon monolayer shows good energetic stability. Figure 2 shows a unit cell relative energy of the predicted hexatetra-carbon monolayer compared with different carbon allotropes under strain conditions.

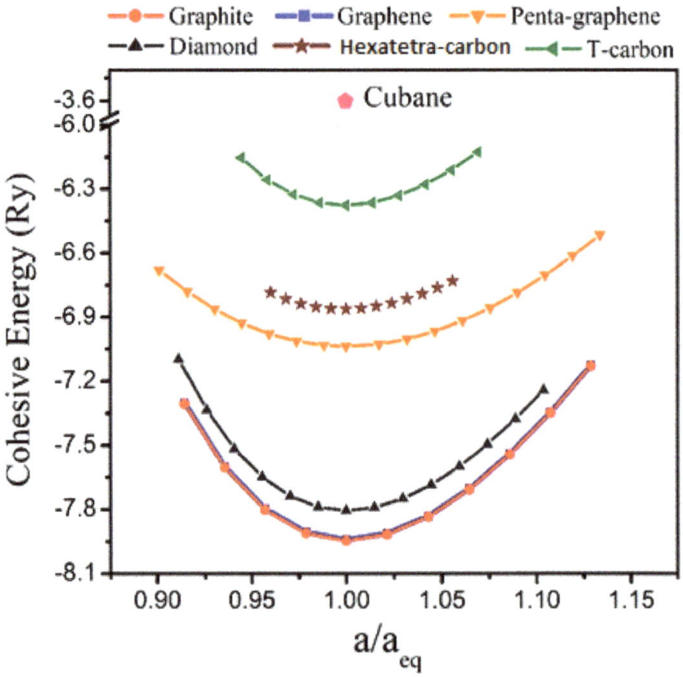

Figure 2. Cohesive energy of different carbon allotropes.

Next, we examined the dynamic stability of the 2D hexatetra-carbon monolayer by calculating its phonon dispersion spectrum. As shown in Figure 3, there are no imaginary phonon modes in the whole Brillouin zone, indicating that the hexatetra-carbon is a local minimum on the potential energy surface and can be considered as a metastable allotrope of carbon compared with the other carbon allotropes. Specifically, we obtained the highest frequency of 1267 cm^{-1}, which is higher than those obtained for silicon [29] (580 cm^{-1}), MoS2 monolayer [30] (473 cm^{-1}), and TiC monolayer [31] (810 cm^{-1}). However, it is lower than the highest phonon frequency of graphene (about 1650 cm^{-1}) and penta-graphene (about 1600 cm^{-1}), indicating robust C–C bonds in the predicted monolayer.

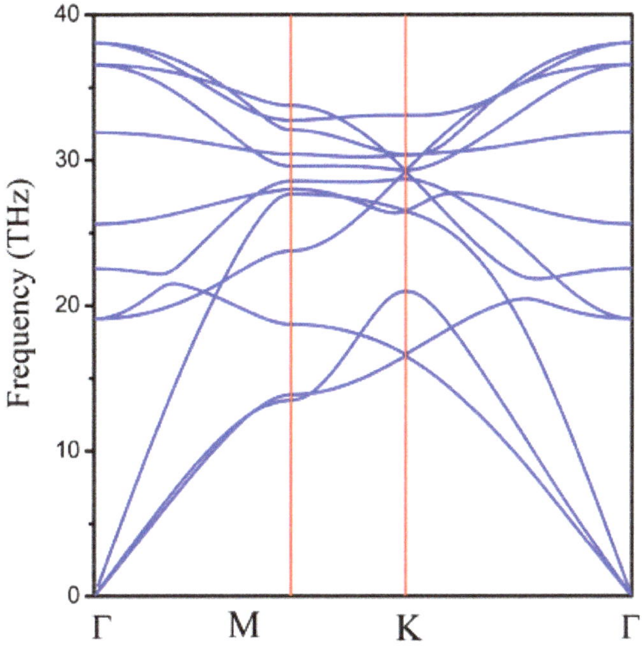

Figure 3. Phonon dispersion spectrum of 2D hexatetra-carbon monolayer.

From the above indications that the 2D hexatetra-carbon may be meta-stable, we next systematically study its mechanical properties. Using Young's modulus (Y), Equation (1) can be calculated [32], where ε is the strain in the vicinity of the optimum lattice vector ($\varepsilon = (a - a_0)/a_0$) and E is the total energy. In addition, V_0 is the equilibrium volume of the 2D material evaluated by $V_0 = \frac{3\sqrt{3}d^2h}{2}$, where d is the adjacent carbon distance in the hexagon ring in the a and b plane, and h is the thickness of the 2D material along the c vector (Figure 1).

$$Y = \frac{1}{V_0}\left(\frac{\partial^2 E}{\partial \varepsilon^2}\right)_{\varepsilon=0} \tag{1}$$

In this work, the estimated Young's modulus value for hexatetra-carbon is 1859.7 GPa. It is smaller than those estimated by Raman spectroscopy [33] for the single- and bilayer graphene, 2400 ± 400 GPa and 2000 ± 500 GPa, respectively. With regards to carbon fibers, the Young's modulus value is 235–427 GPa [34]. Here, the value of hexatetra-graphene is greater than those obtained for other materials, such as imogolites (~320–370 GPa) [35] and GaS (~270 GPa) [36], as well as the nanotubes and MoS_2 monolayer (265 ± 13 GPa), which are similar to the bilayer [37].

4. Electronic Properties

To analyze the electrical properties of the 2D hexatetra-carbon monolayer, we calculated the band structure of the designed 2D monolayer. As shown in Figure 4, an indirect bandgap of about 2.20 eV, which is calculated using the hybrid functional level of theory can be seen in the band structure. Moreover, the valence band maximum (VBM) of this 2D monolayer material is located between K and Γ points, while the conduction band minimum (CBM) is located at the Γ point. Therefore, the 2D hexatetra-carbon monolayer, which is a semiconductor with a moderate band gap, is not similar to the monolayer and bilayer graphene due to the fact that they are both semi-metals.

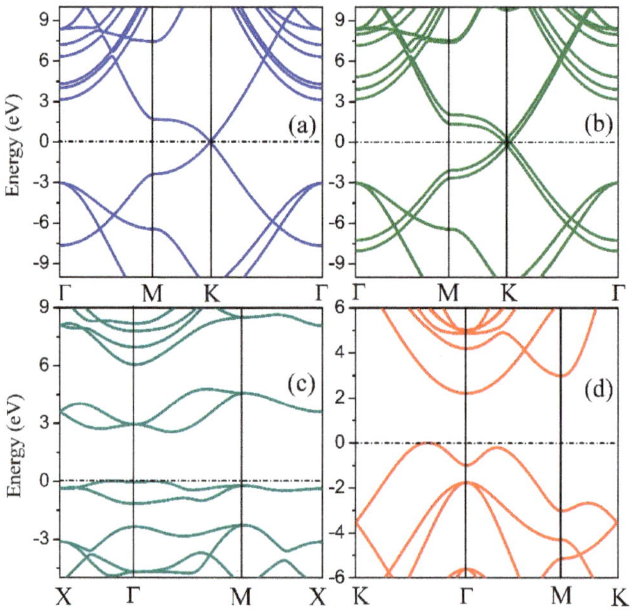

Figure 4. Band structures of (**a**) graphene, (**b**) bilayer graphene, (**c**) penta-graphene, and (**d**) hexatetra-carbon.

To gain a deeper insight into the bonding nature of hexatetra-carbon and its structural analogy with respect to AA-stacked graphene, we investigated the 2D valence charge density distribution of these two nanostructures (see Figure 5d–g). The crystal structure of AA-stacked bilayer graphene, the cubane molecule, and hexatetra-carbon are shown in Figure 5a–c. It is clear that the hexatetra-carbon structure has a hexagonal face similar to graphene (xy plane). In addition, interlayer bonds along the z direction between the two graphene monolayers show a tetragonal side view, which is similar to the cubane molecule.

As shown in Figure 5d, there are three sp^2 sigma covalent bonds (1.43 Å) between the C–C atoms in each monolayer of AA-stacked bilayer graphene, as well as a weak p_z-p_z interaction between the monolayers. In comparison, the in-plane bond length of C–C for hexatetra-carbon is 1.56 Å, which is longer than those obtained in AA-stacked bilayer graphene, i.e., the C–C in-plane orbital overlap decreases for the hexatetra-carbon (see Figure 5f). Therefore, to retain its structural stability, the hexatetra-carbon nanostructure compensates this orbital variation by creating interlayer sigma bonds between the neighboring carbons, which are located in the different planes. Moreover, these interlayer bonds would restrict p_z, resulting in the semiconducting nature of 2D hexatetra-carbon.

Furthermore, we calculated the variation of the cohesive energy of two graphene sheets as they approach each other. As seen in Figure 6, when a graphene sheet moves towards another fixed graphene sheet, they tend to stabilize at the vertical distance of about 3.45 Å and form the AA-stacked bilayer graphene structure, which is the most stable structure in this situation. However, by applying an additional external vertical mechanical strain to the AA-stacked bilayer graphene through decreasing its interlayer distance, its lattice parameter increases. In addition, another local energy minimum occurs for the two graphene sheets with a vertical interlayer distance of about 1.55 Å. In this case, the hexatetra-carbon includes four carbon atoms in its unit cell, and the lattice parameter of about 2.65 Å structure is formed. In other words, by applying the external vertical strain, a type of phase transition occurs.

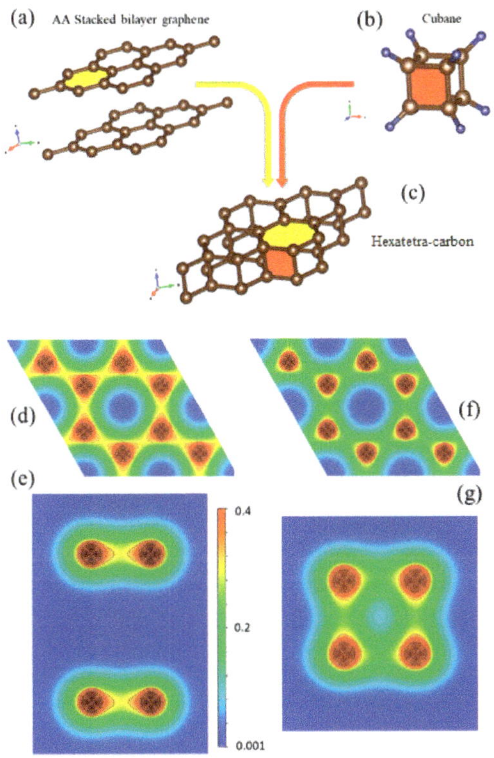

Figure 5. Crystal structure of (**a**) AA-stacked bilayer graphene, (**b**) cubane molecule, and (**c**) hexatetra-carbon (**d**,**e**). Top and side views of valence charge density distribution for AA-stacked graphene and (**f**,**g**) for hexatetra-carbon obtained by WIEN2K code [21].

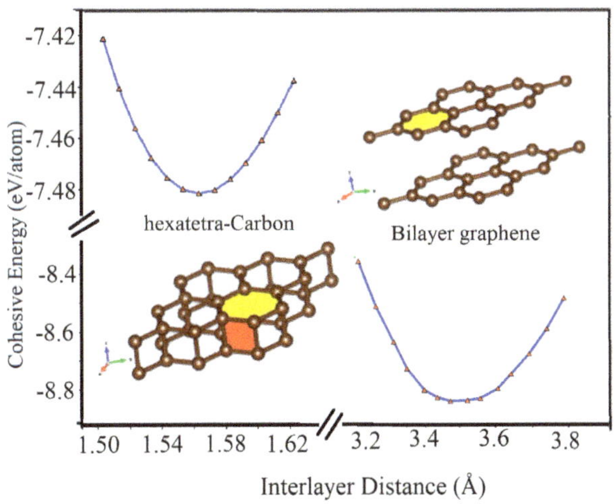

Figure 6. Cohesive energy of AA-stacked bilayer graphene and hexatetra-carbon versus interlayer distance calculated by Quantum Espresso (due to the interruption in accessing WIEN2k while following up on a reviewer's comment, we have used Quantum Espresso for Figure 6).

5. Summary

In conclusion, utilizing first-principles calculations based on DFT, we proposed a new 2D monolayer crystalline allotrope of carbon called hexatetra-carbon, which shows a semiconducting nature with a moderate band gap. The stability of 2D hexatetra-carbon monolayer is confirmed by its rather high cohesive energy, as well as the absence of imaginary phonon modes in its phonon dispersion spectrum. Comparing the cohesive energy of the hexatetra-carbon monolayer with those obtained for other 2D allotropes confirms its high bonding properties. Moreover, our analysis on the charge densities of both bilayer graphene and hexatetra-carbon monolayer indicates that by applying a vertical strain on bilayer graphene, the hexatetra-carbon structure can be obtained. Therefore, one may be optimistic that the 2D hexatetra-carbon monolayer can be experimentally achieved in the foreseeable future.

Author Contributions: All authors contributed equally to this work. All authors have read and agreed to the published version of the manuscript.

Funding: This research was funded by the Natural Sciences and Engineering Research Council of Canada Discovery Grant number RGPIN-2019-03976. Computational resources were provided by Compute Canada/WestGrid.

Institutional Review Board Statement: Not applicable.

Informed Consent Statement: Not applicable.

Data Availability Statement: Not applicable.

Conflicts of Interest: The authors declare no conflict of interest.

References

1. Novoselov, K.S.; Geim, A.K.; Morozov, S.V.; Jiang, D.; Zhang, Y.; Dubonos, S.V.; Grigorieva, I.V.; Firsov, A.A. Electric field effect in atomically thin carbon films. *Science* **2004**, *306*, 666–669. [CrossRef] [PubMed]
2. Butler, S.Z.; Hollen, S.M.; Cao, L.; Cui, Y.; Gupta, J.A.; Gutiérrez, H.R.; Heinz, T.F.; Hong, S.S.; Huang, J.; Ismach, A.F.; et al. Progress, Challenges, and Opportunities in Two-Dimensional Materials Beyond Graphene. *ACS Nano* **2013**, *7*, 2898–2926. [CrossRef] [PubMed]
3. Molle, A.; Goldberger, J.; Houssa, M.; Xu, Y.; Zhang, S.-C.; Akinwande, D. Buckled two-dimensional Xene sheets. *Nat. Mater.* **2017**, *16*, 163–169. [CrossRef] [PubMed]
4. Guzmán-Verri, G.G.; Voon, L.C.L.Y. Electronic structure of silicon-based nanostructures. *Phys. Rev. B* **2007**, *76*, 075131. [CrossRef]
5. Xu, M.; Liang, T.; Shi, M.; Chen, H. Graphene-Like Two-Dimensional Materials. *Chem. Rev.* **2013**, *113*, 3766–3798. [CrossRef]
6. Balendhran, S.; Walia, S.; Nili, H.; Sriram, S.; Bhaskaran, M. Elemental Analogues of Graphene: Silicene, Germanene, Stanene, and Phosphorene. *Small* **2015**, *11*, 640–652. [CrossRef] [PubMed]
7. Zhang, S.; Xie, M.; Li, F.; Yan, Z.; Li, Y.; Kan, E.; Liu, W.; Chen, Z.; Zeng, H. Semiconducting Group 15 Monolayers: A Broad Range of Band Gaps and High Carrier Mobilities. *Angew. Chem. Int. Ed.* **2016**, *55*, 1666–1669. [CrossRef]
8. Kroto, H.W.; Heath, J.R.; Brien, S.C.O.; Curl, R.F.; E Smalley, R. C60: Buckminsterfullerene. *Nature* **1985**, *318*, 162–163. [CrossRef]
9. Iijima, S.; Ichihashi, T. Single-shell carbon nanotubes of 1-nm diameter. *Nature* **1993**, *363*, 603–605. [CrossRef]
10. Charlier, J.-C.; Rignanese, G.-M. Electronic Structure of Carbon Nanocones. *Phys. Rev. Lett.* **2001**, *86*, 5970–5973. [CrossRef]
11. Jin, C.; Lan, H.; Peng, L.; Suenaga, K.; Iijima, S. Deriving Carbon Atomic Chains from Graphene. *Phys. Rev. Lett.* **2009**, *102*, 205501. [CrossRef] [PubMed]
12. Li, Y.; Xu, L.; Liu, H.; Li, Y. Graphdiyne and graphyne: From theoretical predictions to practical construction. *Chem. Soc. Rev.* **2014**, *43*, 2572–2586. [CrossRef] [PubMed]
13. Zhang, S.; Zhou, J.; Wang, Q.; Chen, X.; Kawazoe, Y.; Jena, P. Penta-graphene: A new carbon allotrope. *Proc. Natl. Acad. Sci. USA* **2015**, *112*, 2372–2377. [CrossRef] [PubMed]
14. Wang, Z.; Zhou, X.; Zhang, X.; Zhu, Q.; Dong, H.; Zhao, M.; Oganov, A.R. Phagraphene: A Low-Energy Graphene Allotrope Composed of 5–6–7 Carbon Rings with Distorted Dirac Cones. *Nano Lett.* **2015**, *15*, 6182–6186. [CrossRef]
15. Maruyama, M.; Okada, S. Two-Dimensional sp2Carbon Network of Fused Pentagons: All Carbon Ferromagnetic Sheet. *Appl. Phys. Express* **2013**, *6*, 095101. [CrossRef]
16. Terrones, H.; Hernández, E.; Grobert, N.; Charlier, J.-C.; Ajayan, P.M. New Metallic Allotropes of Planar and Tubular Carbon. *Phys. Rev. Lett.* **2000**, *84*, 1716–1719. [CrossRef]
17. Jia, T.-T.; Fan, X.; Zheng, M.-M.; Chen, G. Silicene nanomeshes: Bandgap opening by bond symmetry breaking and uniaxial strain. *Sci. Rep.* **2016**, *6*, 20971. [CrossRef]

18. Jia, T.-T.; Zheng, M.-M.; Fan, X.; Su, Y.; Li, S.-J.; Liu, H.-Y.; Chen, G.; Kawazoe, Y. Dirac cone move and bandgap on/off switching of graphene superlattice. *Sci. Rep.* **2016**, *6*, 18869. [CrossRef]
19. Yang, D.-C.; Jia, R.; Wang, Y.; Kong, C.-P.; Wang, J.; Ma, Y.; Eglitis, R.I.; Zhang, H.-X. Novel Carbon Nanotubes Rolled from 6, 6, 12-Graphyne: Double Dirac Points in 1D Material. *J. Phys. Chem. C* **2017**, *121*, 14835–14844. [CrossRef]
20. Eaton, P.E.; Cole, T.W. Cubane. *J. Am. Chem. Soc.* **1964**, *86*, 3157–3158. [CrossRef]
21. Abt, R.; Draxl, C.; Knoll, P. Optical response of high temperature superconductors by full potential LAPW band structure calculations. *Phys. B Condens. Matter* **1994**, *194–196*, 1451–1452. [CrossRef]
22. Blaha, P.; Schwarz, K.; Tran, F.; Laskowski, R.; Madsen, G.K.H.; Marks, L.D. An Augmented Plane Wave+ Local Orbitals Program for Calculating Crystal Properties. *J. Chem. Phys* **2020**, *152*, 074101. [CrossRef] [PubMed]
23. Perdew, J.P.; Burke, K.; Ernzerhof, M. Generalized gradient approximation made simple. *Phys. Rev. Lett.* **1996**, *77*, 3865–3868. [CrossRef] [PubMed]
24. Tran, F.; Blaha, P. Implementation of screened hybrid functionals based on the Yukava potential within the LAPW basis set. *Phys. Rev. B* **2011**, *83*, 235118. [CrossRef]
25. Monkhorst, H.J.; Pack, J.D. Special points for Brillouin-zone integrations. *Phys. Rev. B* **1976**, *13*, 5188–5192. [CrossRef]
26. Giannozzi, P.; Baroni, S.; Bonini, N.; Calandra, M.; Car, R.; Cavazzoni, C.; Ceresoli, D.; Chiarotti, G.L.; Cococcioni, M.; Dabo, I.; et al. Quantum espresso: A modular and open-source software project for quantum simulations of materials. *J. Phys. Condens. Matter* **2009**, *21*, 395502. [CrossRef]
27. Troullier, N.; Martins, J.L. Efficient pseudopotentials for plane-wave calculations. *Phys. Rev. B* **1991**, *43*, 1993–2006. [CrossRef]
28. Alborznia, H.; Naseri, M.; Fatahi, N. Pressure effects on the optical and electronic aspects of T-Carbon: A first principles calculation. *Optik* **2019**, *180*, 125–133. [CrossRef]
29. Jahangirov, S.; Topsakal, M.; Akturk, E.; Sahin, H.; Ciraci, S. Two- and One-Dimensional Honeycomb Structures of Silicon and Germanium. *Phys. Rev. Lett.* **2009**, *102*, 236804. [CrossRef]
30. Ganatra, R.; Zhang, Q. Few-Layer MoS2: A Promising Layered Semiconductor. *ACS Nano* **2014**, *8*, 4074–4099. [CrossRef]
31. Li, H.; Zhang, L.; Zeng, Q.; Guan, K.; Li, K.; Ren, H.; Liu, S.; Cheng, L. Structural, elastic and electronic properties of transition metal carbides TMC (TM=Ti, Zr, Hf and Ta) from first-principles calculations. *Solid State Commun.* **2011**, *151*, 602–606. [CrossRef]
32. Lourenço, M.P.; De Oliveira, C.; Oliveira, A.; Guimaraes, L.; Duarte, H. Structural, Electronic, and Mechanical Properties of Single-Walled Chrysotile Nanotube Models. *J. Phys. Chem. C* **2012**, *116*, 9405–9411. [CrossRef]
33. Lee, J.-U.; Yoon, D.; Cheong, H. Estimation of Young's Modulus of Graphene by Raman Spectroscopy. *Nano Lett.* **2012**, *12*, 4444–4448. [CrossRef] [PubMed]
34. Li, Y.; Fleischer, C.M.; Ross, A.E. High Young's modulus carbon fibers are fouling resistant with fast-scan cyclic voltammetry. *Chem. Commun.* **2020**, *56*, 8023–8026. [CrossRef] [PubMed]
35. Lourenço, M.P.; Guimarães, L.; da Silva, M.C.; de Oliveira, C.; Heine, T.; Duarte, H.A. Nanotubes With Well-Defined Structure: Single- and Double-Walled Imogolites. *J. Phys. Chem. C* **2014**, *118*, 5945–5953. [CrossRef]
36. Köhler, T.; Frauenheim, T.; Hajnal, Z.; Seifert, G. Tubular structures of GaS. *Phys. Rev. B* **2004**, *69*, 193403. [CrossRef]
37. Li, Y.; Yu, C.; Gan, Y.; Jiang, P.; Yu, J.; Ou, Y.; Zou, D.-F.; Huang, C.; Wang, J.; Jia, T.; et al. Mapping the elastic properties of two-dimensional MoS2 via bimodal atomic force microscopy and finite element simulation. *NPJ Comput. Mater.* **2018**, *4*, 49. [CrossRef]

computation

MDPI

Article

Length-Gauge Optical Matrix Elements in WIEN2k

Oleg Rubel [1,*] and Peter Blaha [2,*]

[1] Department of Materials Science and Engineering, McMaster University, 1280 Main Street West, Hamilton, ON L8S 4L8, Canada
[2] Institute of Materials Chemistry, Vienna University of Technology, Getreidemarkt 9/165-TC, 1060 Vienna, Austria
* Correspondence: rubelo@mcmaster.ca (O.R.); peter.blaha@tuwien.ac.at (P.B.)

Abstract: Hybrid exchange-correlation functionals provide superior electronic structure and optical properties of semiconductors or insulators as compared to semilocal exchange-correlation potentials due to admixing a portion of the non-local exact exchange potential from a Hartree–Fock theory. Since the non-local potential does not commute with the position operator, the momentum matrix elements do not fully capture the oscillator strength, while the length-gauge velocity matrix elements do. So far, length-gauge velocity matrix elements were not accessible in the all-electron full-potential WIEN2k package. We demonstrate the feasibility of computing length-gauge matrix elements in WIEN2k for a hybrid exchange-correlation functional based on a finite difference approach. To illustrate the implementation we determined matrix elements for optical transitions between the conduction and valence bands in GaAs, GaN, $(CH_3NH_3)PbI_3$ and a monolayer MoS_2. The non-locality of the Hartree–Fock exact exchange potential leads to a strong enhancement of the oscillator strength as noticed recently in calculations employing pseudopotentials (Laurien and Rubel: arXiv:2111.14772 (2021)). We obtained an analytical expression for the enhancement factor for the difference in eigenvalues not captured by the kinetic energy. It is expected that these results can also be extended to other non-local potentials, e.g., a many-body *GW* approximation.

Keywords: semiconductors; oscillator strength; density functional theory; hybrid exchange-correlation functional; non-local potential

Citation: Rubel, O.; Blaha, P. Length-Gauge Optical Matrix Elements in WIEN2k. *Computation* **2022**, *10*, 22. https://doi.org/10.3390/computation10020022

Academic Editor: Henry Chermette

Received: 29 December 2021
Accepted: 21 January 2022
Published: 26 January 2022

This paper is dedicated to the 80th birthday of Professor Karlheinz Schwarz, the founder of the WIEN2k DFT package.

1. Introduction

Calculations of linear optical properties of solids require matrix elements for electric dipole transitions. Momentum matrix elements

$$\mathbf{p}_{mn}(\mathbf{k}) = \langle m,\mathbf{k}| - i\nabla_{\mathbf{r}}|n,\mathbf{k}\rangle \tag{1}$$

are widely used in full-potential codes with periodic boundary conditions [1] when optical properties are computed with local potentials (e.g., LDA (see end of the paper for the full list of abbreviations) or GGA XC functionals) and referred to in the literature as a velocity gauges. (Atomic units will be used throughout the paper.)

Starace [2] emphasised the limitations of Equation (1) when representing matrix elements for electric dipole transitions. Instead, the more general velocity matrix elements should be used

$$\mathbf{v}_{mn}(\mathbf{k}) = \langle m,\mathbf{k}|i[\hat{H},\mathbf{r}]|n,\mathbf{k}\rangle \tag{2}$$

with the velocity operator [2]

$$\hat{\mathbf{v}} = i[\hat{H},\mathbf{r}] = \hat{\mathbf{p}} + i[\hat{V}_{\mathrm{NL}}(\mathbf{r},\mathbf{r}'),\mathbf{r}], \tag{3}$$

which contains an additional commutator term $[\hat{V}_{\rm NL}(\mathbf{r},\mathbf{r}'),\mathbf{r}]$ to account for non-local potentials (e.g., the Hartree-Fock exchange in hybrid XC functionals). With $|n,\mathbf{k}\rangle$ and $E_n(\mathbf{k})$ being eigenstates of the Hamiltonian $\hat{H}$, the alternative (length gauge) matrix elements can be expressed as [2]

$$\langle m,\mathbf{k}|i[\hat{H},\mathbf{r}]|n,\mathbf{k}\rangle = i[E_m(\mathbf{k}) - E_n(\mathbf{k})]\langle m,\mathbf{k}|\mathbf{r}|n,\mathbf{k}\rangle. \tag{4}$$

Since the position operator is not well defined for periodic systems, the following substitution is used instead: [3,4]

$$\mathbf{r} = \lim_{q\to 0}(e^{iq\mathbf{r}} - 1)/iq, \tag{5}$$

which leads to a practical expression for the velocity matrix elements in the long wavelength limit [5,6]

$$v_{mn}^{(\alpha)}(\mathbf{k}) = \lim_{q\to 0}\frac{1}{q}\langle m,\mathbf{k}+\mathbf{q}_\alpha|e^{i\mathbf{q}_\alpha\cdot\mathbf{r}}|n,\mathbf{k}\rangle[E_m(\mathbf{k}+\mathbf{q}_\alpha) - E_n(\mathbf{k})]. \tag{6}$$

Here, $\alpha = x,y,z$ is a Cartesian direction, m and n are band indices, and $\mathbf{q}_x = q\,\hat{\mathbf{x}}$, where $\hat{\mathbf{x}}$ is a unit vector in the direction of the x axis.

WIEN2k [7,8] is one of the most used full-potential all-electron DFT codes for solids. It offers many XC functionals to open the band gap, including hybrids with a non-local Hartree-Fock potential [9]. So far, however, WIEN2k has implemented only momentum matrix elements to compute optical properties as a part of the `optic` module [1]. Laurien and Rubel [10] showed that neglecting the second term in Equation (3) when using hybrid functionals can lead to an underestimation of the squared magnitude of matrix elements for electric dipole transitions between conduction and valence band edges by ca. 30%.

Here, we present a scheme for the calculation of the length-gauge optical matrix elements in WIEN2k based on a finite difference Equation (6) with the help of overlap matrix elements

$$M_{mn}(\mathbf{k},\mathbf{q}) = \langle u_{\mathbf{k},n}|u_{\mathbf{k}+\mathbf{q},m}\rangle \equiv \langle \psi_{\mathbf{k},n}|e^{-i\mathbf{q}\cdot\mathbf{r}}|\psi_{\mathbf{k}+\mathbf{q},m}\rangle \tag{7}$$

that come from the `wien2wannier` module [11]. This development opens an avenue for the calculation of optical properties (frequency-dependent dielectric tensor, absorption spectrum, optical conductivity, refractive index, reflectively, loss function) in the independent particle approximation with hybrid functionals in WIEN2k.

2. Methods

DFT [12,13] calculations were performed with the WIEN2k package (version 21.1) and the Yukawa screened hybrid (YSH) functional [9]. It was shown that by choosing an appropriate screening length λ in the Yukawa potential the YSH functional gives very similar results as the common HSE06 XC functional [14,15]. Important parameters are summarized in Table 1. Experimental structural parameters were used for all solids (Figure 1) with internal atomic positions optimized at the PBE level when permitted by symmetry. Spin-orbit coupling (SOC) was included in all calculations. The structure of $(CH_3NH_3)PbI_3$ was represented by a pseudo-cubic cell taken from Ref. [16], scaled to experimental lattice parameters at 350 K [17,18], followed by a subsequent relaxation of atomic positions while retaining the experimental lattice parameters. The pseudo-cubic structure means that the following constrains $a = b = c$, $\alpha = \beta = \gamma = 90°$ are applied to lattice parameters, while the formal symmetry of the structure (spacegroup P1) is not cubic.

Table 1. Structural and calculation parameters.

Parameters	**GaAs**	**GaN**	**$(CH_3NH_3)PbI_3$**	**1L-MoS_2**
Space group	$F\bar{4}3m$ (216)	$P6_3mc$ (186)	P1 (1)	$P\bar{6}m2$ (187)
Lattice param. (Å)	5.653 [19]	3.18, 5.166 [20]	6.31 [17,18] (pseudo-cubic)	3.16 [21], 29.0
R_{MT} (bohr)	2.23 (Ga) 2.23 (As)	1.90 (Ga) 1.64 (N)	0.68 (H) 1.34 (C) 1.26 (N) 2.50 (Pb) 2.50 (I)	2.36 (Mo) 2.03 (S)
n_{val}	13 (Ga) 15 (As)	19 (Ga) 5 (N)	4 (C) 5 (N) 18 (Pb) 17 (I)	14 (Mo) 6 (S)
$R_{MT_{min}}K_{max}$	8.0	8.0	3.0	8.0
G_{max}	12	12	20	12
l_{max}	10 (all structures)			
$l_{vns_{max}}$	6 (all structures)			
k mesh	$8 \times 8 \times 8$ (Γ centered)	$8 \times 8 \times 4$ (Γ centered)	$3 \times 3 \times 3$ (shifted)	$9 \times 9 \times 1$ (Γ centered)
Energy (Ry) and charge converg.	10^{-4} (all structures) 10^{-3} (all structures)			

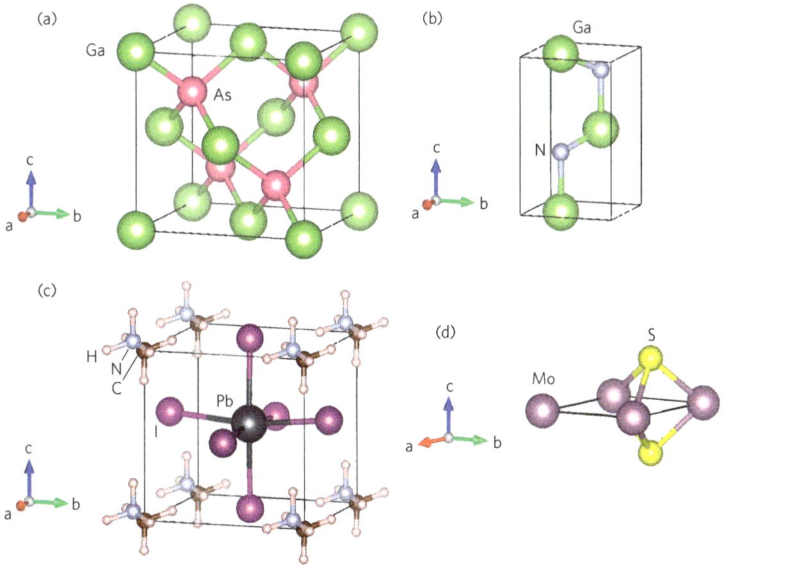

Figure 1. Crystal structures: (**a**) GaAs, (**b**) GaN, (**c**) quasi-cubic $(CH_3NH_3)PbI_3$, and (**d**) monolayer MoS_2.

Velocity-gauge optical matrix elements $\mathbf{p}_{mn}(\mathbf{k})$ were calculated using the `optic` module [1] in WIEN2k. Length-gauge optical matrix elements $\mathbf{v}_{mn}(\mathbf{k})$ were obtained with the forward

$$v_{mn}^{(\alpha)}(\mathbf{k}) \approx \frac{1}{q}\langle u_{\mathbf{k},n}|u_{\mathbf{k}+\mathbf{q}_\alpha,m}\rangle[E_m(\mathbf{k}+\mathbf{q}_\alpha) - E_n(\mathbf{k})] \tag{8}$$

and central

$$v_{mn}^{(\alpha)}(\mathbf{k}) \approx \frac{1}{q}\langle u_{\mathbf{k}-\frac{1}{2}\mathbf{q}_\alpha,n}|u_{\mathbf{k}+\frac{1}{2}\mathbf{q}_\alpha,m}\rangle[E_m(\mathbf{k}+\frac{1}{2}\mathbf{q}_\alpha) - E_n(\mathbf{k}-\frac{1}{2}\mathbf{q}_\alpha)] \tag{9}$$

finite difference method. The overlap matrix elements $\langle u_{\mathbf{k},n}|u_{\mathbf{k}',m}\rangle$ between the cell-periodic parts of the Bloch functions were generated by the `wien2wannier` module [11] (`case.mmn` output file). The length-gauge optical matrix elements for GaAs computed with YSH were additionally verified using VASP [22], HSE06 and projector augmented-wave potentials [22–24]. Sample scripts that illustrate a detailed workflow can be found in the Supporting Information section.

The logarithmic percent change

$$\Delta = \ln\left(\frac{\sum|v_{mn}^{(\alpha)}|^2}{\sum|p_{mn}^{(\alpha)}|^2}\right)100\% \tag{10}$$

was used to evaluate differences between matrix elements. This approach has the following advantages: (i) does not require a reference, (ii) is more suitable for large changes (greater than a few percents), (iii) it has additive properties, and (iv) in the limit of small changes it reduces to the classical ratio of the relative change to the reference.

3. Results

3.1. Finite Difference Calibration and Validation for Local XC

We selected GaAs and the local PBE XC potential [25] to prove the feasibility of computing the length-gauge optical matrix elements using the finite difference methods given by Equations (8) and (9). The local potential is selected here since both the length- and velocity-gauge should lead to identical results under these circumstances. It is also important to get a feeling for the step size q at which the finite difference approximation converges to the accurate result given by the momentum matrix element. Here GaAs serves as an important benchmark since the conduction band and light holes are very sharp and non-parabolic (see Figure 2a).

Results presented in Table 2 suggest that both, the forward and the central finite differences reproduce the values of the momentum matrix element within a 3% error. The central finite difference converges faster (at the wave vector shift of $q \approx 0.003$ rad bohr^{-1}) and will be used to derive matrix elements for the remaining part of this paper. The numerical noise of the finite difference starts to show up at $q < 10^{-6}$ rad bohr^{-1}.

Table 2. Length-gauge velocity matrix elements $|v_{mn}^{(x)}|^2$ (at.u.) in GaAs calculated using the finite difference approximations (forward vs. central) Equations (8) and (9) with various step sizes q. These values are compared with the velocity-gauge momentum matrix elements $|p_{mn}^{(x)}|^2$ (at.u.) from the `optic` module. The local (GGA-PBE) XC functional wa used, which made the velocity and the length gauges identical. The band degeneracy is given as a superscript in parentheses and the meaning of the subscripts is made clear in Figure 2a.

Transition	$\sum\|v_{mn}^{(x)}\|^2$ (q = (16.0/3.5/1.2/0.0006) × 10^{-3} rad bohr^{-1}) Forward FD	Central FD	$\sum\|p_{mn}^{(x)}\|^2$
$\Gamma_{lh,hh}^{(\times 4)} - \Gamma_{c}^{(\times 2)}$	0.264/0.402/0.420/0.402	0.412/0.422/0.422/0.402	0.417
$\Gamma_{so}^{(\times 2)} - \Gamma_{c}^{(\times 2)}$	0.217/0.202/0.200/0.221	0.209/0.201/0.200/0.221	0.206

3.2. Validation for Non-Local XC

After validating our approach with the local potential, we applied it to the non-local YSH XC functional. Again we evaluated the velocity-gauge (momentum) matrix elements and length-gauge (velocity) matrix elements in GaAs. Now we did not expect the two matrix elements to agree given the arguments presented in Section 1. To cross-check our results, we also computed the velocity matrix elements with VASP, which should be comparable with our $v_{mn}^{(\alpha)}$ values.

Our YSH calculations for GaAs gave a band gap of $E_g^{\text{YSH}} = 1.24$ eV vs. $E_g^{\text{exp}} = 1.52$ eV [26] and previously reported $E_g^{\text{HSE06}} = 1.33$ eV [27], which was a significant improvement over PBE. The band structure is shown schematically in Figure 2a where bands are labeled according to the convention. The results presented in Table 3 confirmed the agreement between WIEN2k and VASP for length-gauge matrix elements within less than a 2% deviation. The total length-gauge oscillator strength between valence and conduction band corresponded to $m_0 \sum |v_{cv}^{(x)}|^2 \approx 21$ eV, which agreed well with the 20 eV quoted by (Yu and Cardona [28], Section 2.6) for III-V semiconductors. The momentum (velocity-gauge) matrix elements significantly underestimated the strength of optical transitions, which had previously been reported and quantified in Ref. [10]. The values of $|p_{cv}|^2$ were almost identical to those obtained with PBE (Table 2), even though the momentum matrix elements were derived from YSH wave functions.

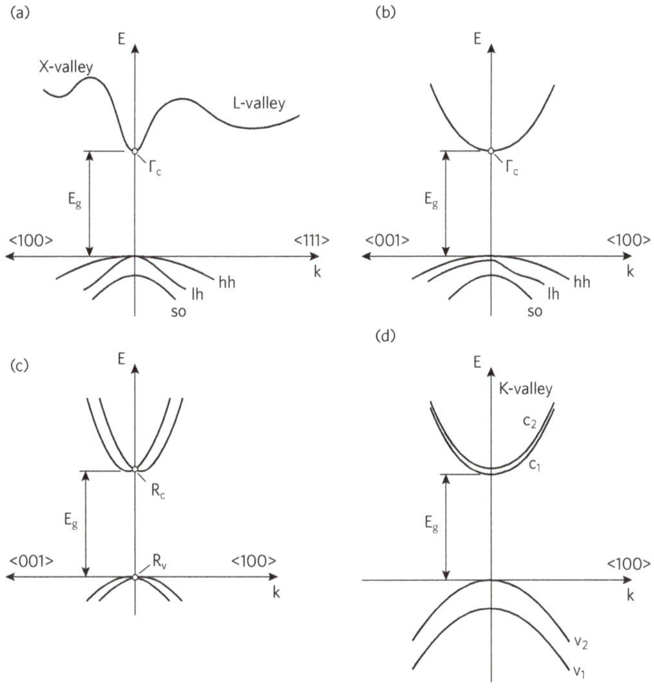

Figure 2. Schematic band structures of materials studied with SOC: (**a**) GaAs, (**b**) GaN, (**c**) quasi-cubic (CH_3NH_3)PbI_3, and (**d**) monolayer MoS_2. The band indices 'c', 'hh', 'lh', and 'so' stand for conduction, valence heavy-hole, light-hole, and split-off bands, respectively. The scale of band splittings is exaggerated.

Table 3. Length-gauge $|v^{(x)}_{mn}|^2$ and velocity-gauge $|p^{(x)}_{mn}|^2$ matrix elements (at.u.) in GaAs calculated using WIEN2k (with YSH) and VASP (with HSE06). Due to the non-local potential the velocity and the length gauges are *not* identical. The band degeneracy is given as a superscript in parentheses and the subscripts are explained in Figure 2a. The logarithmic deviation between $\sum|p^{(x)}_{vc}|^2$ and $\sum|v^{(x)}_{vc}|^2$ is given in parentheses (Δ as per Equation (10)).

Transition	$\sum\lvert v^{(x)}_{mn}\rvert^2$		$\sum\lvert p^{(x)}_{mn}\rvert^2$
	WIEN2k	VASP	
$\Gamma^{(\times 4)}_{lh,hh} - \Gamma^{(\times 2)}_{c}$	0.534	0.541	0.420
$\Gamma^{(\times 2)}_{so} - \Gamma^{(\times 2)}_{c}$	0.255	0.256	0.208
Total	0.789 (+23%)	0.797	0.628

3.3. Illustrative Applications

In the previous subsection we showed that calculations of optical properties for GaAs with the non-local hybrid XC functional (YSH or HSE06) require length-gauge optical matrix elements. If the momentum matrix elements had been used instead, the strength of optical transitions would have been underestimated by 23%. Next, we showed that a similar enhancement of the strength of direct optical transitions was also observed in other semiconductors, such as GaN, $(CH_3NH_3)PbI_3$, and monolayer MoS_2. The corresponding band structures are shown schematically in Figure 2b–d.

The band gap of GaN is well reproduced with YSH: $E_g^{YSH} = 3.19$ eV vs $E_g^{exp} = 3.30$ eV [29]. Compared to GaAs, the optical matrix elements in Table 4 showed an even larger disparity between the length-gauge velocity and the momentum matrix elements.

Table 4. Length-gauge $|v^{(\alpha)}_{mn}|^2$ and velocity-gauge $|p^{(\alpha)}_{mn}|^2$ matrix elements (at.u.) in GaN calculated using the YSH XC functional.

Transition	$\sum\lvert v^{(x)}_{mn}\rvert^2$	$\sum\lvert v^{(z)}_{mn}\rvert^2$	$\sum\lvert p^{(x)}_{mn}\rvert^2$	$\sum\lvert p^{(z)}_{mn}\rvert^2$
$\Gamma^{(\times 2)}_{hh} - \Gamma^{(\times 2)}_{c}$	0.211	0	0.183	0
$\Gamma^{(\times 2)}_{lh} - \Gamma^{(\times 2)}_{c}$	0.256	0.055	0.163	0.042
$\Gamma^{(\times 2)}_{so} - \Gamma^{(\times 2)}_{c}$	0.026	0.507	0.018	0.377
Total	0.493 (+30%)	0.562 (+29%)	0.364	0.419

The monolayer MoS_2 had a direct band gap at the $K = (1/3, 1/3, 1/3)$ point. Due to large excitonic effects [30], direct comparison of the YSH band gap $E_g^{YSH} = 2.22$ eV with experiment was not possible. Thus, we used a many-body result $E_g^{G_0W_0} = 2.53$ eV [31,32] as a reference. Similarly to other materials, the monolayer MoS_2 showed strong enhancement of the matrix elements (Table 5) with the YSH XC functional. Spin selection rules disabled half of the in-plane $v^{(x)}_{cv}$ matrix elements, while the out-of-plane matrix elements $v^{(z)}_{cv}$ were zero for transitions at the band edges due to symmetry arguments.

Table 5. Length-gauge $|v^{(x)}_{mn}|^2$ and velocity-gauge $|p^{(x)}_{mn}|^2$ matrix elements (at.u.) in monolayer MoS_2 calculated using YSH XC functional.

Transition	$\sum\lvert v^{(x)}_{mn}\rvert^2$	$\sum\lvert p^{(x)}_{mn}\rvert^2$
$K_{v_1} - K_{c_1}$	0	0
$K_{v_2} - K_{c_1}$	0.107	0.075
$K_{v_1} - K_{c_2}$	0.106	0.074
$K_{v_2} - K_{c_2}$	0	0
Total	0.213 (+36%)	0.149

The pseudo-cubic ($(CH_3NH_3)PbI_3$) had a direct gap at $R = (1/2, 1/2, 1/2)$. The calculated band gap $E_g^{\text{YSH}} = 1.03$ eV is an improvement relative to the PBE band gap (0.46 eV), but it is still far from the experimental 1.5–1.6 eV for the tetragonal phase [33,34]. This underestimation was due to the lack of stochastic thermal distortions of the PbI_6 octahedra, which further opened the gap by ca. 0.5 eV at room temperature [35,36]. Table 6 captures the matrix elements and their anisotropy caused by the reduced (pseudo-cubic) symmetry of the unit cell. Among all materials studied here, this material showed the lowest enhancement of the velocity matrix elements compared to the momentum matrix elements.

Table 6. Length-gauge $|v_{mn}^{(\alpha)}|^2$ and velocity-gauge $|p_{mn}^{(\alpha)}|^2$ matrix elements (at.u.) in pseudo-cubic $(CH_3NH_3)PbI_3$ calculated using the YSH XC functional.

Transition	$\sum \lvert v_{mn}^{(x,y,z)} \rvert^2$	$\sum \lvert p_{mn}^{(x,y,z)} \rvert^2$
$R_v^{(\times 2)} - R_c^{(\times 2)}$	0.195, 0.150, 0.128 (+11, +11, +8%) [a]	0.174, 0.135, 0.118

[a] The true enhancement should be about 22% due to inaccuracies in $|p_{mn}^{(x,y,z)}|^2$ values. See text below for more details.

It should be mentioned that the momentum matrix elements calculated with the `optic` module in the presence of SOC had an inaccuracy that progressively increased for heavier elements. The discrepancy between $|p_{mn}^{(x,y,z)}|^2$ values calculated at the PBE level (including SOC) with the `optic` module and using the finite difference overlap matrix reached ca. 12% in the case of $(CH_3NH_3)PbI_3$. The discrepancy fully vanished when SOC was excluded. After crosschecking the matrix elements with VASP we concluded that the finite difference results were correct. Since the `optic` module overestimated $|p_{mn}^{(x,y,z)}|^2$ values at PBE with SOC, the same applied to YSH with the SOC results presented in Table 6. After including this error, the true enhancement of YSH matrix elements for $(CH_3NH_3)PbI_3$ were about 22% (10% average enhancement in Table 6 and 12% `optic` error for this material). Additional calculations with VASP and HSE06 XC functional with SOC produced a very similar result (23% enhancement of the matrix elements).

4. Discussion

YSH length-gauge $|v_{m,n}^{\text{YSH}}|^2$ matrix elements were systematically greater than the momentum matrix elements $|p_{m,n}^{\text{YSH}}|^2$. The enhancement ranged from 22 to 36% in the following order: $(CH_3NH_3)PbI_3$, GaAs, GaN, and MoS_2 (from the smaller to higher enhancement). This trend prompted the hypothesis that the enhancement was related to the localization of states involved in the optical transition. $(CH_3NH_3)PbI_3$ has the most extended 5p-I and 6p-Pb states, while MoS_2 had the most localized 4d-Mo and 3p-S states at the band edges.

To gain further insight into the difference between $|p_{m,n}^{\text{YSH}}|^2$ and $|v_{m,n}^{\text{YSH}}|^2$ we wote the momentum matrix element in the length gauge. The corresponding operator was expressed as the commutator

$$\hat{\mathbf{p}} = i[\hat{T}, \mathbf{r}], \tag{11}$$

where $\hat{T}$ is the kinetic energy operator. Following the same logic that leads to Equation (4), we derived an equivalent expression for the momentum matrix element in the length-gauge

$$\mathbf{p}_{m,n}(\mathbf{k}) = i[T_m(\mathbf{k}) - T_n(\mathbf{k})]\langle m, \mathbf{k}|\mathbf{r}|n, \mathbf{k}\rangle. \tag{12}$$

After dividing Equation (4) by (12) we obtain

$$\mathbf{v}_{m,n}(\mathbf{k}) = \mathbf{p}_{m,n}(\mathbf{k}) \frac{E_m(\mathbf{k}) - E_n(\mathbf{k})}{T_m(\mathbf{k}) - T_n(\mathbf{k})}. \tag{13}$$

Thus, the 10 to 36% enhancement of the absolute squared magnitude of velocity matrix elements vs momentum matrix elements in calculations with YSH was directly related to the difference in eigenvalues not captured by the kinetic energy. In contrast, we expected the difference in eigenvalues to be fully captured by the kinetic energy, i.e., $[E_m(\mathbf{k}) - E_n(\mathbf{k})]/[T_m(\mathbf{k}) - T_n(\mathbf{k})] = 1$, when a local potential was employed. Interestingly, Equation (13) predicts an *isotropic* renormalization factor shared by all Cartesian directions ($\alpha = x, y, z$). Indeed, materials with anisotropic $v_{mn}^{(\alpha)}$—GaN (Table 4) and $(CH_3NH_3)PbI_3$ (Table 6)—showed a material-dependent yet isotropic enhancement factor, which was an indirect proof of the validity of Equation (13).

The renormalization of momentum matrix elements should have had implications for optical properties calculated with non-local potentials (hybrid or quasi-particle *GW*). The velocity matrix elements entered the frequency-dependent dielectric tensor (an imaginary part of the inter-band contribution) that took the following form in the independent-particle approximation [5]

$$\epsilon''_{\alpha\beta}(\omega) \propto \sum_{v,c} \int_{\mathbf{k}\in \mathrm{BZ}} \frac{v_{v,c}^{(\alpha)}(\mathbf{k}) v_{c,v}^{(\beta)}(\mathbf{k})}{\omega^2} \delta[E_c(\mathbf{k}) - E_v(\mathbf{k}) - \omega]\, d\mathbf{k}. \tag{14}$$

However, length-gauge $\mathbf{v}_{c,v}$ matrix elements are more difficult to compute than $\mathbf{p}_{c,v}$, especially at the quasi-particle *GW* level of theory where the finite difference method seems the only available technique [3]. Equation (13) opens a convenient possibility to use renormalized momentum matrix elements instead

$$\epsilon''_{\alpha\beta}(\omega) \propto \sum_{v,c} \int_{\mathbf{k}\in \mathrm{BZ}} \frac{p_{v,c}^{(\alpha)}(\mathbf{k}) p_{c,v}^{(\beta)}(\mathbf{k})}{[T_c(\mathbf{k}) - T_v(\mathbf{k})]^2} \delta[E_c(\mathbf{k}) - E_v(\mathbf{k}) - \omega]\, d\mathbf{k}, \tag{15}$$

provided that eigenstates are consistent with the potential, and their kinetic energy is known. The last expression should be valid not only for hybrid XC functionals but also for the quasi-particle *GW* level of theory.

Finally, we would like to comment on the renormalization of optical transition matrix elements proposed by Levine and Allan [37]

$$v_{v,c}^{GW} = v_{v,c}^{\mathrm{LDA/PBE}} \frac{(E_c - E_v)^{GW}}{(E_c - E_v)^{\mathrm{LDA/PBE}}} \tag{16}$$

that is further used in the literature [5,38]. If we apply Equation (16) to the $\Gamma_{so} - \Gamma_c$ transition in GaAs, one would expect the absolute squared magnitude of the velocity matrix element to increase by the ratio of $[(E_c^{\mathrm{YSH}} - E_{so}^{\mathrm{YSH}})/(E_c^{\mathrm{PBE}} - E_{so}^{\mathrm{PBE}})]^2$ which amounts to +151%. This result contradicts the +23% difference between $|p_{so,c}^{\mathrm{PBE}}|^2$ and $|v_{so,c}^{\mathrm{YSH}}|^2$ we observed (compare Tables 2 and 3). At the same time, the dipole matrix element $|\langle\Gamma_{so}|\mathbf{r}|\Gamma_c\rangle|^2$ becomes 127% smaller in YSH relative to PBE and counterbalances (partly) the effect of the gap opening. We further identified contributions of the muffin-tin spheres and of the interstitial volume to the value of the dipole matrix element $\langle\Gamma_{so}|\mathbf{r}|\Gamma_c\rangle$ at PBE and YSH levels of theory: 25% Ga, 44% As, and 31% interstitial. All contributions are in phase with each other, and the proportions remain unchanged from PBE to YSH. Equation (16), in contrast, implies the equality of dipole matrix elements $\langle m,\mathbf{k}|\mathbf{r}|n,\mathbf{k}\rangle^{\mathrm{LDA/PBE}} = \langle m,\mathbf{k}|\mathbf{r}|n,\mathbf{k}\rangle^{GW}$ (see Equation (4). Note that $v_{m,n}^{\mathrm{LDA/PBE}} = p_{m,n}^{\mathrm{LDA/PBE}} \approx p_{m,n}^{GW}$ [10]) leads to a gross overestimation of $v_{m,n}^{GW}$ matrix elements making them inconsistent with the band curvature [10].

5. Conclusions

Strong material-dependent enhancement of the oscillator strength (22–36% in the absolute squared magnitude) was observed in the electronic structure calculations of semiconductors with a hybrid XC functional. The origin of the enhancement was traced to the non-local Hartee–Fock exchange potential. The enhancement of the absolute squared magnitude of velocity matrix elements $|v_{m,n}^{\mathrm{YSH}}|^2$ vs momentum matrix elements $|p_{m,n}^{\mathrm{YSH}}|^2$ in calculations with non-local potentials was directly related to the difference in eigenvalues not captured by the kinetic energy, i.e., $[E_m(\mathbf{k}) - E_n(\mathbf{k})]^2/[T_m(\mathbf{k}) - T_n(\mathbf{k})]^2$. This enhancement is isotropic and can be readily included in a calculation of the dielectric function. Our enhancement factor was much more accurate than that previously proposed by Levine and Allan $(E_g^{GW}/E_g^{\mathrm{LDA}})^2$, which leads to nonphysically large $v_{m,n}^{GW}$ matrix elements.

Supplementary Materials: The following supporting information can be downloaded at: https://www.mdpi.com/article/10.3390/computation10020022/s1, a `SI-Computation.tar.gz` file with WIEN2k workflows (bash scripts), structure files, sample k-point files, and a python script for data processing.

Author Contributions: Conceptualization, O.R. and P.B.; investigation, O.R.; validation, P.B. and O.R.; writing—original draft, O.R.; writing—review and editing, O.R. and P.B. All authors have read and agreed to the published version of the manuscript.

Funding: O.R. acknowledges funding provided by the Natural Sciences and Engineering Research Council of Canada under the Discovery Grant Programs RGPIN-2020-04788. Calculations were performed using the Compute Canada infrastructure supported by the Canada Foundation for Innovation under John R. Evans Leaders Fund.

Institutional Review Board Statement: Not applicable.

Acknowledgments: Authors are thankful to Robert Laskowski (A*STAR, Singapore) and Fabien Tran (VASP Software GmbH, Vienna) for helpful discussions.

Conflicts of Interest: The authors declare no conflict of interest.

Abbreviations

The following abbreviations are used in this manuscript:

BZ	Brillouin zone
DFT	Density functional theory
FD	Finite difference
GGA	Generalized gradient approximations
HSE	Heyd, Scuseria, and Ernzerhof
LDA	Local-density approximation
PBE	Perdew, Burke, and Ernzerhof
SOC	Spin-orbit coupling
VASP	Vienna *ab initio* simulation package
XC	Exchange and correlation
YSH	Yukawa screened hybrid

References

1. Ambrosch-Draxl, C.; Sofo, J.O. Linear optical properties of solids within the full-potential linearized augmented planewave method. *Comput. Phys. Commun.* **2006**, *175*, 1–14. [CrossRef]
2. Starace, A.F. Length and velocity formulas in approximate oscillator-strength calculations. *Phys. Rev. A* **1971**, *3*, 1242–1245. [CrossRef]
3. Deslippe, J.; Samsonidze, G.; Strubbe, D.A.; Jain, M.; Cohen, M.L.; Louie, S.G. BerkeleyGW: A massively parallel computer package for the calculation of the quasiparticle and optical properties of materials and nanostructures. *Comput. Phys. Commun.* **2012**, *183*, 1269–1289. [CrossRef]
4. Laskowski, R.; Blaha, P. Calculations of NMR chemical shifts with APW-based methods. *Phys. Rev. B* **2012**, *85*, 035132. [CrossRef]
5. Rohlfing, M.; Louie, S.G. Electron-hole excitations and optical spectra from first principles. *Phys. Rev. B* **2000**, *62*, 4927–4944. [CrossRef]
6. Rhim, S.H.; Kim, M.; Freeman, A.J.; Asahi, R. Fully first-principles screened-exchange LDA calculations of excited states and optical properties of III-V semiconductors. *Phys. Rev. B* **2005**, *71*, 045202. [CrossRef]
7. Blaha, P.; Schwarz, K.; Madsen, G.K.H.; Kvasnicka, D.; Luitz, J.; Laskowski, R.; Tran, F.; Marks, L.D. *WIEN2k: An Augmented Plane Wave Plus Local Orbitals Program for Calculating Crystal Properties*; Vienna University of Technology: Vienna, Austria, 2018.
8. Blaha, P.; Schwarz, K.; Tran, F.; Laskowski, R.; Madsen, G.K.H.; Marks, L.D. WIEN2k: An APW+lo program for calculating the properties of solids. *J. Chem. Phys.* **2020**, *152*, 074101. [CrossRef]
9. Tran, F.; Blaha, P. Implementation of screened hybrid functionals based on the Yukawa potential within the LAPW basis set. *Phys. Rev. B* **2011**, *83*, 235118. [CrossRef]
10. Laurien, M.; Rubel, O. Importance of the nonlocal exchange potential for effective mass calculations in semiconductors: Benchmarking exchange-correlation potentials with the mstar60 dataset. *arXiv* **2021**, arXiv:2111.14772.

11. Kunes, J.; Arita, R.; Wissgott, P.; Toschi, A.; Ikeda, H.; Held, K. Wien2wannier: From linearized augmented plane waves to maximally localized Wannier functions. *Comput. Phys. Commun.* **2010**, *181*, 1888–1895. [CrossRef]
12. Hohenberg, P.; Kohn, W. Inhomogeneous Electron Gas. *Phys. Rev.* **1964**, *136*, B864–B871. [CrossRef]
13. Kohn, W.; Sham, L.J. Self-Consistent Equations Including Exchange and Correlation Effects. *Phys. Rev.* **1965**, *140*, A1133–A1138. [CrossRef]
14. Heyd, J.; Scuseria, G.E.; Ernzerhof, M. Hybrid functionals based on a screened Coulomb potential. *J. Chem. Phys.* **2003**, *118*, 8207. [CrossRef]
15. Krukau, A.V.; Vydrov, O.A.; Izmaylov, A.F.; Scuseria, G.E. Influence of the exchange screening parameter on the performance of screened hybrid functionals. *J. Chem. Phys.* **2006**, *125*, 224106. [CrossRef]
16. Demchenko, D.O.; Izyumskaya, N.; Feneberg, M.; Avrutin, V.; Özgür, Ü.; Goldhahn, R.; Morkoc, H. Optical properties of the organic-inorganic hybrid perovskite $CH_3NH_3PbI_3$: Theory and experiment. *Phys. Rev. B* **2016**, *94*, 075206. [CrossRef]
17. Dang, Y.; Liu, Y.; Sun, Y.; Yuan, D.; Liu, X.; Lu, W.; Liu, G.; Xia, H.; Tao, X. Bulk crystal growth of hybrid perovskite material $CH_3NH_3PbI_3$. *CrystEngComm* **2015**, *17*, 665–670. [CrossRef]
18. Whitfield, P.S.; Herron, N.; Guise, W.E.; Page, K.; Cheng, Y.Q.; Milas, I.; Crawford, M.K. Structures, Phase Transitions and Tricritical Behavior of the Hybrid Perovskite Methyl Ammonium Lead Iodide. *Sci. Rep.* **2016**, *6*, 35685. [CrossRef]
19. Averkieva, G.K.; Vaipolin, A.A.; Goryunova, N.A.; Prochukhan, V.D. Quinary Tetrahedral Semiconductors. In *Chemical Bonds in Solids*; Springer: New York, NY, USA, 1972; pp. 81–85. [CrossRef]
20. Juza, R.; Hahn, H. Über die Kristallstrukturen von Cu_3N, GaN und InN Metallamide und Metallnitride. *Z. Anorg. Allg. Chem.* **1938**, *239*, 282–287. [CrossRef]
21. Opalovskii, A.A.; Fedorov, V.E. Mixed molybdenum chalcogenides. *Dokl. Akad. Nauk SSSR* **1965**, *163*, 1163–1164.
22. Kresse, G.; Furthmüller, J. Efficient iterative schemes for ab initio total-energy calculations using a plane-wave basis set. *Phys. Rev. B* **1996**, *54*, 11169–11186. [CrossRef]
23. Kresse, G.; Joubert, D. From ultrasoft pseudopotentials to the projector augmented-wave method. *Phys. Rev. B* **1999**, *59*, 1758–1775. [CrossRef]
24. Blöchl, P.E. Projector augmented-wave method. *Phys. Rev. B* **1994**, *50*, 17953–17979. [CrossRef] [PubMed]
25. Perdew, J.P.; Burke, K.; Ernzerhof, M. Generalized Gradient Approximation Made Simple. *Phys. Rev. Lett.* **1996**, *77*, 3865–3868. [CrossRef] [PubMed]
26. *Band Gap of Gallium Arsenide (GaAs): Data Extracted from the Landolt-Börnstein Book Series and Associated Databases*; Springer: Berlin/Heidelberg, Germany, 2017.
27. Kim, Y.S.; Marsman, M.; Kresse, G.; Tran, F.; Blaha, P. Towards efficient band structure and effective mass calculations for III-V direct band-gap semiconductors. *Phys. Rev. B* **2010**, *82*, 205212. [CrossRef]
28. Yu, P.Y.; Cardona, M. *Fundamentals of Semiconductors: Physics and Materials Properties*; Springer: Berlin/Heidelberg, Germany, 2005.
29. *Band Gap of Gallium Nitride (GaN): Data Extracted from the Landolt-Börnstein Book Series and Associated Databases*; Springer: Berlin/Heidelberg, Germany, 2017.
30. Hill, H.M.; Rigosi, A.F.; Roquelet, C.; Chernikov, A.; Berkelbach, T.C.; Reichman, D.R.; Hybertsen, M.S.; Brus, L.E.; Heinz, T.F. Observation of Excitonic Rydberg States in Monolayer MoS_2 and WS_2 by Photoluminescence Excitation Spectroscopy. *Nano Lett.* **2015**, *15*, 2992–2997. [CrossRef]
31. Haastrup, S.; Strange, M.; Pandey, M.; Deilmann, T.; Schmidt, P.S.; Hinsche, N.F.; Gjerding, M.N.; Torelli, D.; Larsen, P.M.; Riis-Jensen, A.C.; et al. The Computational 2D Materials Database: High-throughput modeling and discovery of atomically thin crystals. *2D Mater.* **2018**, *5*, 042002. [CrossRef]
32. Gjerding, M.N.; Taghizadeh, A.; Rasmussen, A.; Ali, S.; Bertoldo, F.; Deilmann, T.; Knøsgaard, N.R.; Kruse, M.; Larsen, A.H.; Manti, S.; et al. Recent progress of the Computational 2D Materials Database (C2DB). *2D Mater.* **2021**, *8*, 044002. [CrossRef]
33. Ishihara, T. Optical properties of PbI-based perovskite structures. *J. Lumin.* **1994**, *60–61*, 269–274. [CrossRef]
34. Dittrich, T.; Awino, C.; Prajongtat, P.; Rech, B.; Lux-Steiner, M.C. Temperature Dependence of the Band Gap of $CH_3NH_3PbI_3$ Stabilized with PMMA: A Modulated Surface Photovoltage Study. *J. Phys. Chem. C* **2015**, *119*, 23968–23972. [CrossRef]
35. Wiktor, J.; Rothlisberger, U.; Pasquarello, A. Predictive Determination of Band Gaps of Inorganic Halide Perovskites. *J. Phys. Chem. Lett.* **2017**, *8*, 5507–5512. [CrossRef]
36. Zheng, C.; Yu, S.; Rubel, O. Structural dynamics in hybrid halide perovskites: Bulk Rashba splitting, spin texture, and carrier localization. *Phys. Rev. Mater.* **2018**, *2*, 114604. [CrossRef]
37. Levine, Z.H.; Allan, D.C. Linear optical response in silicon and germanium including self-energy effects. *Phys. Rev. Lett.* **1989**, *63*, 1719–1722. [CrossRef] [PubMed]
38. Adolph, B.; Gavrilenko, V.I.; Tenelsen, K.; Bechstedt, F.; Sole, R.D. Nonlocality and many-body effects in the optical properties of semiconductors. *Phys. Rev. B* **1996**, *53*, 9797–9808. [CrossRef] [PubMed]

computation

MDPI

Article

Performance Enhancement of APW+*lo* Calculations by Simplest Separation of Concerns

Long Zhang [1,2,3], Anton Kozhevnikov [4], Thomas Schulthess [4], Hai-Ping Cheng [1,2,3] and Samuel B. Trickey [1,2,3,*]

1 Department of Physics, University of Florida, Gainesville, FL 32611, USA; longzhang2008@gmail.com (L.Z.); hping@ufl.edu (H.-P.C.)
2 Quantum Theory Project, University of Florida, Gainesville, FL 32611, USA
3 Center for Molecular Magnetic Quantum Materials, University of Florida, Gainesville, FL 32611, USA
4 Swiss National Supercomputing Centre, 6900 Lugano, Switzerland; kozhevnikov@cscs.ch (A.K.); schulthess@cscs.ch (T.S.)
* Correspondence: trickey@qtp.ufl.edu

Abstract: Full-potential linearized augmented plane wave (LAPW) and APW plus local orbital (APW+*lo*) codes differ widely in both their user interfaces and in capabilities for calculations and analysis beyond their common central task of all-electron solution of the Kohn–Sham equations. However, that common central task opens a possible route to performance enhancement, namely to offload the basic LAPW/APW+*lo* algorithms to a library optimized purely for that purpose. To explore that opportunity, we have interfaced the Exciting-Plus ("EP") LAPW/APW+*lo* DFT code with the highly optimized SIRIUS multi-functional DFT package. This simplest realization of the separation of concerns approach yields substantial performance over the base EP code via additional task parallelism *without* significant change in the EP source code or user interface. We provide benchmarks of the interfaced code against the original EP using small bulk systems, and demonstrate performance on a spin-crossover molecule and magnetic molecule that are of size and complexity at the margins of the capability of the EP code itself.

Keywords: LAPW method; APW+*lo* method; all-electron DFT

Citation: Zhang, L.; Kozhevnikov, A.; Schulthess, T.; Cheng, H.-P.; Trickey, S.B. Performance Enhancement of APW+*lo* Calculations by Simplest Separation of Concerns. *Computation* **2022**, *10*, 43. https://doi.org/10.3390/computation10030043

Academic Editor: Henry Chermette

Received: 7 February 2022
Accepted: 4 March 2022
Published: 16 March 2022

1. Dedication

Much credit for the widespread use of full-potential linearized augmented plane wave (LAPW) methodology to solve the Kohn–Sham (KS) [1] equation for solids goes to Karlheinz Schwarz. The history of that contribution is related in Section 3 of Ref. [2]. It suffices to say here that Heinz started using the original APW in his thesis work, then came to Gainesville and Quantum Theory Project (QTP) in 1969 to work with Prof. J.C. Slater, the inventor of the APW method. That is how the last author of this paper became a collaborator with Heinz and a friend.

Years later, when linearized methods removed the explicit energy dependence difficulty in the APW basis, Heinz undertook development of the code that became WIEN [3]. Again there was a collaboration involving QTP, University of Florida, and the last author. Apparently that was the first publicly available FLAPW code. By now, it has evolved to WIEN2k [2,4]. During that evolutionary period, methodological developments led to revival of the APW scheme via the APW plus local orbitals (APW+*lo*) combination (summarized below). That has been instantiated in several other codes as well as WIEN2k, e.g., ELK [5], FlEUR [6], *exciting* [7], and Exciting-plus [8]. Here, we are pleased to contribute to the further advance of this important methodology and particularly pleased to be able to do so in honor of Heinz' birthday.

2. Motivating Physical Systems

The materials physics problem class driving our effort is molecular magnetism, in particular, the contriving of condensed aggregates of molecular magnets into materials of relevance to quantum information systems, notably, quantum computing [9,10]. The computational challenge is to predict promising molecular magnets [11,12], and promising aggregates of them as well as parametrizing spin Hamiltonians and aiding interpretation of experimental data. The molecules themselves are large and complicated.

A pertinent example is the molecular magnet $NiCl_2$-$[SC(NH_2)_2]_4$, dichloro-tetrakis-thiourea-nickel, commonly called "DTN" and its Co analogue, DTC [13]. DTN is important in this context because of its multi-ferroicity, the coexistence of ferromagnetism and ferroelectricity [14], in contrast with the absence of multi-ferroicity in DTC. The DTN cubic molecular crystal structure has two Ni atoms as magnetic centers in the unit cell. Each Ni has four S atoms and two Cl atoms as nearest neighbors, forming an octahedral structure (like the BO_6 octahedra in ABO_3 perovskites). The unit cell has 70 atoms and 444 electrons,

Spin-crossover systems are a closely related, highly relevant class, as they are candidate linkers for quantum information systems [15–17]. The electronic structure challenge is to calculate the low-spin, high-spin energy difference and provide the potential surface to calculate the vibrational entropic contributions. A particular significant example is the so-called [Mn(taa)] molecule ($[Mn^{3+}(pyrol)_3(tren)]$), a meridional pseudo-octahedral chelate complex of a single Mn as the magnetic center and the hexadentate tris[(E)-1-(2-azolyl)-2-azabut-1-en-4-yl]amine ligand. It has 53 atoms and 224 electrons. Calculating its spin-crossover energy with low-computational-cost, commonly used density functional methodology without user intervention and tuning has proven to be a formidable task [18].

While these two examples are convenient for the demonstration of capacity focus of this paper, they actually are a bit on the small side for the investigation of molecular magnetic materials in general. An illustration of that challenge is a molecule of current interest, the $[Mn_{12}O_{12}(O_2CPh)_{16}(H_2O)_4]$ complex [19,20]. It has spin $S = 10$ from 176 atoms and 1210 electrons.

Essential computational issues are made evident by these examples. The individual molecules are structurally and electronically intricate. Typically they have complicated spin manifolds that are strongly structurally dependent. Their condensed aggregates are correspondingly complicated and demanding. Moreover, the presence of heavy nuclei and the importance of anisotropy both implicate the significance of relativistic effects, including spin-orbit coupling. In sum, predictive, materials-specific simulations of condensed magnetic molecule systems and spin-crossover systems are extremely challenging tasks.

In the remainder of this paper, we describe the context and need for all-electron computational methods with emphasis on LAPW and APW+*lo* methodology, then discuss impediments to use of existing codes on the physical systems of interest, introduce separation of concerns as identifying and off-loading algorithmic elements common to any LAPW/APW+*lo* code, and the SIRIUS package used as a library for that off-loading, show how interfacing between the Exciting-Plus code and SIRIUS can be achieved, and give numerical examples and timings for the combination.

3. Predictive Computational Approaches

Balance of computational cost-effectiveness and accuracy in treatment of electronic structure of challengingly complicated systems is the pragmatic reason for prevalent use of density functional theory (DFT) [21] in its KS variational form [21–24]. In the context motivating this work, accuracy is crucial for predicting both structural properties and characterizing spin manifolds. The primary choices regarding accuracy are the selection of an exchange-correlation (E_{xc}) approximation and selection of a method for solving the resulting KS equations. We address the second. The first is an arena of intense effort that has provided many options.

Most "electronic structure methods" come down to the choice of a basis set (and its truncation) by which to reduce the KS equation to a linear algebra problem. The obvious, naïve basis for periodic systems is plane waves. It provides systematic enrichment and is unbiased with respect to ionic charge. The long-known limitation is that the basis becomes unmanageably large if the oscillations of near-nucleus orbitals caused by the bare Coulomb potential are included [25]. This burden is removed by use of a pseudo-potential instead [26] or, more recently, use of projector augmented waves (PAWs) [27]. Widely used examples of such "plane-wave pseudo-potential-PAW" (PW-PP-PAW) codes include VASP [28], QuantumEspresso [29], and ABINIT [30–32].

Such calculations intrinsically are not truly all-electron. Pseudo-potentials eliminate core states, while PAWs reconstruct them. There is a need therefore to test and cross-check plane-wave pseudo-potential calculations against truly all-electron calculations. For only three examples of many, see Refs. [33–35]. Another example is a comparatively early use of all-electron calculations for materials-by-design [36]. That was a study of Li-ion battery formulations with the WIEN2k code. Though nontrivial (especially at the time), at 14 atoms per cell with 170–178 electrons, those systems were smaller than the motivating examples discussed above. Cross-validation is particularly important in the context of molecular magnetic quantum materials because of core contributions to spin manifolds and spin-orbit interaction effects.

The all-electron methodology of choice is the LAPW method or its close kin, APW+*lo* [37]. Basis set construction is by use of the "muffin-tin" potential, the spherical average of the KS potential in non-overlapping, nuclear-centered spheres and a constant average elsewhere (the "interstitial" region). "Full-potential" denotes use of the whole KS crystalline potential, not just the so-called muffin-tin (MT) part. Historically that was an important distinction but today the MT potential is used only for basis set construction. Both LAPW and APW+*lo* are rooted in Slater's original APW scheme [38–43]. Within the MT spheres, all three sets have basis functions that are atomic-like solutions of spherical potentials. Those are matched with plane waves in the interstitial region.

Original LAPW literature is extensive, see Refs. [44–58]. Subsequently there was a particular kind of local orbital ("LO", not "*lo*") added [59], and then the closely related APW+*lo* scheme [60]. These are covered in at least two other books [37,61] as well as various review chapters (e.g., Refs. [2,62]). Therefore here we display only those equations directly relevant to our discussion of codes and algorithmic libraries.

The original APW basis function for Bloch wave-vector $\boldsymbol{k}$ and plane-wave vector $\boldsymbol{G}$ is

$$\begin{aligned} \varphi_{\boldsymbol{k}}^{\boldsymbol{G}}(\boldsymbol{r}) &:= \sum_{\ell,m}\sum_{\nu} A_{\ell m\nu}^{\alpha,\boldsymbol{k}}(\boldsymbol{G}) u_{\ell\nu}^{\alpha}(r) Y_{\ell m}(\hat{\boldsymbol{r}}) \quad \boldsymbol{r} \in \alpha \\ &= (1/\sqrt{\Omega}) e^{i(\boldsymbol{G}+\boldsymbol{k})\cdot\boldsymbol{r}} \quad \boldsymbol{r} \notin \alpha \end{aligned} \tag{1}$$

Here $u_{\ell\nu}^{\alpha}(r)$ is the solution of the (energy-dependent, ϵ) radial Schrodinger equation in the MT sphere labeled α, that is regular at the origin with principle quantum number ν, $Y_{\ell m}(\hat{\boldsymbol{r}})$ are spherical harmonics, $A_{\ell m\nu}^{\alpha,\mathbf{k}}(\mathbf{G})$ are the coefficients for matching with the interstitial plane wave, ℓ and m are the azimuthal and magnetic quantum numbers in a particular sphere. (The APW basis does not have continuous radial first derivatives at the sphere surfaces). Since the radial functions are ϵ-dependent, continuity at the sphere surface requires that energy to correspond to a KS eigenvalue. The APW secular equation thus is highly non-linear in the one-electron energies. That non-linearity induces both an important computational computational cost and difficult-to-manage singularities whenever a radial basis function has a node on a sphere surface.

The LAPW basis addresses those difficulties by using both the radial functions $u(r;\epsilon_\ell)$ at reference energies ε_ℓ and their energy derivatives

$$\dot{u}_{\ell\nu} := \frac{\partial u_{\ell\nu}(r;\epsilon)}{\partial\epsilon}\Big|_{\epsilon_\ell} . \tag{2}$$

(The "dotted" notation for the energy derivative is conventional in LAPW literature). Thus the basis functions become

$$\begin{aligned}\varphi_{k}^{G}(r) &:= \sum_{\ell,m}\sum_{\nu}[A_{\ell m\nu}^{\alpha,k}(G)u_{\ell\nu}^{\alpha}(r;\epsilon_{\ell}) + B_{\ell m\nu}^{\alpha,k}(G)\dot{u}_{\ell\nu}^{\alpha}(r,\epsilon_{\ell})]Y_{\ell m}(\hat{r}) \quad r \in \alpha \\ &:= \frac{1}{\sqrt{\Omega}}\exp^{i(G+k)\cdot r} \quad r \notin \alpha\,. \end{aligned} \tag{3}$$

The coefficients follow from making each basis function continuous with continuous first radial derivative at each sphere boundary. Unlike the APW case, the KS secular equation in the LAPW basis is of the ordinary linear variational form. The only user-dependent choices are the reference energies ϵ_{ℓ} and muffin-tin radii.

As the LAPW linearization is not unique, exploration of options eventually led [60] to the recognition that a more efficient linearization combines the original APW basis functions inside the sphere *at fixed reference energies* with a set of linearized (in energy) radial functions inside the sphere, each of which vanishes at the sphere surface. This "APW+*lo*" basis consists of Equation (1) enhanced with different local orbitals ("*lo*"),

$$\varphi_{lo}(r) := \sum_{\ell,m}\sum_{\nu}[A_{\ell m\nu}^{\alpha}u_{\ell\nu}^{\alpha}(r;\epsilon_{\ell}) + B_{\ell m\nu}^{\alpha}\dot{u}_{\ell\nu}^{\alpha}(r,\epsilon_{\ell})]Y_{\ell m}(\hat{r})\,, \quad r \in \alpha\,. \tag{4}$$

These localized basis functions do not have continuous derivatives at the sphere boundary, so surface terms arise in the kinetic energy and in any gradient-dependent exchange correlation functional.

The LAPW and APW+*lo* basis sets can be used together with suitable reference energy choices and consideration of the atomic structure differences among spheres. Observe that both basis sets are adaptive in that the radial functions evolve as the KS potential evolves from SCF iteration to iteration.

The forms of the electron number density $n(r)$ and the KS effective potential $v_{KS}(r)$ matrix elements in these basis sets are given in Appendix A for completeness.

4. Codes and Libraries

4.1. Base Code

The present work focuses on the *Exciting-Plus* code, hereafter "EP" for brevity [8]. EP was developed from an early version of the ELK/*exciting* code, that was branched at the time when independent evolution of *exciting* and ELK had just begun. EP was developed with emphasis on post-ground-state calculations such as for the density response function [63] and RPA [8] and GW [64] calculations. Ground state KS calculations are done in EP with k-point task distribution and LAPACK [65] diagonalization support. EP also implemented a convenient *mpi-grid* task parallelization in several independent variable dimensions, e.g., k-points, i-j index pairs of KS states, and q points in the calculation of the KS density response function.

EP was constructed conscientiously in terms of coding practices. However, its design did not focus on high performance for multi-atom unit cells. Our context makes that important. Our goal is to retain the features and capabilities of EP while making it fast enough for routine all-electron DFT calculations to be feasible for large, complicated systems such as the magnetic molecules, spin-crossover molecules, and aggregates discussed at the outset.

4.2. Separation of Concerns and the SIRIUS Package

LAPW/APW+*lo* codes evidently share their central formalism. Because their basis sets start from plane waves, those codes also share significant procedural elements with PW-PP-PAW codes. Shared tasks include unit cell setup, atomic configurations, definition and generation of reciprocal lattice vectors G, combinations with Bloch vectors $G + k$, definition of basis functions on regular grids as Fourier expansion coefficients, construction of the plane wave contributions to the KS Hamiltonian matrix, generation of the charge density, effective potential, and magnetization on a regular grid, iteration-to-iteration mixing schemes for density and potential, and diagonalization of the secular equation. Compared to PW-PP-PAW codes, LAPW/APW+*lo* codes additionally have spatially decomposed basis sets as outlined above.

These extensive commonalities constitute an opportunity for performance enhancement via separation of concerns. Computer scientists can bring their skills to bear on the shared algorithmic core of LAPW/APW+*lo* methodology while computational materials physicists can focus on implementation of analysis, post-processing, better exchange-correlation functionals, etc.

With achievement of the benefits of this separation in mind, an optimized package, SIRIUS [66], was created by some of us. It has explicit, focused, highly refined implementation of LAPW/APW+*lo* commonalities (and PW-PP-PAW to the extent of the broader commonality) as the goal. That is, abstracting and encapsulating objects common to LAPW and APW+*lo* as the design objective for SIRIUS. By concept, it had both task parallelization and data parallelization. It has been optimized for multiple MPI levels as well as OpenMP parallelization and for GPU utilization.

SIRIUS can be used two ways, as a library or as a simple LAPW/APW+*lo* code. Elsewhere, we will report on its use in the latter way [67]. In that case the compromise involved is to accept the functionality limits of SIRIUS in return for being able to handle very large systems by both task and data parallelism. Here we report on exploitation of SIRIUS purely as a DFT library by construction of an EP-SIRIUS interface using the SIRIUS API. The expected gain is speed-up while retaining the familiar user-interface and post-processing functionalities of EP. Figure 1 illustrates the scheme. The intrinsic limitation of separation of concerns is that the resulting package has limitations that, in essence, are the union of the limitations of the host code and of the library. We discuss that briefly at the end.

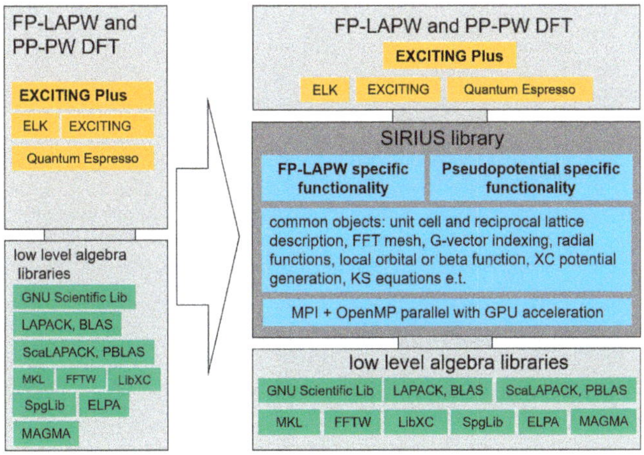

Figure 1. General scheme for utilization of SIRIUS as a library to enhance performance of a host code.

4.3. SIRIUS Characteristics and Features

SIRIUS is written in C++ in combination with the CUDA [68] back-end to provide (1) low-level support (e.g., pointer arithmetic, type casting) as well as high-level abstractions (e.g., classes and template meta-programming); (2) easy interoperability between C++ and widely used Fortran90; (3) full support from the standard template library (STL) [69]; and (4) easy integration with the CUDA nvcc compiler [70]. The SIRIUS code provides dedicated API functions to interface to *exciting* and to QuantumEspresso [29,71,72].

Virtually all KS electronic structure calculations rely at minimum on two basic functionalities: distributed complex matrix-matrix multiplication (e.g., `pzgemm` in LAPACK) and a distributed generalized eigenvalue solver (e.g., `pzhegvx` also in LAPACK). SIRIUS handles these two major tasks with data distribution and multiple task distribution levels.

The eigenvalue solver deserves particular attention. Development of *exciting* led to significant code facilities to scale the calculation to larger numbers of distributed tasks than originally envisioned by making the code switchable from LAPACK to ScaLAPACK. This can be verified by comparing the task distribution and data distribution of the base ground state subroutine in recent versions of *exciting* (version Nitrogen for example) and ELK (version 5.2.14 or earlier for example). ELK development appears to have emphasized physics features and functionalities rather than adding ScaLAPACK support. The EP situation is similar. It has only LAPACK support and does not have data distribution of large arrays. Table 1 summarizes the diagonalization methods available in these codes.

Table 1. Eigensolver options.

	Full Diagonalization		**Iterative Diagonalization**
	LAPACK	ScaLAPACK	Davidson algorithm
Exciting-Plus	Yes	No	No
ELK	Yes	No	Yes
exciting	Yes	Yes	Yes
SIRIUS	Yes	Yes	Yes

Eigenvalue solver performance depends strongly upon the algorithm type. Widely used linear algebra libraries (e.g., LAPACK, ScaLAPACK) implement robust full diagonalization. They can handle system size up to about 10^6. Unfortunately for LAPW/APW+*lo* calculations on systems as large as 100+ atoms, the eigensystem often is several times larger. A Davidson-type iterative diagonalization algorithm is appropriate in that case because it typically suffices to solve for the lowest 10–20 percent of all occupied eigenvalues and associated eigenvectors up through and somewhat above the Fermi energy.

Davidson-type diagonalization algorithms are available in some APW+*lo* and PW-PP-PAW codes, e.g., WIEN2k [73] and PWscf [74] respectively. They are not offered in standard linear algebra libraries however. At least in part that is because such algorithms repeatedly apply the Hamiltonian to a sub-space of the system. Therefore the algorithm depends upon details of the Hamiltonian matrix, hence upon the specific basis-set formalism. By virtue of focus on tasks central to LAPW/APW+*lo* and PW-PP calculations, the SIRIUS package can provide an efficient implementation of Davidson-type diagonalization [75] for LAPW/APW+*lo* and PW-PP-PAW codes.

4.4. *Interfacing Exciting-Plus with SIRIUS*

Despite its many attractive features, especially for important post-ground state calculations, EP has some significant limitations in regard to ground state calculations on large systems such as magnetic and spin-crossover molecules. Those limits include: (1) provision of only the LAPACK eigensolver; and (2) $\boldsymbol{k}$-point-only MPI parallelization. This second limit renders the code completely serial for single $\boldsymbol{k}$-point calculations, e.g., on an isolated molecule in a big cell.

We frame the task therefore as straight-forward interfacing to SIRIUS as an unaltered library with comparatively minimal modification of EP. This black-box approach is pure separation of concerns, since it is the simplest route an experienced EP user could take to try to gain advantage from SIRIUS without investing effort in learning its inner workings. A benefit is that the user interface to EP+SIRIUS is essentially unaltered EP, yet the combined system provides (a) ScaLAPACK support, (b) Davidson iterative eigensolver, (c) band MPI parallelization for one $\boldsymbol{k}$-point, and (d) thread-level OMP parallelization per k-point per band. It also exposes some oddities introduced by the black-box strategy.

Interface implementation benefits from the FORTRAN API functionalities provided by SIRIUS. Listing 1 displays the FORTRAN API function calls for parsing the atomic configuration, the APW and *lo* basis from EP, and passing them to SIRIUS.

Listing 1: Setting up atomic configuration.

```
call sirius_set_atom_type_coniguration(sctx, string(trim(label)),&
& spn(ist,is), spl(ist,is),spk(ist,is), spocc(ist,is),&
& logical(spcore(ist,is),kind=c_bool))
enddo
```

The code segment in Listing 1 loops over the number of states of a single atom type atom (`spnst`: species' number of states). For each state, the API call provides to SIRIUS the quantum numbers n, l and k for each state (`spn`, `spl` and `spk`), the occupation of that state (`spocc`), and whether that state is treated as a core state (`spcore`).

The first of the two double loops in the code chunk shown in Listing 2 goes over the APWs of one atom type and the ℓ-channels of each APW. For each ℓ-channel, the API call passes the following information to SIRIUS: principle quantum numbers n (`apwpqn`), value of ℓ (`l`), value of the initial linearization energy (`apwe0`), the order of energy derivative of that APW (`apwdm`), and whether the linearization energy is allowed to be adjusted automatically (`autoenu`). The second loop is over the total number of local orbitals (`nlorb`) of one atom type and the *orders* (`lorbord`) of each local orbital (i.e., number of $u(r)$ or $\dot{u}(r)$ terms in that local orbital). The API call passes the following information to SIRIUS: quantum numbers n and l (`lopqn` and `lorbl`), initial linearization energy (`lorbe0`), order of energy derivative (`lorbdm`), and whether the linearization energy is allowed to be adjusted automatically (`autoenu`).

Listing 2: Fortran API for basis description.

```
! parsing APW descriptions from host code to SIRIUS
do l = 0, lmaxapw
do io = 1, apword(l, is)
autoenu = .false.
if (use_sirius_autoenu.and.apwve(io,l,is)) autoenu = .true.
call sirius_add_atom_type_aw_descriptor(sctx, string(trim(label)),&
&apwpqn(l,is), l, apwe0(io, l, is), apwdm(io, l, is),&
&logical(autoenu,kind=c_bool))
enddo
enddo
! parsing LO/lo description from host code to SIRIUS
do ilo = 1, nlorb(is)
do io = 1, lorbord(ilo, is)
autoenu = .false.
if (use_sirius_autoenu.and.lorbve(io, ilo, is)) autoenu = .true.
call sirius_add_atom_type_lo_descriptor(sctx, string(trim(label)),&
&ilo, lopqn(ilo,is), lorbl(ilo, is),lorbe0(io, ilo, is),&
&lorbdm(io, ilo, is), logical(autoenu,kind=c_bool))
enddo
enddo
```

General input parameters such as the plane-wave cutoff, ℓ cutoff for the APWs and for density and potential expansion, $\boldsymbol{k}$-points, lattice vectors and atom positions, etc., all are set as usual in the EP input file. Then they are passed to SIRIUS via its built-in `import` and `set parameter` functionalities. Other important parameters such as the fast Fourier transform grid, radial function grid inside each MT sphere, and number of first variational states [37] often are not set in EP input files but defaulted. For EP+SIRIUS, however, those also must be passed to SIRIUS in the initialization step to ensure that the Hamiltonian matrix and eigenvectors are precisely the same in EP and SIRIUS. Other information such as specification of core states, linearization energy values and MT radii defined in the so-called species files of EP is passed to SIRIUS at the beginning of the calculation to overwrite the corresponding SIRIUS default values. Consider Listing 3 therefore.

Listing 3: Fortran API for setting inputs for SIRIUS.

```
call sirius_set_parameters( sctx,&
&use_symmetry=bool(.true.),&
&valence_rel=string('zora'),&
&core_rel=string('none'),&
&auto_rmt=0,&
&fft_grid_size=ngrid(1),&
&num_mag_dims=ndmag,&
&num_fv_states=nstfv,&
&pw_cutoff=gmaxvr,&
&gk_cutoff=gkmax,&
&lmax_apw=lmaxapw,&
&lmax_rho=lmaxvr,&
&lmax_pot=lmaxvr )
```

The code chunk shown in Listing 3 is an example of basic inputs that are added to EP in the initialization step, in the piece of code named init0.f90. Most of the meanings are explicit in the name. `zora` means zero-order relativistic approximation. `ngrid` is the FFT grid set up in EP and passed to SIRIUS. Plane wave cutoff and $|G + k|$ cutoff values are `gmaxvr` and `gkmax` in EP. The `lmaxapw` and `lmaxvr` are the angular momentum cutoff for APW and for charge density (and potential) inside the MT.

The inserted code shown in Listing 4 supplies SIRIUS with additional parameters for the Davidson method if it is used. After ensuring that the setup of input quantities is identical between the host code (EP) and SIRIUS, the ground state calculation is done solely by SIRIUS. The results, eigenvalues and eigenvectors, are passed back to EP for further calculation.

Listing 4: Eigen-solver selection and Davidson solver parameter setup.

```
if (sirius_davidson_eigen_solver) then
call sirius_import_parameters(sctx, string('{''iterative_solver'' : {''type'' :
    ''davidson'' }}'))
call sirius_import_parameters(sctx, string('{''iterative_solver'' :
    {''energy_tolerance'' : 1e-13}}'))
call sirius_import_parameters(sctx, string('{''iterative_solver'' :
    {''residual_tolerance'' : 1e-6}}'))
call sirius_import_parameters(sctx, string('{''iterative_solver'' : {''num_steps''
    : 32}}'))
call sirius_import_parameters(sctx, string('{''iterative_solver'' :
    {''subspace_size'' : 8 }}'))
call sirius_import_parameters(sctx, string('{''iterative_solver'' :
    {''converge_by_energy'' : 1 }}'))
call sirius_import_parameters(sctx, string('{''iterative_solver'' :
    {''num_singular'' : 20 }}'))
else
! otherwise use full eigen solver from LAPACK
call sirius_set_parameters(sctx, iter_solver_type=string('exact'))
endif
```

Next we display, in Listing 5, a code segment with the typical API calls from EP to retrieve the resulting eigenvalues and eigenvectors. It is inserted in the ground state subroutine, the piece of code named gndstate.f90.

Listing 5: Fortran API for retrieving eigenvalues and eigenvectors from SIRIUS.

```
! get local fraction of eigen-vectors
do ikloc=1,nkptloc
ik=mpi_grid_map(nkpt,dim_k,loc=ikloc)
call sirius_get_fv_eigen_vectors(ks_handler, ik, evecfvloc(1, 1, 1, ikloc),
    nmatmax, nstfv)
call sirius_get_sv_eigen_vectors(ks_handler, ik, evecsvloc(1, 1, ikloc), nstsv)
enddo !ikloc
! get all eigen-values and band occupancies
do ik = 1, nkpt
if (ndmag.eq.0.or.ndmag.eq.3) then
call sirius_get_band_energies(ks_handler, ik, 0, evalsv(1, ik))
call sirius_get_band_occupancies(ks_handler, ik, 0, occsv(1, ik))
else
call sirius_get_band_energies(ks_handler, ik, 0, evalsv(1, ik))
call sirius_get_band_energies(ks_handler, ik, 1, evalsv(nstfv+1, ik))
call sirius_get_band_occupancies(ks_handler, ik, 0, occsv(1, ik))
call sirius_get_band_occupancies(ks_handler, ik, 1, occsv(nstfv+1, ik))
endif
enddo
```

Care is needed in dealing with the MPI task schedules when interfacing to SIRIUS as a library because typically the host code will have an MPI implementation that differs from that in SIRIUS. For EP as the host, the task is simplified because EP has only *k*-point parallelization in the ground-state calculation. In the initialization step, we set the SIRIUS MPI communicator to be derived from the global MPI communicator (MPI_COMM_WORLD) of the host code so that all MPI ranks will be used by SIRIUS. Then the user needs to specify how SIRIUS will carry out the *k*-point distribution, how to plan further band parallelization within a *k*-point, and thread-level parallelization. The schedules of *k*-point parallelization and band parallelization are required additional inputs. Thread-level parallelization also has additional inputs which are specified in the run job script.

If band parallelization is used in SIRIUS, the eigenvalues and eigenvectors associated with a single *k*-point are distributed in multiple MPI tasks. It therefore is necessary to combine the band subset results before transmitting the eigenvalues and eigenvectors back to EP. Thus, after SIRIUS finishes the ground state calculation but before calling the API to return the eigenvalues and eigenvectors to EP, SIRIUS will do *mpi_reduce* in the MPI band dimension and prepare full eigenvalues and eigenvectors labeled by *k*-points and by the global band index at each *k*-point.

The last piece of the interface provides the additional inputs for the SIRIUS Davidson diagonalization algorithm. These are adjustable numerical parameters passed directly to SIRIUS by EP.

As anticipated, the MPI parallelization in the band degree of freedom is one major gain from interfacing EP to SIRIUS. We noted above that EP runs entirely in non-parallel mode for a single *k*-point calculation (often a "Gamma-point calculation" or "Balderschi-point calculation"), such as is typical for isolated molecule calculations. Hence the SIRIUS-enhanced-EP has the same scaling as SIRIUS alone in the case of single *k*-point calculations. This is an example of the antithesis of the union of limitations that is inherent in separation of concerns. Here, separation of concerns actually avoids a limitation of the host code.

To illustrate, Figure 2 displays the benchmark of band-parallelization on the DTN molecule (brief details about the molecule are below). It is placed in a $10 \times 10 \times 10$ Å cubic unit cell, with plane-wave cutoff $20\,a_0^{-1}$ (inverse Bohr radius) and angular momentum cutoff $= 7$. All jobs were set to 16 multi-threads in one task in accord with the hardware configuration. The recorded time is for the first 100 SCF iterations using the Davidson diagonalization eigensolver. Note that the figure also shows that employment of EP as a front-end to SIRIUS does not introduce any significant overhead. The timings for EP+SIRIUS are almost identical to those for SIRIUS alone. Timings compared to PW-PP-PAW codes are in the next section.

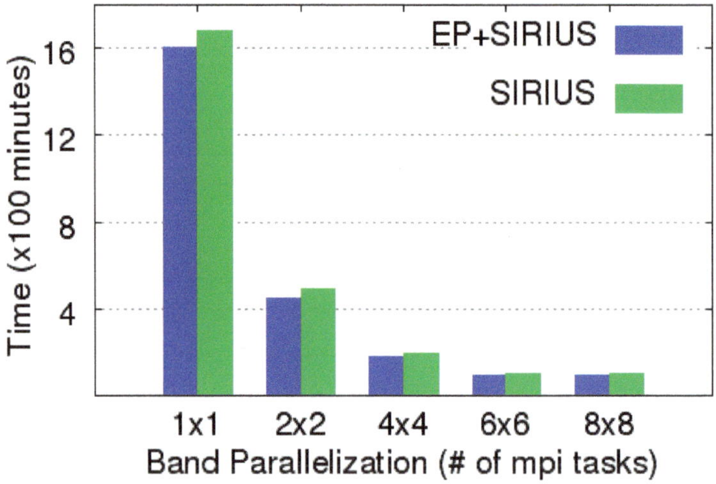

Figure 2. Benchmark of band parallelization in single k-point jobs. $n \times n$ ranks are used for one k-point.

5. EP+SIRIUS: Verification Tests on Small Solid State Systems

The EP+SIRIUS combination was bench-marked first against SIRIUS standalone on ground-state calculations of the total energy (and magnetization for magnetic systems) for the simple bulk materials Al, Ni, Fe, NiO, C, Si, Ge, and GaAs. For each system, identical input parameters were used for the SIRIUS and EP+SIRIUS runs. To be systematic, we adopted the experimental lattice parameters for all systems. The APW+*lo* and LAPW bases were used. Both local density approximation (LDA) and generalized gradient approximation (GGA; PBE [76]) exchange-correlation functionals were used. In the interstitial potential and charge density expansions, the maximum length of the reciprocal lattice vector $|G|$ used as plane wave cut-off for the APW was set to $12\,a_0^{-1}$ for all systems. The angular momentum truncation was taken as $\ell_{\max} = 8$ for APW, with the same value used for the charge density, potential, and orbital inside the MT sphere. The linearization energy associated with each APW radial function was chosen at the center of the corresponding band with ℓ-like character for all systems. Sampling of the first Brillouin zone was by a dense $16\,16 \times 16$ k-mesh for all systems. All parameters were tested carefully to achieve total energy convergence (tolerance $= 10^{-6}$ Hartree). For the EP+SIRIUS calculations, diagonalization always was done with the Davidson iterative eigensolver.

Table 2 summarizes the APW+*lo* basis configuration, settings other than those already stated, and the converged total energy and magnetization of these small systems. Table 3 summarizes the same calculation setup with the LAPW basis. The good agreements in total energy and magnetization in these tests validate the assumption that the identical basis setup was invoked and that the constructed interface linked the SIRIUS calculation properly with the EP host code.

Table 2. EP+SIRIUS vs. SIRIUS, using APW+*lo*.

	Al (Non-Mag.)	**NiO (Non-Mag.)**	**Ni (FM)**	**Fe (FM)**
crystal structure	fcc	rock-salt	fcc	bcc
latt. const. (Å)	4.05	4.17	3.52	2.87
R_{mt} (a_0)	1.8	1.8, 1.6	2.0	2.0
valence relativity	z.o.r.a.	z.o.r.a.	z.o.r.a.	z.o.r.a.
lo config.	Al: s, p	Ni: s, p, d O: s, p, d	Ni: s, p, d	Fe: s, p, d
LO for semi-core	$\epsilon'_{2p} = -2.55$	$\epsilon'_{Ni,3d} = -0.33$ $\epsilon'_{O,2s} = -0.87$	$\epsilon'_{Ni,3d} = -0.33$ $\epsilon'_{Ni,3p} = -2.59$	$\epsilon'_{Fe,3d} = -0.28$ $\epsilon'_{Fe,3p} = -2.18$ $\epsilon'_{Fe,3s} = -3.43$
treated as core state	1s, 2s	Ni: 1s, 2s, 2p, 3s O: 1s	1s, 2s, 2p, 3s	1s, 2s, 2p
LDA: (unit: Ha, μ_B) E_{tot}, SIRIUS E_{tot}, EP+SIRIUS μ_{tot}, SIRIUS μ_{tot}, EP+SIRIUS	 −241.40085447 −241.40085447	 −1593.13659104 −1593.13659102	 −1518.09194282 −1518.09194282 0.564822 0.564825	 −1270.11766996 −1270.11766997 2.308247 2.308245
GGA-PBE: (unit: Ha, μ_B) E_{tot}, SIRIUS E_{tot}, EP+SIRIUS μ_{tot}, SIRIUS μ_{tot}, EP+SIRIUS	 −241.54245824 −241.54245824	 −1593.27058366 −1593.27058365	 −1518.15356943 −1518.15356942 0.563466 0.563467	 −1270.18442575 −1270.18442575 2.327534 2.327530

Table 3. EP+SIRIUS vs. SIRIUS, using LAPW.

	Al (Non-Mag.)	**NiO (Non-Mag.)**	**Ni (FM)**	**Fe (FM)**
R_{mt} (a_0)	1.8	1.8, 1.6	2.0	2.0
	LAPW has same linearization energy as APW. No more *lo* configurations. *LO* configuration is same as Table 2. other parameters are also same as in Table 2			
LDA: (unit: Ha, μ_B) E_{tot}, SIRIUS E_{tot}, EP+SIRIUS μ_{tot}, SIRIUS μ_{tot}, EP+SIRIUS	 −241.40085321 −241.40085422	 −1593.13659761 −1593.13659902	 −1518.09194596 −1518.09194752 0.564830 0.564827	 −1270.11766882 −1270.11766822 2.308243 2.308240
GGA-PBE: (unit: Ha, μ_B) E_{tot}, SIRIUS E_{tot}, EP+SIRIUS μ_{tot}, SIRIUS μ_{tot}, EP+SIRIUS	 −241.54245124 −241.54245372	 −1593.27058546 −1593.27058701	 −1518.15356717 −1518.15356932 0.563460 0.563466	 −1270.18442211 −1270.18442394 2.327533 2.327528

6. EP+SIRIUS: Two Molecular Examples

6.1. [Mn(taa)] Molecule

As the first known example of a manganese(III) d^4 spin-crossover system [77], [Mn(taa)] is a system of long-standing interest. Experiment shows that the Mn^{3+} cation goes from a low-spin state (LS) to a high-spin state (HS) at a transition temperature of about 45 K.

The [Mn(taa)] structure (see Figure 3) is sufficiently large that it has non-negligible intra-molecular dispersion interactions with significant HS-LS dependence. The HS ground state involves anti-bonding molecular orbital occupation, hence the octahedral HS complex tends to have weaker and therefore longer metal-ligand bonds than in the LS ground state.

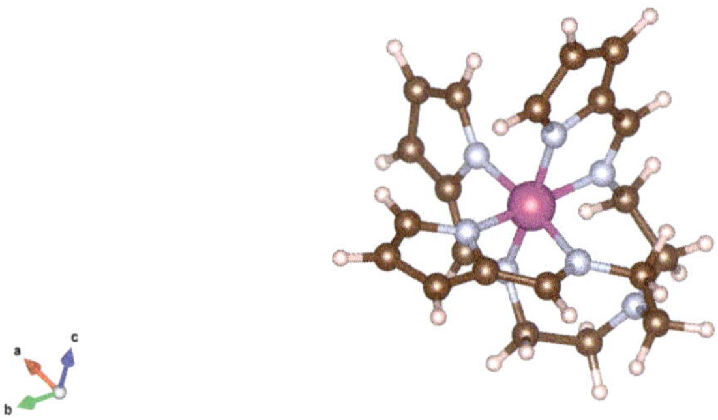

Figure 3. [Mn(taa)] molecule.

This combination of spin- and structural dependence makes [Mn(taa)] a significant challenge to the computational determination of the two ground states. The purely molecular (non-thermal) $\Delta E_{HL} := E_{HS} - E_{LS}$ is small compared to the total energies. Estimates are about 50 ± 30 meV but as high as a few hundred meV. Extensive details of studies with various codes are in Ref. [18]. Several factors can affect a DFT calculation of the molecular ΔE_{HL}. For consistency with condensed phase calculations, it is appropriate to study the isolated molecule in a large, periodically bounded box. Appropriate accuracy necessitates a rather large plane wave cutoff, a need that is worsened by the amount of vacuum in the unit cell. (We remark that the self-interaction error of the usual GGA exchange-correlation functions (e.g., PBE) tends to cause the the LS state to be favored, hence cause overestimated ΔE_{HL} values. That is not of concern here since what we are testing is algorithmic efficiency. Similarly we did not use Hubbard U).

For the test of EP + SIRIUS, we used the experimentally determined HS and LS [Mn(taa)] structures and did PBE calculations for a single molecule in a $10 \times 10 \times 10$ Å^3 box. Comparison data are from VASP calculations on optimized structures, also with PBE and without U. Notice, however, that the VASP calculations used a $20 \times 20 \times 20$ Å^3 box. Table 4 gives the parameters and results for the LS state. Its total energy is determined to be about 412 meV below that of the HS state. In contrast, the VASP results are 458-497 mev (at the optimized geometry) with the variation arising from whether the Mn pseudo-potential has 7, 13, or 15 Mn valence electrons. This illustrates the kind of assessment that all-electron calculations facilitate. Regarding timing, observe that the EP + SIRIUS timing is for 16 (4×4) MPI tasks with 8 cores per task.

Table 5 compares timing for the EP only and EP+SIRIUS all-electron calculations and VASP PW-PP-PAW calculations. Evidently EP-only is not competitive but EP+SIRIUS is, at least on a per iteration basis.

There is a difficulty hidden in these results however. The lesser aspect is that we cannot run EP alone at all in a $20 \times 20 \times 20\ ^3$ box. The appropriate cutoffs for such a large vacuum region cause out-of-memory problems with EP because of the way its arrays are structured. The more severe consequence is that we also cannot do a full EP+SIRIUS run, in the sense of returning solutions from SIRIUS to EP for post-processing, on that size box. In effect, EP+SIRIUS is limited in this situation to being an EP user interface for input and control of SIRIUS. The work goes to SIRIUS from EP but the results cannot be returned to EP. Examination of EP suggests that it would take some significant restructuring to remedy the problem, a task well outside the scope of this work or of the separation of concerns approach.

Table 4. Parameters and results for the APW+*lo* calculation of the [Mn(taa)] LS state, with comparison with VASP timings.

structure	[Mn(taa)], LS state structure
unit cell	$10 \times 10 \times 10$ Å^3 box
number of atoms in unit cell	55
R_{mt} (a_0)	Mn: 2.0; C/N: 1.2; H: 1.0;
G_{max} (a_0^{-1}) for APW	20
l_{max} for APW and ρ, ($l_{\text{max}}^{\text{APW}}$ and l_{max}^{ρ})	8
l_{max} for V_{eff}, ($l_{\text{max}}^{\text{pot}}$)	8
k-points grid	$1 \times 1 \times 1$
(L)APW configuration for $l \leq l_{\text{max}}^{\text{APW}}$	$\epsilon_l = -0.15\,\text{eV}; \partial_E = 0;$
lo configuration	Mn: s, p, d; O/C: s, p; H: s;
LO for semi-core	$\epsilon'_{Mn,3d} = -0.32; \epsilon'_{Mn,3p} = -2.45$
treated as core state	Mn: 1s, 2s, 2p, 3s; O/C: 1s
μ_{tot} (μ_B)	total: 2.00 Mn atom: 1.65
Minutes per SCF cycle EP+SIRIUS	4.5
Minutes per SCF cycle VASP	3.03–3.71

Table 5. For $|G + k|_{max} \cdot R_{MT}$ = 4, the average time (seconds) consumed per SCF iteration of EP-SIRIUS for single [Mn(taa)] in $10 \times 10 \times 10$ Å^3 box, single k-point calculation over 60 min of iterations. Comparison is to VASP for three different pseudo-potentials (see text) in a $20 \times 20 \times 20$ Å^3 box.

	$\|G^{\rho,v}\|_{max}$ (a_0^{-1}) = 12	14	16	18	20
EP-only (1 MPI task)	1140 (s)	1180	1227	1275	1323
EP+SIRIUS (1 MPI task)	156	190	178	181	318
EP+SIRIUS (4 MPI task)	45	60	42	46	72
VASP (1 MPI task)	181–223				

6.2. EP+SIRIUS: DTN Molecule

The challenges and opportunities posed by the DTN molecule were summarized in Section 2. In essence one has two transition metals in a complicated structure reminiscent of the perovskites such that the system is both ferromagnetic and ferroelectric. See Figure 4. Recall that the molecule has 70 atoms and 444 electrons.

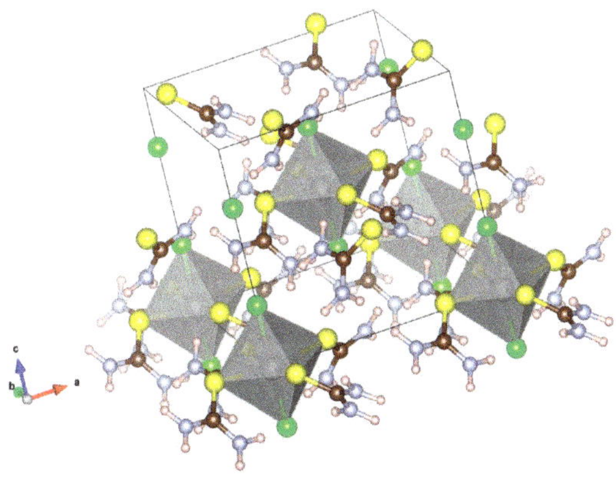

Figure 4. DTN molecule crystal.

We used EP+SIRIUS to calculate the AFM ground state of DTN. We make no attempt at a thorough study, but simply use DTN to show the speed of an all-electron APW+*lo* calculation done with EP+SIRIUS versus with the conventional implementation in EP. Table 6 shows the parameters used and the basic results.

Table 6. Input parameters and outputs of DTN.

	DTN
unit cell	$\approx 10 \times 10 \times 10$ Å^3 box
number of atoms in unit cell	70
R_{mt} (a_0)	Ni: 2.0; Cl/S: 1.2; C/N: 1.0; H: 0.75;
$G_{\max}$ (a_0^{-1}) for APW	20
l_{max} for APW, ($l_{\max}^{\rm APW}$)	8
$l_{\max}$ for ρ, ($l_{\max}^{\rho}$)	8
l_{max} for $V_{\rm eff}$, ($l_{\max}^{\rm pot}$)	8
k-points grid	$2 \times 2 \times 2$
(L)APW configuration for $l \leq l_{\max}^{\rm APW}$	$\epsilon_l = -0.15$ eV; $\partial_E = 0$;
lo configuration	Na: s, p, d Cl/S/C/N: s, p; H: s
LO for semi-core	$\epsilon'_{Ni,3d} = -0.28$ $\epsilon'_{Ni,3p} = -2.18$
treated as core state	Ni: 1s, 2s, 2p, 3s Cl/S/C/N: 1s
$\mu_{\rm tot}$ (μ_B)	total: 0.0 Ni atom: +/−0.72

Table 7 shows the average time per scf iteration as a function of the longest expansion vector $G^{\rho,v}|_{max}$ for the density and potential. Notice that the main gain from EP+SIRIUS over EP alone at the level of one MPI task per k-vector is that the iteration time is almost independent of that vector magnitude. The bigger gain comes from the multiple MPI tasks.

Table 7. For the DTN MOF structure, with $|G + k|_{max} \cdot R_{MT} = 4$, the average time (seconds) consumed per SCF iteration as a function of longest expansion vector for the potential and density. $2 \times 2 \times 2$ k-points, run of 60 min.

DTN	$G_{max}^{\rho,v}$ (a_0^{-1}) = 12	14	16	18	20
EP-only (1 MPI task per k-point)	420 (s)	420	450	515	515
EP+SIRIUS (1 MPI task per k-point)	420	440	430	440	440
EP+SIRIUS (4 MPI task per k-point)	171	184	171	195	180

7. Summary and Conclusions

To summarize, we have implemented a performance enhancement strategy for the Exciting-Plus LAPW/APW+*lo* code by interfacing it with SIRIUS used as a library. We have explored the simplest possible approach to exploiting the separation-of-concerns design philosophy of SIRIUS, namely to interface to it as a black box. The interface outsources the central tasks of the ground-state KS problem from EP to SIRIUS. The objective is to embed a SIRIUS SCF loop inside EP. The implementation effort involved is moderate, benefiting from the similarity of the data structures between EP and the LAPW/LAPW+*lo* elements of SIRIUS.

The EP+SIRIUS combination provides performance gains through diagonalization and parallelization improvements while retaining the user interface and post-processing functionalities of EP. The result is a major advance in capability for treating large, complex molecular aggregates. From the user perspective, only small modifications to the original EP input files are needed. A few lines to select use of SIRIUS and to specify the additional parameters for the Davidson eigensolver are the only changes.

This simplest separation of concerns implementation resolves the eigenvalue solver bottleneck in EP that comes from use of LAPACK full diagonalization. (It cannot handle Hamiltonian matrices larger than $\approx 10^6$). The hand-off to SIRIUS provides the option to use diverse diagonalization algorithms (Davidson, ScaLAPACK, or LAPACK). Use of Davidson-type diagonalization of the Hamiltonian in the self-consistent loop thus benefits from multiple level parallelization within k-points and bands. The eigenvalues and eigenvectors resulting from the SIRIUS calculation have the same structure as those of EP. The design intent therefore is to transfer them back to EP. However, the array structure design of EP inhibits this, as we found with [Mn(taa)]. We return to that below.

For testing and validation, we showed results from small bulk systems calculated in both the APW+*lo* basis and the LAPW basis. The resulting total energy and magnetization show no meaningful deviation from the SIRIUS standalone runs. Two very much larger molecular systems were calculated using the APW+*lo* basis using both EP alone and EP+SIRIUS. Good scaling in band parallelization for a single k-point is observed, The parallelization of the interfaced code works well on high-performance computers, and the computational time is drastically reduced in comparison with the original EP.

The main advantage of the interfaced code is the ease of its construction and the support from advanced eigensolvers. We expect similar interface construction can be done with the ELK or *exciting* codes without unreasonable effort.

Looking ahead, we have found that the non-distributed large arrays defined in EP have become the new bottleneck. That is especially the case when dealing with molecular systems containing more than ≈ 100 atoms in a large unit cell. The primary cause of the bottleneck is the high plane wave G cutoff for large systems, and the fact that some fundamental multi-dimensional arrays are defined with one dimension containing all indices of the G vector or $G + k$ vector. Examples of such fundamental quantities include the augmentation wave part ($u \cdot Y_{lm}$ part) of the APW basis and the so-called structure factor, the form $\exp[i(G + k) \cdot r]$. These basic quantities are used in many places in the host code. It is not an easy job to change them to be distributed data in all occurrences. Although that system-size limitation remains, calculations based on the current EP+SIRIUS can handle larger systems than the original Exciting-Plus and offer a significantly improved foundation for examining the validity of the results from calculations based on various pseudo-potentials.

Author Contributions: Conceptualization, T.S. and H.-P.C.; Formal analysis, A.K. and S.B.T.; Methodology, L.Z., A.K., T.S., H.-P.C. and S.B.T.; Resources, H.-P.C.; Software, L.Z., A.K. and T.S.; Supervision, H.-P.C. and S.B.T.; Validation, L.Z.; Writing—original draft, L.Z.; Writing—review & editing, S.B.T. All authors have read and agreed to the published version of the manuscript.

Funding: This research received no external funding.

Acknowledgments: The work of L.Z., H.-P.C., and S.B.T. was supported as part of the Center for Molecular Magnetic Quantum Materials, an Energy Frontier Research Center funded by the U.S. Department of Energy, Office of Science, Basic Energy Sciences under Award No. DE-SC0019330. Computations were performed at the U.S. National Energy Research Supercomputer Center and the University of Florida Research Computer Center. SBT thanks Angel M. Albavera Mata for providing some [Mn(taa)] calculation timing data from VASP and a helpful remark.

Conflicts of Interest: The authors declare no conflict of interest.

Appendix A. Expressions for Number Density and KS Potential

In the LAPW and APW+*lo* basis sets, the number density and KS potential obviously are adapted, through their matrix elements, to the MT subdivision of the unit cell. In the interstitial region they are expanded in plane waves and inside MT spheres in real spherical harmonics $R_{\ell m}(\boldsymbol{r})$:

$$n(\boldsymbol{r}) = \begin{cases} \sum_{\ell m} n^{\alpha}_{\ell m}(r) R_{\ell m}(\hat{\boldsymbol{r}}), & \mathbf{r} \in \alpha \\ \sum_{\mathbf{G}} \tilde{n}(\boldsymbol{G}) e^{i\boldsymbol{G}\cdot\boldsymbol{r}}, & \mathbf{r} \notin \alpha \end{cases} \quad \text{(A1)}$$

and

$$v_{KS}(\boldsymbol{r}) = \begin{cases} \sum_{\ell m} v^{\alpha}_{\ell m}(r) R_{\ell m}(\hat{\boldsymbol{r}}), & \mathbf{r} \in \alpha \\ \sum_{\boldsymbol{G}} \tilde{v}(\boldsymbol{G}) e^{i\boldsymbol{G}\cdot\boldsymbol{r}}, & \mathbf{r} \notin \alpha\,. \end{cases} \quad \text{(A2)}$$

Here $n^{\alpha}_{\ell m}(r)$, $\tilde{n}(\boldsymbol{G})$, $v^{\alpha}_{\ell m}(r)$, and $\tilde{v}(\boldsymbol{G})$ are expansion coefficients determined through the self-consistent solution of the KS equation.

References

1. Kohn, W.; Sham, L.J. Self-Consistent Equations Including Exchange and Correlation Effects. *Phys. Rev.* **1965**, *140*, A1133–A1138. [CrossRef]
2. Blaha, P.; Schwarz, K.; Trickey, S. Electronic structure of solids with WIEN2k. *Mol. Phys.* **2010**, *108*, 3147–3166.
3. Blaha, P.; Schwarz, K.; Sorantin, P.; Trickey, S. Full-potential, Linearized Augmented Plane Wave Programs for Crystalline Systems. *Comput. Phys. Commun.* **1990**, *59*, 399–415. [CrossRef]
4. Blaha, P.; Schwarz, K.; Tran, F.; Laskowski, R.; Madsen, G.; Marks, L. WIEN2k: An APW+lo program for calculating the properties of solids. *J. Chem. Phys.* **2020**, *152*, 074101. [CrossRef] [PubMed]
5. The Elk Code. Available online: http://elk.sourceforge.net (accessed on 16 October 2020).
6. The Fleur Code. Available online: http://www.flapw.de (accessed on 11 December 2020).
7. Gulans, A.; Kontur, S.; Meisenbichler, C.; Nabok, D.; Pavone, P.; Rigamonti, S.; Sagmeister, S.; Werner, U.; Draxl, C. Exciting—A full-potential all-electron package implementing density-functional theory and many-body perturbation theory. *J. Phys. Condens. Matter* **2014**, *26*, 363202. [CrossRef] [PubMed]
8. Kozhevnikov, A.; Eguiluz, A.G.; Schulthess, T.C. Toward First Principles Electronic Structure Simulations of Excited States and Strong Correlations in Nano- and Materials Science. In Proceedings of the 2010 ACM/IEEE International Conference for High Performance Computing, Networking, Storage and Analysis, New Orleans, LA, USA, 13–19 November 2010; pp. 1–10. [CrossRef]
9. Wasielewski, M.R.; Forbes, M.D.E.; Frank, N.L.; Kowalski, K.; Scholes, G.D.; Yuen-Zhou, J.; Baldo, M.A.; Freedman, D.E.; Goldsmith, R.H.; Goodson, T., III; et al. Exploiting chemistry and molecular systems for quantum information science. *Nat. Rev. Chem.* **2020**, *4*, 490–504. [CrossRef]
10. Gaita-Ari no, A.; Luis, F.; Hill, S.; Coronado, E. Molecular spins for quantum computation. *Nat. Chem.* **2019**, *11*, 301–309. [CrossRef] [PubMed]
11. Castro, S.L.; Sun, Z.; Grant, C.M.; Bollinger, J.C.; Hendrickson, D.N.; Christou, G. Single-Molecule Magnets: Tetranuclear Vanadium(III) Complexes with a Butterfly Structure and an S = 3 Ground State. *J. Am. Chem. Soc.* **1998**, *120*, 2365–2375. [CrossRef]
12. Bagai, R.; Christou, G. The Drosophila of single-molecule magnetism: [$Mn_{12}O_{12}(O_2CR)_{16}(H_2O)_4$]. *Chem. Soc. Rev.* **2009**, *38*, 1011–1026. [CrossRef]
13. Mun, E.; Wilcox, J.; Manson, J.L.; Scott, B.; Tobash, P.; Zapf, V.S. The Origin and Coupling Mechanism of the Magnetoelectric Effect in $TMCl_2$-4SC$(NH_2)_2$ (TM = Ni and Co). *Adv. Condens. Matter Phys.* **2014**, *2014*, 512621. [CrossRef]
14. Jain, P.; Stroppa, A.; Nabok, D.; Marino, A.; Rubano, A.; Paparo, D.; Matsubara, M.; Nakotte, H.; Fiebig, M.; Picozzi, S.; et al. Switchable electric polarization and ferroelectric domains in a metal-organic-framework. *NPJ Quantum Mater.* **2016**, *1*, 16012. [CrossRef]
15. Kepp, K.P. Consistent descriptions of metal–ligand bonds and spin-crossover in inorganic chemistry. *Coord. Chem. Rev.* **2013**, *257*, 196–209. [CrossRef]
16. Shatruk, M.; Phan, H.; Chrisostomo, B.A.; Suleimenova, A. Symmetry-breaking structural phase transitions in spin crossover complexes. *Coord. Chem. Rev.* **2015**, *289–290*, 62. [CrossRef]
17. Collet, E.; Guionneau, P. Structural analysis of spin-crossover materials: From molecules to materials-Etudes structurales des materiaux a conversion de spin: De la molecule aux materiaux. *Comptes Rendu Chim.* **2018**, *21*, 1133–1151. [CrossRef]
18. Mejía-Rodríguez, D.; Albavera-Mata, A.; Fonseca, E.; Chen, D.T.; Chen, H.P.; Hennig, R.G.; Trickey, S. Barriers to Predictive High-throughput Screening for Spin-crossover. *Comput. Mater. Phys.* **2022**, *206*, 111161. [CrossRef]

19. Caneschi, A.; Gatteschi, D.; Sessoli, R.; Barra, A.L.; Brunel, L.C.; Guillot, M. Alternating current susceptibility, high field magnetization, and millimeter band EPR evidence for a ground S = 10 state in $[Mn_{12}O_{12}(Ch_3COO)_{16}(H_2O)_4]\cdot 2CH_3COOH\cdot 4H_2O$. *J. Am. Chem. Soc.* **1991**, *113*, 5873–5874. [CrossRef]
20. Sessoli, R.; Tsai, H.L.; Schake, A.R.; Wang, S.; Vincent, J.B.; Folting, K.; Gatteschi, D.; Christou, G.; Hendrickson, D.N. High-spin molecules: $[Mn_{12}O_{12}(O_2CR)_{16}(H_2O)_4]$. *J. Am. Chem. Soc.* **1993**, *115*, 1804–1816. [CrossRef]
21. Hohenberg, P.; Kohn, W. Inhomogeneous electron gas. *Phys. Rev.* **1964**, *136*, B864. [CrossRef]
22. Parr, R.G.; Yang, R.G.P.W. *Density-Functional Theory of Atoms and Molecules*; Oxford University Press: Oxford UK, 1989.
23. Dreizler, R.; Gross, E. *Density Functional Theory: An Approach to the Quantum Many-Body Problem*; Springer: Berlin/Heidelberg, Germany, 1990.
24. Engel, E.; Dreizler, R.M. *Density Functional Theory*; Springer: Berlin/Heidelberg, Germany, 2013.
25. Schlosser, H.; Marcus, P.M. Composite Wave Variational Method for Solution of the Energy-Band Problem in Solids. *Phys. Rev.* **1963**, *131*, 2529–2546. [CrossRef]
26. Vanderbilt, D. Soft self-consistent pseudopotentials in a generalized eigenvalue formalism. *Phys. Rev. B* **1990**, *41*, 7892. [CrossRef] [PubMed]
27. Blöchl, P. Projector augmented-wave method. *Phys. Rev. B* **1994**, *50*, 17953–17979. [CrossRef] [PubMed]
28. Kresse, G.; Furthmüller, J. Efficient iterative schemes for ab initio total-energy calculations using a plane-wave basis set. *Phys. Rev. B* **1996**, *54*, 11169–11186. [CrossRef] [PubMed]
29. Giannozzi, P.; Baroni, S.; Bonini, N.; Calandra, M.; Car, R.; Cavazzoni, C.; Ceresoli, D.; Chiarotti, G.L.; Cococcioni, M.; Dabo, I.; et al. QUANTUM ESPRESSO: A modular and open-source software project for quantum simulations of materials. *J. Phys. Condens. Matter* **2009**, *21*, 395502. [CrossRef] [PubMed]
30. Gonze, X.; Amadon, B.; Anglade, P.M.; Beuken, J.M.; Bottin, F.; Boulanger, P.; Bruneval, F.; Caliste, D.; Caracas, R.; Côté, M.; et al. ABINIT: First-principles approach to material and nanosystem properties. *Comput. Phys. Commun.* **2009**, *180*, 2582–2615. [CrossRef]
31. Gonze, X.; Beuken, J.M.; Caracas, R.; Detraux, F.; Fuchs, M.; Rignanese, G.M.; Sindic, L.; Verstraete, M.; Zerah, G.; Jollet, F.; et al. First-principles computation of material properties: The ABINIT software project. *Comput. Mater. Sci.* **2002**, *25*, 478–492. [CrossRef]
32. Torrent, M.; Jollet, F.; Bottin, F.; Zèrah, G.; Gonze, X. Implementation of the projector augmented-wave method in the ABINIT code: Application to the study of iron under pressure. *Comput. Mater. Sci.* **2008**, *42*, 337–351. [CrossRef]
33. Paier, J.; Marsman, M.; Hummer, K.; Kresse, G.; Gerber, I.C.; Ángyán, J.G. Screened hybrid density functionals applied to solids. *J. Chem. Phys.* **2006**, *124*, 154709. [CrossRef] [PubMed]
34. Haas, P.; Tran, F.; Blaha, P. Calculation of the lattice constant of solids with semilocal functionals. *Phys. Rev. B* **2009**, *79*, 085104. [CrossRef]
35. Fu, Y.; Singh, D. Applicability of the Strongly Constrained and Appropriately Normed Density Functional to Transition-Metal Magnetism. *Phys. Rev. Lett.* **2018**, *121*, 207201. [CrossRef] [PubMed]
36. Eglitis, R.I.; Borstel, G. Towards a practical rechargeable 5 V Li ion battery. *Phys. Status Solidi A* **2005**, *202*, R13–R15. [CrossRef]
37. Singh, D.J.; Nordstrom, L. *Plane Waves, Pseudopotentials, and the LAPW Method*, 2nd ed.; Springer: New York, NY, USA, 2006.
38. Slater, J. Wave Functions in a Periodic Potential. *Phys. Rev.* **1937**, *51*, 846–851. [CrossRef]
39. Slater, J. An Augmented Plane Wave Method for the Periodic Potential Problem. *Phys. Rev.* **1953**, *92*, 603–608. [CrossRef]
40. Leigh, R.S. The Augmented Plane Wave and Related Methods for Crystal Eigenvalue Problems. *Proc. Phys. Soc. A* **1956**, *69*, 388–400. [CrossRef]
41. Loucks, T. *The Augmented Plane Wave Method*; Benjamin: New York, NY, USA, 1967.
42. Slater, J. Energy Bands and the Theory of Solids. In *Methods of Computational Physics*; Alder, B., Fernbach, S., Rotenberg, M., Eds.; Academic: New York, NY, USA, 1968; Volume 8, pp. 1–20.
43. Mattheiss, L.; Wood, J.; Switendick, A. A Procedure for Calculating Electronic Energy Bands Using Symmetrized Augmented Plane Waves. In *Methods of Computational Physics*; Alder, B., Fernbach, S., Rotenberg, M., Eds.; Academic: New York, NY, USA, 1968; Volume 8, pp. 63–147.
44. Marcus, P. Variational methods in the computation of energy bands. *Int. J. Quantum Chem.* **1967**, *1*, 567–588. [CrossRef]
45. Koelling, D. Linearized form of the APW method. *J. Phys. Chem. Solids* **1972**, *33*, 1335–1338. [CrossRef]
46. Andersen, O. Simple Approach to the Band-Structure Problem. *Solid State Commun.* **1973**, *13*, 133–136. [CrossRef]
47. Andersen, O. Linear methods in band theory. *Phys. Rev. B* **1975**, *12*, 3060–3083. [CrossRef]
48. Koelling, D.; Arbman, G. Use of energy derivative of the radial solution in an augmented plane wave method: Application to copper. *J. Phys. F* **1975**, *5*, 2041–2054. [CrossRef]
49. Koelling, D.; Harmon, B. A technique for relativistic spin-polarised calculations. *J. Phys. C* **1977**, *10*, 3107–3114. [CrossRef]
50. Wimmer, E.; Krakauer, H.; Weinert, M.; Freeman, A. Full-potential self-consistent linearized-augmented-plane-wave method for calculating the electronic structure of molecules and surfaces: O_2 molecule. *Phys. Rev. B* **1981**, *24*, 864–875. [CrossRef]
51. Weinert, M. Solution of Poisson's equation: Beyond Ewald-type methods. *J. Math. Phys.* **1981**, *22*, 2433–2439. [CrossRef]
52. Weinert, M.; Wimmer, E.; Freeman, A. Total-energy all-electron density functional method for bulk solids and surfaces. *Phys. Rev. B* **1982**, *26*, 4571–4578. [CrossRef]

53. Blaha, P.; Schwarz, K. Electron densities and chemical bonding in TiC, TiN, and TiO derived from energy band calculations. *Int. J. Quantum Chem.* **1983**, *XXIII*, 1535–1552. [CrossRef]
54. Jansen, H.; Freeman, A. Total-energy full-potential linearized augmented-plane-wave method for bulk solids: Electronic and structural properties of tungsten. *Phys. Rev. B* **1984**, *30*, 561–569. [CrossRef]
55. Blaha, P.; Schwarz, K.; Herzig, P. First-Principles Calculation of the Electric Field Gradient of Li_3N. *Phys. Rev. Lett.* **1985**, *54*, 1192–1195. [CrossRef]
56. Wei, S.; Krakauer, H.; Weinert, M. Linearized augmented-plane-wave calculation of the electronic structure and total energy of tungsten. *Phys. Rev. B* **1985**, *32*, 7792–7797. [CrossRef] [PubMed]
57. Mattheiss, L.; Hamann, D. Linear augmented-plane-wave calculation of the structural properties of bulk Cr, Mo, and W. *Phys. Rev. B* **1986**, *33*, 823–840. [CrossRef] [PubMed]
58. Goedecker, S.; Maschke, K. Operator approach in the linearized augmented-plane-wave method: Efficient electronic-structure calculations including forces. *Phys. Rev. B* **1992**, *45*, 1597–1604. [CrossRef] [PubMed]
59. Singh, D.J. Ground-state properties of lanthanum: Treatment of extended-core states. *Phys. Rev. B* **1991**, *43*, 6388–6392. [CrossRef] [PubMed]
60. Sjöstedt, E.; Nordstrom, L.; Singh, D. An alternative way of linearizing the augmented plane-wave method. *Solid State Commun.* **2000**, *114*, 15–20. [CrossRef]
61. Cottenier, S. *Density Functional Theory and the Family of (L)APW-Methods: A Step-by-Step Introduction*; Self-Published: Zwijnaarde, Belgium, 2013; ISBN 978-90-807215-1-7.
62. Schwarz, K. Computation of Materials Properties at Atomic Scale. In *Selected Topics in Applications of Quantum Mechanics*; Pahlavani, M.R., Ed.; IntechOpen: London, UK, 2015; pp. 275–310. [CrossRef]
63. Petersilka, M.; Gossmann, U.J.; Gross, E.K.U. Excitation Energies from Time-Dependent Density-Functional Theory. *Phys. Rev. Lett.* **1996**, *76*, 1212–1215. [CrossRef]
64. Chu, I.H.; Trinastic, J.P.; Wang, Y.P.; Eguiluz, A.G.; Kozhevnikov, A.; Schulthess, T.C.; Cheng, H.P. All-electron self-consistent *GW* in the Matsubara-time domain: Implementation and benchmarks of semiconductors and insulators. *Phys. Rev. B* **2016**, *93*, 125210. [CrossRef]
65. Anderson, E.; Bai, Z.; Bischof, C.; Blackford, S.; Demmel, J.; Dongarra, J.; Du Croz, J.; Greenbaum, A.; Hammarling, S.; McKenney, A.; et al. *LAPACK Users' Guide*, 2rd ed.; Society for Industrial and Applied Mathematics: Philadelphia, PA, USA, 1999.
66. Solcá, R.; Kozhevnikov, A.; Haidar, A.; Tomov, S.; Dongarra, J.; Schulthess, T.C. Efficient Implementation of Quantum Materials Simulations on Distributed CPU-GPU Systems. In Proceedings of the International Conference for High Performance Computing, Networking, Storage and Analysis, ser. SC'15, Austin, TX, USA, 15–20 November 2015; ACM: New York, NY, USA, 2015; pp. 1–12. [CrossRef]
67. Zhang, L.; Kozhevnikov, A.; Schulthess, T.C.; Trickey, S.B.; Cheng, H.P. Large-scale All Electron APW+*lo* calculations using SIRIUS. 2022, *unpublished*.
68. CUDA ToolKit Documentation. Version: v11.2.1. Available online: https://docs.nvidia.com/cuda/index.html (accessed on 16 February 2021).
69. Josuttis, N.M. *The C++ Standard Library: A Tutorial and Reference*, 3rd ed.; Addison-Wesley: Philadelphia, PA, USA, 2000.
70. NVCC Compiler. Available online: https://docs.nvidia.com/cuda/cuda-compiler-driver-nvcc/ (accessed on 16 February 2021).
71. Giannozzi, P.; Andreussi, O.; Brumme, T.; Bunau, O.; Nardelli, M.B.; Calandra, M.; Car, R.; Cavazzoni, C.; Ceresoli, D.; Cococcioni, M.; et al. Advanced capabilities for materials modelling with QUANTUM ESPRESSO. *J. Phys. Condens. Matter* **2017**, *29*, 465901. [CrossRef] [PubMed]
72. Giannozzi, P.; Baseggio, O.; Bonfà, P.; Brunato, D.; Car, R.; Carnimeo, I.; Cavazzoni, C.; de Gironcoli, S.; Delugas, P.; Ferrari Ruffino, F.; et al. Quantum ESPRESSO toward the exascale. *J. Chem. Phys.* **2020**, *152*, 154105. [CrossRef] [PubMed]
73. Blaha, P.; Hofstätter, H.; Koch, O.; Laskowski, R.; Schwarz, K. Iterative diagonalization in augmented plane wave based methods in electronic structure calculations. *J. Comput. Phys.* **2010**, *229*, 453–460. [CrossRef]
74. Chandran, A.K. A Performance Study of Quantum ESPRESSO's Diagonalization Methods on Cutting Edge Computer Technology for High-Performance Computing. Master's Thesis, Scuola Internazionale Superiore di Studi Avanzati (SISSA), Trieste, Italy, 2017.
75. Gulans, A. Implementation of Davidson Iterative Eigen-solver for LAPW. *to be published*.
76. Perdew, J.P.; Burke, K.; Ernzerhof, M. Generalized Gradient Approximation Made Simple. *Phys. Rev. Lett.* **1996**, *77*, 3865–3868. [CrossRef]
77. Sim, P.G.; Sinn, E. First manganese(III) spin crossover, first d4 crossover. Comment on cytochrome oxidase. *J. Am. Chem. Soc.* **1981**, *103*, 241–243. [CrossRef]

computation MDPI

Article

Influence of the Chemical Pressure on the Magnetic Properties of the Mixed Anion Cuprates Cu_2OX_2 (X = Cl, Br, I)

William Lafargue-Dit-Hauret *,†,‡ and Xavier Rocquefelte *,‡

Univ Rennes, CNRS, ISCR (Institut des Sciences Chimiques de Rennes) UMR 6226, F-35000 Rennes, France
* Correspondence: william.lafargue-dit-hauret@univ-pau.fr (W.L.-D.-H.); xavier.rocquefelte@univ-rennes1.fr (X.R.)
† Current address: Universite de Pau et des Pays de l'Adour, E2S UPPA, CNRS, IPREM, F-64053 Pau, France.
‡ These authors contributed equally to this work.

Abstract: In this study, we theoretically investigate the structural, electronic and magnetic properties of the Cu_2OX_2 (X = Cl, Br, I) compounds. Previous studies reported potential spin-driven ferroelectricity in Cu_2OCl_2, originating from a non-collinear magnetic phase existing below $T_N \sim$70 K. However, the nature of this low-temperature magnetic phase is still under debate. Here, we focus on the calculation of J exchange couplings and enhance knowledge in the field by (i) characterizing the low-temperature magnetic order for Cu_2OCl_2 and (ii) evaluating the impact of the chemical pressure on the magnetic interactions, which leads us to consider the two new phases Cu_2OBr_2 and Cu_2OI_2. Our *ab initio* simulations notably demonstrate the coexistence of strong antiferromagnetic and ferromagnetic interactions, leading to spin frustration. The T_N Néel temperatures were estimated on the basis of a quasi-1D AFM model using the *ab initio* J couplings. It nicely reproduces the T_N value for Cu_2OCl_2 and allows us to predict an increase of T_N under chemical pressure, with $T_N = 120$ K for the dynamically stable phase Cu_2OBr_2. This investigation suggests that chemical pressure is an effective key factor to open the door of room-temperature multiferroicity.

Keywords: Cu_2OCl_2; Cu_2OBr_2; Cu_2OI_2; oxyhalides; density functional theory; magnetic couplings; Néel temperature; chemical pressure

Citation: Lafargue-Dit-Hauret, W.; Rocquefelte, X. Influence of the Chemical Pressure on the Magnetic Properties of the Mixed Anion Cuprates Cu_2OX_2 (X = Cl, Br, I). *Computation* **2022**, *10*, 73. https://doi.org/10.3390/computation10050073

Academic Editor: Henry Chermette

Received: 7 April 2022
Accepted: 24 April 2022
Published: 12 May 2022

1. Introduction

The design of multiferroics (MF) in which magnetic and ferroelectric orders are coupled is one of the hottest current topics in materials science [1]. These compounds appear as a fantastic playground to deeply investigate the origin and the interactions of the related ferroic properties and to design materials with improved or novel properties. Such multi-functional materials are also extremely appealing for the development of spintronic devices. For example, the control of a magnetic order by an electric field is targeted for several applications ranging from magnetic sensors to memory technologies. At the present time, very few MF compounds exhibit such properties at room temperature (RT), which dramatically limits potential industrial interests. In that context, we embarked on the quest for discovering high-temperature range MF materials, which ideally also show strong electric polarization.

Among the prospective strategies to tune MF stability domain(s) is the application of an hydrostatic (physical) pressure [2]. For instance, in 2013 [3], we predicted theoretically that under high pressure values of about 20 GPa, CuO would become MF at RT. This prediction has been confirmed through dielectric constants measurements up to 7 GPa in 2021 [4] and more recently by neutron diffraction up to 18.5 GPa [5]. An alternative route is to apply a chemical pressure through chemical substitutions, leading to create internal constraints. For instance, the two multiferroic compounds $CuCl_2$ [6] and $CuBr_2$ [7] are characterized by Néel temperatures of 23.9 K and 73.5 K, respectively. Such an increase of T_N is the direct consequence of the chemical substitution of Cl by Br.

Herein, we propose to study the mixed anion system Cu_2OX_2 (with X = Cl, Br, I). It derives from the CuO compound in which half of the O^{2-} ions have been replaced by two X^- ions. Magnetic measurements performed on Cu_2OCl_2 evidenced an antiferromagnetic order with a Néel temperature of ~70 K [8,9], which was firstly attributed to an all-in-all-out arrangement of magnetic moments carried by Cu^{2+} centers forming a pyrochlore-like lattice [10]. Then, powder neutron diffraction analyses [11] revealed an incommensurate magnetic order, with a propagation vector $\vec{q} = [0.827(7), 0, 0]$ and concomitant with a ferroelectric phase. It has also been evidenced that the incommensurate magnetic structure induces a spin–phonon coupling similar to the ones observed in perovskite compounds [12]. In 2019, two contradictory neutron-scattering investigations were published. In the first investigation [13], the spin-driven nature of the multiferroicity of Cu_2OCl_2 was demonstrated, leading to the proposition of a cycloidal non-collinear magnetic order with competing magnetic exchange couplings and driven by an inverse Dzyaloshinskii–Moriya mechanism. In the second article [14], the authors claimed the simultaneous existence, at low temperature, of a collinear antiferromagnetic order and an antiferroelectric phase resulting from a Cl→O charge transfer.

In terms of J magnetic exchange values, only one experimental estimation from magnetic susceptibilty [8] has been proposed, one conference's abstract gave contradicting values deduced from Density Functional Theory (DFT) calculations [15] without providing any details, and one last article [13] focused on the sign of J couplings by studying Wannier functions with DFT. The present study aims to clarify the understanding of the magnetic interactions in Cu_2OCl_2 by doing DFT calculations at ambient pressure but also under chemical pressure, by predicting the magnetic properties of Cu_2OBr_2 and Cu_2OI_2, for the very first time.

2. Materials and Methods

Geometry optimizations on the isostructural Cu_2OX_2 (X = Cl, Br, I) systems were performed within the Density Functional Theory (DFT) framework using the PAW method, as implemented in the VASP code [16–18]. Valence electrons were treated using the following electronic configurations for the different species: $3p^24s^23d^9$ for Cu, $3s^23p^5$ for Cl, $4s^24p^5$ for Br, $5s^25p^5$ for I and $2s^22p^4$ for O. The PBE approach was considered for the exchange–correlation functional [19]. To improve the description of the electronic properties for Cu-3d orbitals, a Hubbard effective correction following Dudarev's scheme [20] was used. We previously showed that U_{eff} = 6.5 eV enables us to recover the experimental volume variation under hydrostatic pressure for the CuO compound [3]. Here, an identical value was chosen. Long-distance van der Waals interactions were taken into account by introducing the DFT-D3 correction proposed by Grimme et al. [21]. Spin-polarized simulations were done with a $5 \times 4 \times 4$ Γ-centered k-mesh and a cutoff energy of 550 eV. The dynamical stability of all optimized structures was checked by computing the phonon modes with the finite-differences method for symmetrically non-equivalent displacements on a $2 \times 1 \times 1$ supercell and a $2 \times 4 \times 4$ Γ-centered k-mesh.

For each compound, the exchange couplings were estimated based on the optimized structures with the Wien2K code, using DFT within the FP-LAPW approach [22]. The PBE0 on-site functional was chosen for the treatment of the exchange correlation part in the lines of our previous works [3,4]. The RK_{max} parameter was set to 7. Muffin-Tin radii were fixed to 1.94, 2.10, 2.20 and 1.71 a_0 for Cu, Cl, Br, I and O species, respectively. The first Brillouin zone was sampled with 52–58 irreducible k-points. The magnetic interactions were extracted from a least-squares fitting procedure applied between the DFT energy related to a given collinear magnetic state and the energy of an Ising Hamiltonian used to describe the magnetic structure:

$$E_{Ising} = \sum_{i \neq j} S_i \cdot J_{ij} \cdot S_j \tag{1}$$

where S_i and S_j are the spins carried by the Cu_i and Cu_j centers, respectively. The J_{ij} parameter represents the magnetic exchange interactions between S_i and S_j. Hereafter, $J > 0$ indicates an antiferromagnetic (AFM) coupling, and $J < 0$ indicates a ferromagnetic (FM) coupling.

3. Results and Discussion

3.1. Crystal Structure of Cu_2OCl_2

The melanothallite Cu_2OCl_2 crystallizes in the orthorhombic space group *Fddd* with a = 7.4477 Å, b = 9.5989 Å, c = 9.6888 Å, V = 692.65 Å^3, and Z = 8 [11]. The structure shown in Figure 1a is characterized by one symmetrically non-equivalent Cu^{2+} site located in a strongly distorted octahedral environment due to a Jahn–Teller effect. It can be viewed as a CuO_2Cl_2 square-planar environment (also called plaquette), in which the bond lengths are d_{Cu-O} = 1.943 Å and d_{Cu-Cl} = 2.283 Å. Two longer bonds are formed with chlorine ions in apical positions with d_{Cu-Cl} = 3.123 Å. These CuO_2Cl_2 *plaquettes* are sharing edges to form ribbons which are inter-connected together by the oxygen ions in a three-dimensional network, which has many common features with the CuO atomic structure. For instance, both Cu_2OCl_2 and CuO [23–25] exhibit unusual thermal variations of lattice parameters due to the so-called hinge mechanism [26]. The ribbons are parallel to the [110] and $[\bar{1}10]$ directions. As shown in Figure 1b, another vision of the structure is a pyrochlore-like lattice where the tridimensional network is based on O-centered OCu_4 tetrahedra which are sharing corners. Here, the chlorine species are pointing toward the center of cavities.

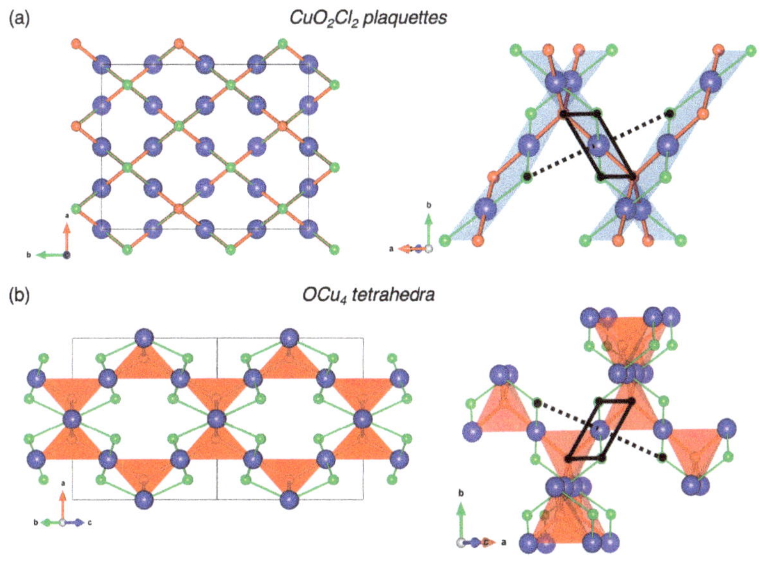

Figure 1. Crystallographic structure of Cu_2OCl_2 described by (**a**) Cu-centered CuO_2Cl_2 square planar environments and (**b**) O-centered OCu_4 tetrahedral environments. The blue, red and green spheres represent copper, oxygen and chlorine ions, respectively. Square planar and tetrahedral environments are evidenced in transparent blue and red, respectively. The CuO_2Cl_4 distorted octahedral environment is highlighted with black lines.

3.2. Geometry Optimization

Geometry relaxations have been performed at ambient pressure in order to validate our approach by comparing with the experimental data of Cu_2OCl_2 and to predict the atomic structure of Cu_2OBr_2 and Cu_2OI_2. The experimental Cu_2OCl_2 structure reported in Ref. [11] was used as an initial guess for all compounds. The main structural parameters of the optimized geometries are summarized in Table 1, including the experimental data of Cu_2OCl_2.

Table 1. Optimized structural parameters of the Cu_2OX_2 compounds (X = Cl, Br, I) deduced from DFT+U calculations ($U_{eff}[Cu(3d)]$ = 6.5 eV) and experimental data of Cu_2OCl_2 Ref. [11]. For Cu_2OCl_2, the deviation (in percentage) with respect to the experimental data is given in parentheses.

	Cu_2OCl_2 (Exp.)	Cu_2OCl_2	Cu_2OBr_2	Cu_2OI_2
a (Å)	7.4477	7.4675 (+0.3)	7.7203	8.1680
b (Å)	9.5989	9.6448 (+0.5)	9.9962	10.5578
c (Å)	9.6888	9.7337 (+0.5)	9.7661	10.1226
V (Å^3)	692.65	701.05 (+1.2)	753.69	872.93
z_X	0.3241	0.3237 (−0.1)	0.3138	0.3104
Cu-O (Å)	1.943	1.951 (+0.4)	1.996	2.094
Cu-X_{eq} (Å)	2.283	2.295 (+0.5)	2.408	2.543
Cu-X_{ap} (Å)	3.123	3.132 (+0.3)	3.214	3.391
Cu-O-Cu (°)	102.86	102.82 (0.0)	104.58	105.65
Cu-X_{eq}-Cu (°)	83.42	83.25 (−0.2)	81.93	82.01
hinge angle (°)	63.20	63.13 (−0.1)	62.42	62.10

Regarding Cu_2OCl_2, an excellent agreement is found between the experimental and the theoretical structures. The optimized lattice parameters deviate from the experimental ones by less than 0.5%. Considering the first neighbor's shell, the Cu-O, Cu-Cl_{eq} and Cu-Cl_{ap} bonds of the optimized geometry are 1.95, 2.30 and 3.13 Å, respectively (deviation smaller than 0.5%). Inside a ribbon, the Cu-O-Cu and Cu-Cl_{eq}-Cu bond angles are computed at 102.8° and 85.25° (less than 0.2% deviation), respectively. Finally, the hinge angle that corresponds to the Cu-O-Cu angle between two ribbons is 63.1° (0.1% deviation).

Focusing now on the new Cu_2OBr_2 compound, one may notice that the lattice parameters are larger than the optimized ones of Cu_2OCl_2, as expected from the larger atomic size of bromine. It leads to $a = 7.720$ Å (3.4% larger compared to Cu_2OCl_2), $b = 9.996$ Å (3.6% larger), $c = 9.766$ Å (0.3% larger), and $V = 753.69$ Å^3 (7.5% larger). This steric effect mainly affects a and b parameters, which are defining the directions of the ribbons. The optimized Cu-O, Cu-Br_{eq} and Cu-Br_{ap} bond lengths are 2.00, 2.41 and 3.21 Å, respectively. The angles within a ribbon, i.e., Cu-O-Cu and Cu-Br_{eq}-Cu, are about 2° larger and 1.3° smaller than in Cu_2OCl_2, respectively. The hinge angle appears slightly smaller than in Cu_2OCl_2 at 62.42°.

For the second new compound Cu_2OI_2, the lattice parameters are still increasing compared to the two former cases, following the chemical sense directed by the atomic radii ($r_{Cl} < r_{Br} < r_I$). In detail, we found a = 8.168 Å (9.4% larger compared to Cu_2OCl_2), b = 10.558 Å (9.5% larger), c = 10.123 Å (4.0% larger) and V = 872.93 Å^3 (24.5% larger). Once more, the significant variation of volume results from more important changes reported for a and b lattice parameters. Regarding the distorted CuO_2I_4 environment, the optimized Cu-O, Cu-I_{eq} and Cu-I_{ap} bond lengths are respectively measured at 2.09, 2.54 and 3.39 Å. Focusing on the ribbons, the Cu-O-Cu angle equals 105.7° (2.3° larger than in Cu_2OCl_2), while the Cu-I_{eq}-Cu angle is reduced to 82.0° (1.2% smaller).

No imaginary phonon frequencies were retrieved for both Cu_2OCl_2 and Cu_2OBr_2 systems. First, this result confirms that the present modeling reproduces properly both the atomic parameters and the dynamical stability of Cu_2OCl_2. Experimentally, the crystals are grown using the chemical vapor transport techniques with a stoichiometric mixture of CuO and $CuCl_2$ compounds [11]. Second, Cu_2OBr_2 is predicted to be dynamically stable. For Cu_2OI_2, one imaginary phonon frequency was calculated at $\sim$20i cm^{-1}, suggesting that such a phase does not exist. However, Cu_2OI_2 was kept in our investigation as a virtual compound allowing us to probe the chemical pressure effect.

3.3. Electronic Properties

Based on the optimized structures, the electronic properties were investigated using the Wien2K code. The projected density of states (pDOS) of the three Cu_2OX_2 systems are represented in Figure 2. The present PBE0 calculations lead to band gap values of 2.18 eV, 1.85 eV and 0.95 eV for X = Cl, Br and I, respectively, and rather similar pDOS. The valence band (VB) is based on the O-2*p*, Cl-3*p* (or Br-4*p* or I-5*p*) and Cu-3*d* states (from −7 to 0 eV with respect to E_F). While the top of the VB is mainly composed of O-2*p* states in Cu_2OCl_2, it is mainly based on Br-4*p* and I-5*p* states in Cu_2OBr_2 and Cu_2OI_2, respectively. Such a point is emphasized in Figure 3 by the electronic densities calculated in the energy range from −0.5 to 0 eV (with respect to E_F). The magnetically active orbital, i.e., the one carrying the magnetic moment, is the Cu-3$d_{x^2-y^2}$, as expected from the d^9 electronic configuration and the square planar environment. The calculated magnetic moment of copper is 0.6–0.7 μ_B in Cu_2OX_2, which is in good agreement with the available experimental data for Cu_2OCl_2. Indeed, the ordered magnetic moment was found to be 0.64(5) μ_B and 0.66(2) μ_B from powder and single-crystal neutron refinements, respectively [13].

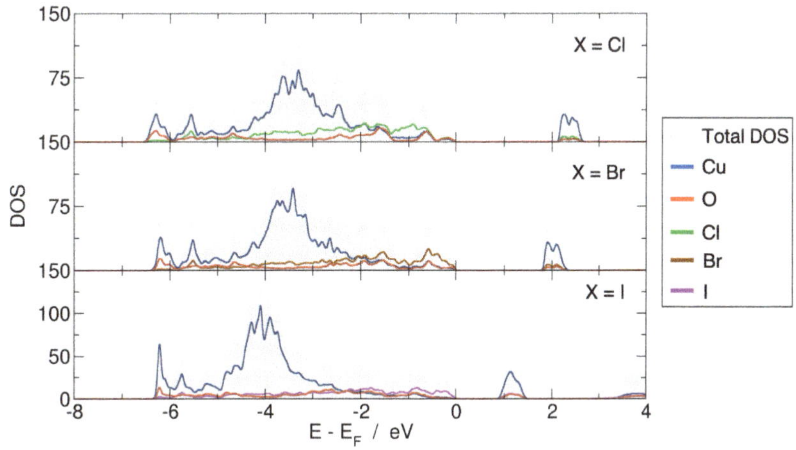

Figure 2. Total and projected densities of states for the Cu_2OX_2 compounds (X = Cl, Br, I) deduced from PBE0 on-site calculations on the ground-state AFM collinear order. Energies for pDOS are given with respect to E_F.

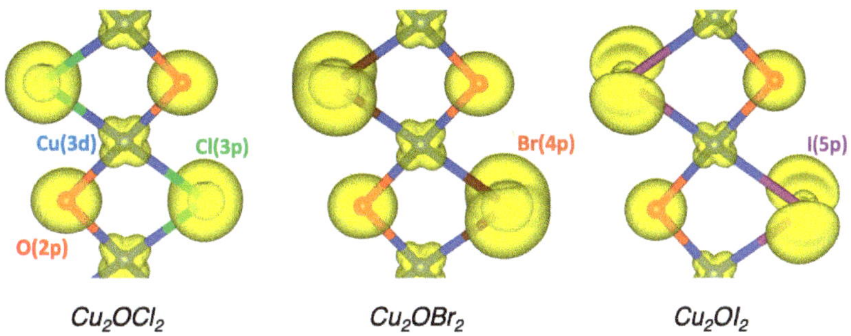

Figure 3. Electronic density corresponding to the top of the valence band (from −0.5 to 0 eV with respect to E_F) for the Cu_2OX_2 compounds (X = Cl, Br, I) deduced from PBE0 on-site calculations on the ground state AFM collinear order. Isovalue is set to 0.1 electron/Å^3. Blue, red, green, brown and violet spheres indicate Cu, O, Cl, Br and I atoms, respectively.

3.4. Magnetic Exchange Interactions

Magnetic susceptibiltity measurements on Cu_2OCl_2 provided a Néel temperature T_N of ∼70 K followed by a broad maximum at ∼140 K [8,9]. Many investigations were carried out to determine the nature of the magnetic order at lower temperatures, leading to several propositions such as an *all-in-all-out* model [10], an incommensurate spin spiral phase [11], an incommensurate spin cycloidal phase [13] and a collinear antiferromagnetic (AFM) phase [14]. However, only one of these experimental investigations proposed magnetic exchange coupling (J) values extracted from magnetic susceptibility fits [8], which are in disagreement with the reported J values estimated from DFT calculations [15]. However, these latter values were found in a conference abstract and no related publication is available, to our knowledge.

Based on the optimized geometries, we have thus calculated the J exchange values using PBE0 on-site functional as we did for CuO [3,4]. Examining the atomic structure leads to defining four magnetic interactions depicted in Figure 4 for both ribbons- and pyrochlore-like lattices. The related d_{Cu-Cu} bond lengths and Cu-O-Cu angles are specified in Table 2 for Cu_2OCl_2, Cu_2OBr_2 and Cu_2OI_2. More specifically, considering the ribbons-like structure, we targeted two intra-ribbon (J_1 and J_4) and two inter-ribbon (J_2 and J_3) couplings. It should be noticed that among these four interactions, only J_4 is based on a super-superexchange (SSE) path, i.e., the magnetic moments are mediated through an orbital overlap implying two ligand atoms (namely Cu-L-L-Cu, with L = O, Cl, Br or I), the others being based on a superexchange (SE) path implying only one ligand atom (namely Cu-L-Cu, with L = O, Cl, Br or I).

The J couplings were estimated using the *mapping analysis* procedure. The strategy consists of (1) determining the DFT energies of a set of magnetic states, (2) defining the related energy expressions in terms of J parameters using an Ising Hamiltonian, and (3) refining the J parameters on top of the DFT results using a least squares method. One may underline that the accuracy of such an approach strongly depends on the size and the quality of the set of magnetic states [27]. In our case, eight collinear magnetic structures were selected to obtain four J values. The results are summarized in Table 3.

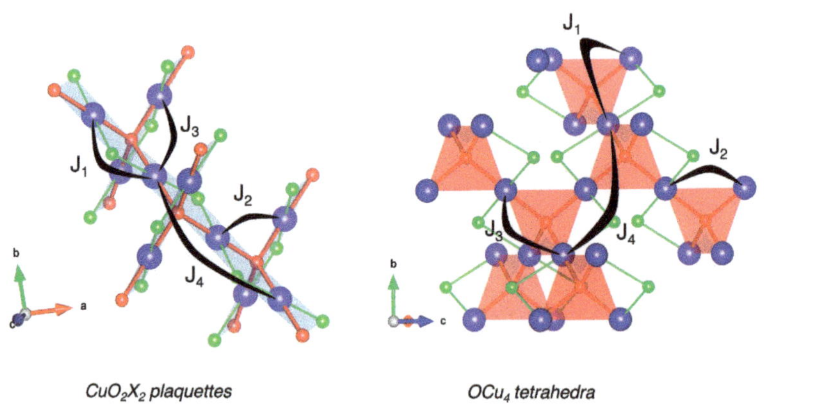

Figure 4. Representation of J exchange couplings between copper sites in both CuO_2X_2- and OCu_4-based frameworks. Cu, O and X atoms are depicted by blue, red and green spheres, respectively. CuO_2X_2 *plaquettes* and OCu_4 tetrahedra are shown in transparent blue and red, respectively.

Table 2. Geometrical parameters related to the four magnetic couplings for the Cu_2OX_2 compounds (X = Cl, Br, I) deduced from DFT+U relaxation ($U_eff[Cu(3d)] = 6.5$ eV). The experimental data of Cu_2OCl_2 taken from Ref. [11] are given in parentheses.

		J_1	J_2	J_3	J_4
Cu_2OCl_2	Cu-Cu (Å)	3.049	3.067	3.426	6.099
		(3.037)	(3.055)	(3.410)	(6.075)
	Cu-O-Cu (°)	102.82	103.65	122.82	-
		(102.86)	(103.70)	(122.72)	-
Cu_2OBr_2	Cu-Cu (Å)	3.158	3.112	3.494	6.315
	Cu-O-Cu (°)	104.58	102.47	122.16	-
Cu_2OI_2	Cu-Cu (Å)	3.337	3.252	3.657	6.674
	Cu-O-Cu (°)	105.65	101.87	121.64	-

Table 3. Magnetic exchange interactions (in meV) calculated at the PBE0 on-site level for the Cu_2OX_2 compounds (X = Cl, Br, I). The available experimental and theoretical data are given for comparison. $J > 0$ indicates an AFM coupling, and $J < 0$ indicates an FM coupling.

		J_1	J_2	J_3	J_4
Cu_2OCl_2	This work	−14.0	−2.2	19.1	8.7
	Theo. [15]	−15.5	19.0 [1]		8.6
	Exp. [8]	9.7 [2]	9.3 [2]		-
Cu_2OBr_2	This work	−13.1	1.9	25.2	9.2
Cu_2OI_2	This work	−8.1	24.2	41.0	15.1

[1] This value has been defined as an interchain coupling without giving more details. It could be alternatively J_2, J_3 or a combined effective value. [2] These two parameters correspond to mean field estimation of $J_{intra-ribbon} = J_1$ and $J_{inter-ribbon} = J_2 + J_3$.

Before discussing our results, let us on comment the experimental values extracted by Okabe et al. [8] considering the ribbons-like picture. They considered two effective J parameters which were found to be both AFM and very close to each other, i.e., J_{intra} = 9.7 meV and J_{inter} = 9.3 meV, thus leading to strong magnetic frustrations. These mean field values can be compared to our J parameters considering that $J_{intra} = J_1 + J_4$ and $J_{inter} = J_2 + J_3$. It leads to DFT effective intra- and inter-ribbon couplings, which are respectively FM and AFM, i.e., J_{intra} = −5.3 meV and J_{inter} = 16.8 meV, in disagreement with Okabe et al. [8]. In contrast, in Ref. [15], three DFT J values have been proposed, i.e., two intra-ribbon (J_1 = −15.5 meV and J_4 = 8.6 meV) and one inter-ribbon ($J_{inter} = J_2 + J_3$ = 19.0 meV), which are in good agreement with our values, i.e., J_1 = −14.0 meV and J_4 = 8.7 meV for the intra-ribbon interactions and J_2 = −2.2 meV and J_3 = 19.1 meV for the inter-ribbon ones.

In other words, the present calculations evidenced that the ribbons are mainly FM ordered due to the first-neighbor (SE) interaction J_1, and the spins are frustrated due to the AFM second-neighbor (SSE) interaction J_4, which is 1.6 times smaller than J_1 in amplitude. Between the ribbons, the interaction is AFM due to J_3 (J_2 appears negligible). In the pyrochlore-like lattice, each OCu_4 tetrahedron is globally AFM with two up and two down magnetic centers, which further emphasizes the spin frustration character of the magnetic structure. The resulting most stable collinear magnetic order is depicted in Figure 5.

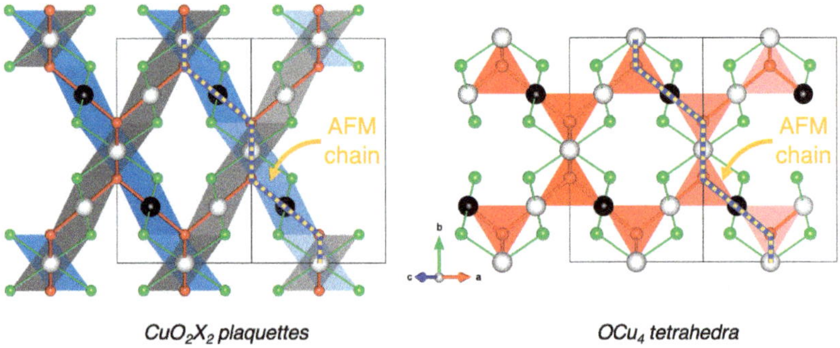

Figure 5. Schematic representation of the ground-state AFM collinear magnetic structure within the (**left**) ribbons- and (**right**) pyrochlore-like lattices. Black and white spheres indicate up and down spins carried by Cu sites, respectively. Red and green spheres represent oxygen and halogen species, respectively. The OCu_4 tetrahedral coordinations are represented in transparent red. The CuO_2X_2 *plaquettes* are shown in transparent blue and gray for an up or down spin carried by the central Cu, respectively. One AFM spin chain, formed by the $J = J_3$ coupling and considered within the quasi-1D AFM model, is evidenced by a yellow–blue dotted line.

A similar picture is obtained for Cu_2OBr_2 and Cu_2OI_2 compounds. Regarding the intra-ribbon interactions, one may notice that the FM J_1 coupling decreases to -13.1 meV and -8.1 meV, respectively, while the AFM J_4 coupling is enhanced to 9.2 meV and 15.1 meV, respectively. For the inter-ribbon interactions, the FM J_2 coupling becomes AFM at 1.9 meV for Cu_2OBr_2 and strongly increases up to 24.2 meV for Cu_2OI_2. J_3 stays the strongest (AFM) coupling and is increased to 25.2 meV and 41.0 meV, respectively. The larger values for J_3 cannot be explained only based on geometrical arguments. Indeed, the related Cu-O bond length and Cu-O-Cu angle, respectively, increases and decreases, when Cl is substituted by Br or I. It is the signature that the halogen element indirectly participates in this interaction by setting the size and the shape of the magnetic orbital due to the nephelauxetic effect, which increases in the series $Cl^- < Br^- < I^-$, leading to an expansion of the copper 3d orbitals and thus larger interactions. Figure 6 shows that the spin densities of the three compounds are quite similar, with spin-up densities (in yellow) along the FM J_1 coupling for both copper and halogen elements involved in this interaction. For oxygen atoms, which are also involved in the AFM couplings (J_3 and J_4), two lobes are found with one up- and one down-spin densities in yellow and cyan, respectively. These observations have a direct consequence on the magnetic moments carried by the ligands, which are 0.11, 0.10, 0.08 and 0.00 μ_B for Cl, Br, I and O, respectively. Indeed, the magnetic moment of the oxygen is null due to its participation in AFM couplings.

To go further, we focused more deeply on the magnetically active orbital, i.e., Cu-$3d_{x^2-y^2}$, and its interaction with the surrounding ligands. The spin-polarized pDOS of Cu-$3d_{x^2-y^2}$, O-2p and Cl-3p states of one *plaquette* are given in Figure 7a for Cu_2OCl_2. It appears that the occupied Cu-$3d_{x^2-y^2}$ states are mainly positioned within the energy range from -7 to -5 eV (highlighted by the gray dotted line rectangle). The electronic charge density (ρ) and spin density related to this energy window are shown in Figure 7b,c, respectively. First of all, from the analysis of both pDOS and ρ, it appears that the Cu-$3d_{x^2-y^2}$ orbital interaction is larger with O-2p than Cl-3p states. It confirms that the magnetic moment of the magnetically active orbital is mainly mediated by oxygen atoms and thus justifies that the largest coupling is along a Cu-O-Cu superexchange path. In addition, the up- and down-pDOS of O-2p states are perfectly symmetric, leading to a net magnetic moment of zero. Similarly, the spin-density around oxygen is constituted of two equivalent lobes with opposite spins (as previously discussed). Looking in more detail, Figure 7c evidenced the pivotal role of oxygen that connects two ribbons (with opposite spins), which are nearly perpendicular. To summarize the results of the present analysis, the magnetically active orbital results from an overlap of Cu-3$d_{x^2-y^2}$ with O-2p and to a less extent with Cl-3p states in Cu_2OCl_2. One may notice that in this energy range, the pDOS of the magnetically active orbital is constituted by two peaks: (1) the lower in energy originating from the Cu-O interaction, while (2) the higher in energy is associated to the Cu-Cl interaction. A similar picture can be retrieved for Cu_2OBr_2 and Cu_2OI_2, except that the first peak (Cu-O interaction) increases and the second peak (Cu-X interaction) decreases (see Figure 2). This observation explains the enhancement of J_3 from Cu_2OCl_2 to Cu_2OBr_2 and to Cu_2OI_2.

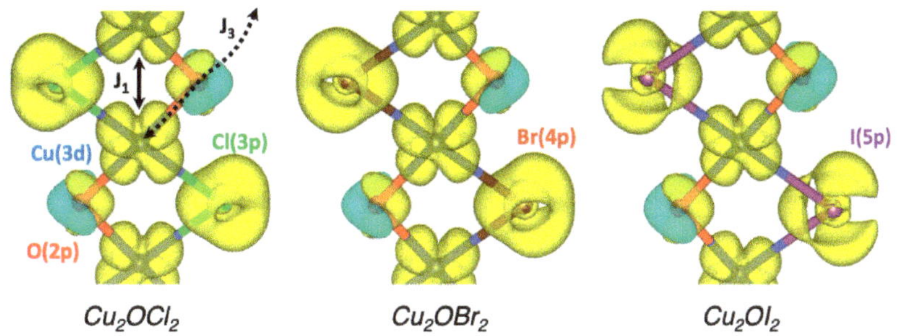

Figure 6. Spin density for the AFM ground state of Cu_2OCl_2, Cu_2OBr_2 and Cu_2OI_2. The isovalue is set to 0.017 electron/Å^3, up and down regions of the spin density are evidenced in transparent yellow and cyan, respectively. Blue, red, green, brown and violet spheres indicate Cu, O, Cl, Br and I atoms, respectively. The FM J_1 and AFM J_3 couplings are indicated.

Interestingly, these three systems can be compared to their parent compound CuO, for which the atomic and magnetic structures are described using different descriptors. While the atomic structure can be viewed as based on ribbons of edge-sharing CuO_4 *plaquettes*, the magnetic order is mainly governed by AFM spin chains of corner-sharing CuO_4 *plaquettes*. Here also, it appears that the strongest magnetic coupling, i.e., J_3, is at the origin of AFM spin chains of corner-sharing CuO_2X_2 *plaquettes*, highlighted by a yellow–blue dotted line in Figure 5. The presence of such AFM spin chains can be more clearly seen and understood in the OCu_4-based vision, where these AFM spin chains propagate along tetrahedra.

As we did previously for CuO [3,4], we can thus estimate T_N based on the random phase approximation model of a quasi-1D AFM Heisenberg cubic lattice [28]:

$$J' = \frac{T_N}{4c\sqrt{\ln(\alpha J/T_N) + 0.5\ln(\ln(\alpha J/T_N))}} \tag{2}$$

where $\alpha = 2.6$ and $c = 0.233$ are numerical parameters, J' is an effective inter-chain coupling and J is the intra-chain coupling. In the present case, half of the spin chains are 90° rotated, leading to half as many inter-chain couplings. The resulting ground-state expression is $E_{GS}(model) = J + J'$, instead of $E_{GS}(model) = J + 2J'$ in the CuO model where all chains are oriented in the same direction. Such an expression has to be compared to the collinear ground-state (Figure 5) energy defined as:

$$E_{GS} = J_3 - J_1 + J_2 - J_4 \tag{3}$$

Thus, one may define:

$$J = J_3 \tag{4}$$

$$J' = -J_1 + J_2 - J_4 \tag{5}$$

Using the aforementioned J values and the original c and α parameters, we obtain T_N = 65, 111 and 294 K for Cu_2OCl_2, Cu_2OBr_2 and Cu_2OI_2, respectively (see Figure 8). Experimentally, $T_N \sim$70 K for Cu_2OCl_2. Such a good agreement validates the present magnetic model, which consists of considering AFM spin chains based on $J = J_3$ interacting through an effective coupling defined as $J' = -J_1 + J_2 - J_4$. It should be noticed that if we use the refined value for CuO c = 0.284 [3], we obtain T_N = 77, 131 and 344 K for Cu_2OCl_2, Cu_2OBr_2 and Cu_2OI_2, respectively (see Figure 8). The agreement for Cu_2OCl_2 is still very good with the experiment, and whatever the c value is, we predict a significant increase of T_N when Cl is replaced by either Br or I in the compound. A similar trend has been reported experimentally for $CuCl_2$ and $CuBr_2$ [6,7], for which T_N = 23.9 and 77 K, respectively.

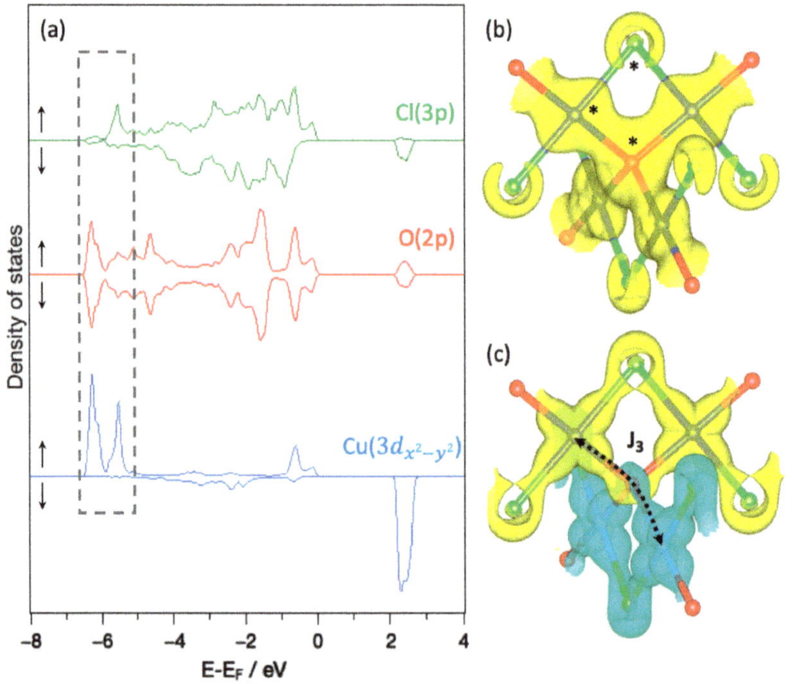

Figure 7. (**a**) Spin-polarized projected density of states (pDOS) of one *plaquette* in Cu_2OCl_2. The energy window of Cu-3$d_{x^2-y^2}$ occupied states is evidenced by a gray dotted rectangle on the pDOS. The (**b**) electronic charge density and (**c**) the spin density calculated in this energy range are also displayed. The atoms considered for the pDOS are shown by black stars on the electronic charge density. Up and down spin channels of the pDOS are evidenced by ↑ and ↓, respectively. Isovalues are set to 0.1 and 0.017 electron/Å^3 for electronic charge density and the spin density, respectively. Up and down regions of the spin density are highlighted in transparent yellow and cyan, respectively. Blue, red and green spheres indicate Cu, O and Cl atoms, respectively.

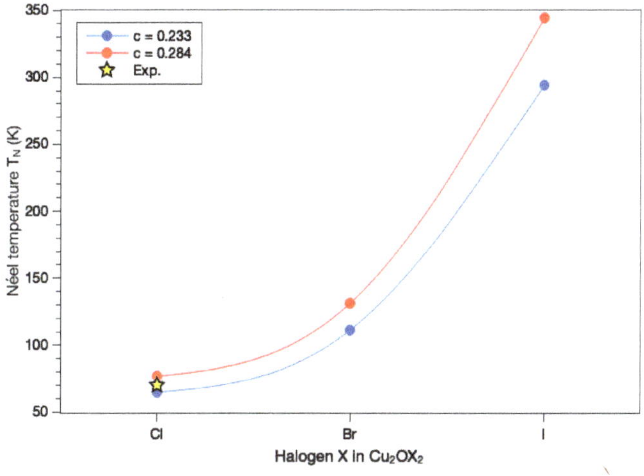

Figure 8. Estimation of the T_N Néel temperature depending on the nature of halogen species and considering two sets of parameters for the magnetic model presented in Equation (2): (1, in blue) $\alpha = 2.6$ and $c = 0.233$, and (2, in red) $\alpha = 2.6$ and $c = 0.284$. The experimental value for Cu_2OCl_2 is represented by a yellow star.

4. Conclusions

In summary, we have investigated the magnetic properties of Cu_2OX_2 (X = Cl, Br, I) systems based on first-principles calculations. The atomic structure of Cu_2OX_2 can be viewed either as based on ribbons of CuO_2X_2 edge-sharing *plaquettes* or as a pyrochlore-like lattice constituted of OCu_4 corner-sharing tetrahedra. The present study shows that the magnetic order is governed by AFM spin chains of CuO_2X_2 corner-sharing *plaquettes*, which defines an AFM intra-chain coupling $J = J_3$. These spin chains interact with each other through three inter-chain couplings, which leads to an effective interaction $J' = -J_1 + J_2 - J_4$. Using these two parameters J and J' in an analytical expression for a quasi-1D AFM Heisenberg system, one may recover the Néel temperature of the known multiferroic compound Cu_2OCl_2 at $T_N = 70$ K. We also predict that Cu_2OBr_2 could be a stable phase and is expected to have a larger T_N value, which is evaluated to be about 120 K. Using chemical pressure to enhance T_N is thus an effective and appealing technique to design new mixed anion compounds in order to reach RT operating MF. Ongoing investigations in our group consist of combining chemical and physical pressures to reach this goal.

Author Contributions: Investigation, W.L.-D.-H.; supervision, X.R.; formal analysis, writing—original draft preparation, review and editing, W.L.-D.-H., X.R. All authors have read and agreed to the published version of the manuscript.

Funding: Research funded by Agence Nationale de la Recherche (ANR-19-CE08-0013).

Institutional Review Board Statement: Not applicable.

Informed Consent Statement: Not applicable.

Data Availability Statement: The data presented in this study are available in insert article.

Acknowledgments: The theoretical work was granted access to the HPC resources of TGCC/CINES /IDRIS under Allocation No. 2019-A0050907682 made by GENCI. The authors also acknowledge the CCIPL (Centre de Calcul Intensif des Pays de la Loire) for computational resources. Crystal structures were represented using the VESTA software [29].

Conflicts of Interest: The authors declare no conflict of interest.

References

1. Fiebig, M.; Lottermoser, T.; Meier, D.; Trassin, M. The evolution of multiferroics. *Nat. Rev. Mater.* **2016**, *1*, 16046. [CrossRef]
2. Gilioli, E.; Ehm, L. High pressure and multiferroics materials: A happy marriage. *Int. Union Crystallogr. J.* **2014**, *1*, 590–603. [CrossRef] [PubMed]
3. Rocquefelte, X.; Schwarz, K.; Blaha, P.; Kumar, S.; van den Brink, J. Room-temperature spin-spiral multiferroicity in high-pressure cupric oxide. *Nat. Commun.* **2013**, *4*, 2511. [CrossRef] [PubMed]
4. Lafargue-Dit-Hauret, W.; Braithwaite, D.; Huxley, A.D.; Kimura, T.; Saúl, A.; Rocquefelte, X. Potential room-temperature multiferroicity in cupric oxide under high pressure. *Phys. Rev. B* **2021**, *103*, 214432. [CrossRef]
5. Manuel, P. Private communication. 2022.
6. Banks, M.G.; Kremer, R.K.; Hoch, C.; Simon, A.; Ouladdiaf, B.; Broto, J.M.; Rakoto, H.; Lee, C.; Whangbo, M.H. Magnetic ordering in the frustrated Heisenberg chain system cupric chloride $CuCl_2$. *Phys. Rev. B* **2009**, *80*, 024404. [CrossRef]
7. Zhao, L.; Hung, T.L.; Li, C.C.; Chen, Y.Y.; Wu, M.K.; Kremer, R.K.; Banks, M.G.; Simon, A.; Whangbo, M.H.; Lee, C.; et al. $CuBr_2$—A New Multiferroic Material with High Critical Temperature. *Adv. Mater.* **2012**, *24*, 2469–2473. [CrossRef]
8. Okabe, H.; Suzuki, K.; Kawashima, K.; Muranaka, T.; Akimitsu, J. New Pyrochlore-like Compound Cu_2OCl_2 with S = 1/2. *J. Phys. Soc. Jpn.* **2006**, *75*, 123705. [CrossRef]
9. Kawashima, K.; Okabe, H.; Suzuki, K.; Kuroiwa, S.; Akimitsu, J.; Sato, K.H.; Koda, A.; Kadono, R. Antiferromagnetic ordering in Cu_2OCl_2 studied by the muon spin rotation/relaxation technique. *J. Phys. Condens. Matter* **2007**, *19*, 145275. [CrossRef]
10. Nishiyama, M.; Oyamada, A.; Itou, T.; Maegawa, S.; Okabe, H.; Akimitsu, J. NMR study of pyrochlore lattice antiferromagnet, melanothallite Cu_2OCl_2. *J. Phys. Conf. Ser.* **2011**, *320*, 012030. [CrossRef]
11. Zhao, L.; Fernández-Díaz, M.T.; Tjeng, L.H.; Komarek, A.C. Oxyhalides: A new class of high-T_c multiferroic materials. *Sci. Adv.* **2016**, *2*, e1600353. [CrossRef]
12. Araújo, B.S.; Arévalo-López, A.M.; Attfield, J.P.; Paschoal, C.W.A.; Ayala, A.P. Spin-phonon coupling in melanothallite Cu_2OCl_2. *Appl. Phys. Lett.* **2018**, *113*, 222901. [CrossRef]
13. Guo, H.; Zhao, L.; Schmidt, W.; Fernández-Díaz, M.T.; Becker, C.; Melendez-Sans, A.; Peng, W.; Zbiri, M.; Hansmann, P.; Komarek, A.C. Multiferroic properties of melanothallite Cu_2OCl_2. *Phys. Rev. Mater.* **2019**, *3*, 124405. [CrossRef]
14. Wu, H.C.; Yuan, J.K.; Chandrasekhar, K.D.; Lee, C.H.; Li, W.H.; Wang, C.W.; Chen, J.M.; Lin, J.Y.; Berger, H.; Yen, T.W.; et al. Observation of charge–transfer–driven antiferroelectricity in 3d-pyrochlore multiferroic Cu_2OCl_2. *Mater. Today Phys.* **2019**, *8*, 34–42. [CrossRef]
15. Tsirlin, A.A.; Janson, O.; Rosner, H. *Electronic Structure and Magnetic Properties of Melanothallite (Cu_2OCl_2), a Gateway to Understanding Copper Oxychlorides*; Russian Mineralogical Society: Kirovsk, Russia, 2010; Volume 1, p. 61.
16. Kresse, G.; Furthmüller, J. Efficiency of ab-initio total energy calculations for metals and semiconductors using a plane-wave basis set. *Comput. Mater. Sci.* **1996**, *6*, 15–50. [CrossRef]
17. Kresse, G.; Furthmüller, J. Efficient iterative schemes for ab initio total-energy calculations using a plane-wave basis set. *Phys. Rev. B* **1996**, *54*, 11169–11186. [CrossRef]
18. Kresse, G.; Joubert, D. From ultrasoft pseudopotentials to the projector augmented-wave method. *Phys. Rev. B* **1999**, *59*, 1758–1775. [CrossRef]
19. Perdew, J.P.; Burke, K.; Ernzerhof, M. Generalized Gradient Approximation Made Simple. *Phys. Rev. Lett.* **1996**, *77*, 3865–3868. [CrossRef]
20. Dudarev, S.L.; Botton, G.A.; Savrasov, S.Y.; Humphreys, C.J.; Sutton, A.P. Electron-energy-loss spectra and the structural stability of nickel oxide: An LSDA+U study. *Phys. Rev. B* **1998**, *57*, 1505–1509. [CrossRef]
21. Grimme, S.; Antony, J.; Ehrlich, S.; Krieg, H. A consistent and accurate ab initio parametrization of density functional dispersion correction (DFT-D) for the 94 elements H-Pu. *J. Chem. Phys.* **2010**, *132*, 154104. [CrossRef]
22. Blaha, P.; Schwarz, K.; Madsen, G.K.H.; Kvasnicka, D.; Luitz, J.; Laskowsk, R.; Tran, F.; Marks, L.; Marks, L. *WIEN2k: An Augmented Plane Wave Plus Local Orbitals Program for Calculating Crystal Properties*; Vienna University of Technology Institute of Materials Chemistry Getreidemarkt 9/165-TC A-1060: Vienna, Austria, 2019. ISBN 3-9501031-1-2.
23. Krivovichev, S.V.; Filatov, S.K.; Burns, P.C. The cuprite-like framework of OCu_4 tetrahedra in the crystal structure of synthetic melanothallite, Cu_2OCl_2, and its negative thermal expansion. *Can. Mineral.* **2002**, *40*, 1185–1190. [CrossRef]
24. Filatov, S.K.; Bubnova, R.S. The nature of special points on unit cell parameters temperature dependences for crystal substances. In *Tenth European Powder Diffraction Conference: Geneva, Switzerland, 1–4 September 2006*; Für Kristallographie, D.G., Ed.; Number 26; Oldenbourg Wissenschaftsverlag: Munchen, Germany, 2007; pp. 447–452.
25. Zheng, X.G.; Kubozono, H.; Yamada, H.; Kato, K.; Ishiwata, Y.; Xu, C.N. Giant negative thermal expansion in magnetic nanocrystals. *Nat. Nanotechnol.* **2008**, *3*, 724–726. [CrossRef] [PubMed]
26. Sleight, A.W. Compounds That Contract on Heating. *Inorg. Chem.* **1998**, *37*, 2854–2860. [CrossRef]
27. Rocquefelte, X.; Schwarz, K.; Blaha, P. Comment on High-T_c Ferroelectricity Emerging from Magnetic Degeneracy in Cupric Oxide. *Phys. Rev. Lett.* **2011**, *107*, 239701. [CrossRef] [PubMed]
28. Yasuda, C.; Todo, S.; Hukushima, K.; Alet, F.; Keller, M.; Troyer, M.; Takayama, H. Néel Temperature of Quasi-Low-Dimensional Heisenberg Antiferromagnets. *Phys. Rev. Lett.* **2005**, *94*, 217201. [CrossRef]
29. Momma, K.; Izumi, F. VESTA 3 for three-dimensional visualization of crystal, volumetric and morphology data. *J. Appl. Crystallogr.* **2011**, *44*, 1272–1276. [CrossRef]

computation

MDPI

Article

Regression Machine Learning Models Used to Predict DFT-Computed NMR Parameters of Zeolites

Robin Gaumard [1], Dominik Dragún [2], Jesús N. Pedroza-Montero [1], Bruno Alonso [1], Hazar Guesmi [1], Irina Malkin Ondík [2,3] and Tzonka Mineva [1,*]

1 ICGM, CNRS, ENSCM, Universite de Montpellier, 34296 Montpellier, France; robin.gaumard@enscm.fr (R.G.); jesus-nain.pedroza-montero@umontpellier.fr (J.N.P.-M.); bruno.alonso@enscm.fr (B.A.); hazar.guesmi@enscm.fr (H.G.)

2 FIIT STU in Bratislava, Ilkovičova 2, 84216 Bratislava, Slovakia; domco.dragun@gmail.com (D.D.); malkin.ondik@gmail.com (I.M.O.)

3 MicroStep-MIS, Spol. S.R.O., Čavojského 1, 84104 Bratislava, Slovakia

* Correspondence: tzonka.mineva@enscm.fr

Abstract: Machine learning approaches can drastically decrease the computational time for the predictions of spectroscopic properties in materials, while preserving the quality of the computational approaches. We studied the performance of kernel-ridge regression (KRR) and gradient boosting regressor (GBR) models trained on the isotropic shielding values, computed with density-functional theory (DFT), in a series of different known zeolites containing out-of-frame metal cations or fluorine anion and organic structure-directing cations. The smooth overlap of atomic position descriptors were computed from the DFT-optimised Cartesian coordinates of each atoms in the zeolite crystal cells. The use of these descriptors as inputs in both machine learning regression methods led to the prediction of the DFT isotropic shielding values with mean errors within 0.6 ppm. The results showed that the GBR model scales better than the KRR model.

Keywords: NMR; machine learning; zeolites

Citation: Gaumard, R.; Dragún, D.; Pedroza-Montero, J.N.; Alonso, B.; Guesmi, H.; Malkin Ondík, I.; Mineva, T. Regression Machine Learning Models Used to Predict DFT-Computed NMR Parameters of Zeolites. *Computation* **2022**, *10*, 74. https://doi.org/10.3390/computation10050074

Academic Editor: Henry Chermette

Received: 30 March 2022
Accepted: 10 May 2022
Published: 13 May 2022

1. Introduction

Machine learning (ML) coupled with density functional theory (DFT) calculations has been rapidly emerging for predictions of nuclear magnetic resonance (NMR) isotropic shielding values [1–9]. The role of the experimental NMR investigations to recognise the local atomic environment in chemical and biological systems has been established for decades. Theoretical DFT calculations, using either the gauge-invariant atomic orbital (GIAO) or gauge invariant-projector augmented wave (GIPAW), have been widely employed to improve the NMR signal assignments and/or identify the local structural environment and molecular interactions of the targeted nucleus [10,11]. The interest in the last few years in developing and applying ML models for the prediction of NMR parameters thus originates in the importance of the rapid achievement of accurate theoretical NMR parameters.

Hitherto, several ML models [12] have been built and applied for predicting NMR isotropic shielding (σ_{iso}) or, respectively, the chemical shift ($\delta = \sigma_{ref} - \sigma_{iso}$) of ^{1}H, ^{13}C, ^{13}O, and ^{13}N nuclei in small organic, aromatic molecules or molecular crystals [2,6,13–20]. These ML models comprise deep neural networks (DNNs) [15], convolutional neural networks (CNNs) [16], the IMPRESSION model based on kernel-ridge regression (KRR) [6,19,20], linear-ridge regression [2], gradient boosting regression (GBR) [21,22], graph neural networks (GNNs) [23,24], and the Δ-ML method [7]. Chemical shifts of proteins have been predicted using random forest regression (RFR) [13,14,17,18]. Despite the strong decrease of the computational time to train the model and predict the NMR parameters, in comparison to the GIAO and GIPAW calculations, most of the ML models yielded somewhat less accurate results in comparison to the experimental data than the DFT σ_{iso} with PBE exchange–correlation functionals [7]. Significantly lesser is the amount of works devoted to NMR property calculations in silicates [1–3]. The ML precision in predicting ^{29}Si and ^{17}O chemical shifts in these amorphous solids is found more accurate than in the organic compounds. For example, the ML-predicted deviation from DFT-GIPAW calculations is obtained to be only 0.7 ppm for ^{29}Si and 1.5 ppm for ^{17}O in SiO_2 glasses [2]. The supervised feed-forward neural network representation yielded mean absolute errors (MAEs) of $\delta_{iso} < 1$ ppm for ^{29}Si in ZSM-11 and a-cristobalite [1]. The same NN model also performed very well for the ^{17}O quadrupolar coupling constant predictions, giving MAEs ($C_q(^{17}O)$) of 0.07 MHz in cristobalite and 0.06 MHz in ZSM-11 zeolite.

One of the most significant tasks to take into account in the ML applications is the choice of the descriptors, representing the local chemical environment of each atom in the system. This choice is not trivial because it greatly depends on the shape of the molecular system (simple organic molecules or crystalline materials) and on the considered data set [17]. The most widely used descriptor for predicting the NMR properties in organic molecules and materials is the smooth overlap of the atomic positions (SOAP) descriptor. This descriptor can also be used as a kernel when it is coupled with the kernel-ridge regression methods. Indeed the SOAP descriptor has been already found very efficient to describe the local chemical environment of a large range of chemical compounds, and in particular, it allows obtaining the accurate prediction of NMR properties [1,2]. Furthermore, the symmetry functions are widely used for describing the chemical environment in the neural network representation [1]. Molecular descriptors and fragment descriptors [25] led to predicting with a great accuracy the J-coupling constants in small organic molecules. The ML combination with DFT is therefore a promising tool, and further validations are of high interest.

In this work, we apply two simple state-of-the-art regression ML methods, namely KRR and GBR, to predict σ_{iso} in a set of crystalline zeolite structures, selected from the International Zeolite Association's (IZA) structure database [26]. The zeolites are the crystalline alumino-silicate porous materials with waste industrial applications as catalysts or molecular sieves. The three-dimensional zeolite structure is composed by tetrahedron units with Si atoms in the centre and four oxygen atoms at the vertices, which can organise in a variety of porous frameworks, with pores of sizes varying between 2 and 10 nm [27,28]. ML methods coupled to DFT computations have already emerged for predicting mechanical properties [29], nitrogen adsorption [30], molar volumes, and cohesive energies [31] in zeolites. The success rate of these ML applications to zeolites vary according to the predicted properties and the proposed ML approach [32]. Among the spectroscopy techniques, used to study zeolite structures and chemical compositions, most of the NMR techniques can today be routinely applied to the as-synthesised zeolitic materials. We therefore found it of interest to examine and report in this work the performance of simple ML methods trained on the computed DFT σ_{iso} values in a series of known zeolite structures.

2. Methods and Computational Details

2.1. Kernel-Ridge Regression

The first ML approach used by us is KRR [6], which consists of a combination of the ridge regression and the kernel method. The KRR model is suitable for complex continuous data, which cannot be described by a linear regression. Unlike the linear regression, the kernel-ridge regression method offers larger flexibility by transforming the input with a regression function.

Below, we briefly illustrate the KRR scheme. In the case of the ML linear regression algorithm, the goal is to minimise a function Ω called the quadratic cost [33], which is defined as

$$\Omega(\mathbf{w}) = \frac{1}{2}\sum_{i=1}^{N}\left(Y_i - \mathbf{w}^T\mathbf{X}_i\right)^2, \tag{1}$$

where $\mathbf{X}_i$ represents the vector of the input data, Y_i are the scalar output data, N corresponds to the dimension of the input data, and the vector $\mathbf{w}$ is the vector of weights that will be optimised during the training process. In the case of the ridge regression algorithm, an additional term is implemented to the previous quadratic cost in order to prevent over-fitting problems during the training stage by regularising its value. Hence, the form of the quadratic cost becomes

$$\Omega(\mathbf{w}) = \frac{1}{2}\sum_{i=1}^{N}\left(Y_i - \mathbf{w}^T\mathbf{X}_i\right)^2 + \frac{1}{2}\lambda||\mathbf{w}||_2^2, \tag{2}$$

where λ is a positive parameter that controls the value of the vector norm $\mathbf{w}$. This step is called L2-regularisation because of the use of the L2-norm of the vector $\mathbf{w}$. In order to determine the parameter λ, a cross-validation algorithm is widely used [34]. Thus, by minimising the function $\Omega(\mathbf{w})$, it leads to a simple linear problem to be solved for the set of weights as follows:

$$\sum_{i=1}^{N}\left(Y_i - \mathbf{w}^T\mathbf{X}_i\right)\mathbf{X}_i = \lambda\mathbf{X}_i. \tag{3}$$

These optimised weights are thus obtained as

$$\mathbf{w} = \left(\lambda\mathbf{I} + \sum_{i=1}^{N}\mathbf{X}_i\mathbf{X}_i^T\right)^{-1}\left(\sum_{j=1}^{N}Y_j\mathbf{X}_j\right), \tag{4}$$

where $\mathbf{I}$ is the identity matrix.

This linear regression method is limited to problems that can be described as a linear function; thus, to overcome this limit, a non-linear kernel function is introduced in order to measure the similarity between two samples of a high-dimensional space. The most widely used kernel function is the Gaussian kernel function. In the KRR method, the vector of the input data, $\mathbf{X}_i$, is substituted by the non-linear kernel function $\varphi(\mathbf{X}_i)$. Therefore, we can rewrite the expression of the optimised weight parameters as a function of $\varphi(\mathbf{X}_i)$:

$$\begin{aligned}\mathbf{w} &= \left(\lambda\mathbf{I} + \sum_{i=1}^{N}\varphi(\mathbf{X}_i)\varphi(\mathbf{X}_i^T)\right)^{-1}\left(\sum_{j=1}^{N}Y_j\varphi(\mathbf{X}_j)\right)\\ &= \left(\lambda\mathbf{I} + \varphi(\mathbf{X}_i)\varphi(\mathbf{X}_i^T)\right)^{-1}\varphi(\mathbf{X}_i)Y_i\\ &= \varphi(\mathbf{X}_i)\left(\varphi(\mathbf{X}_i^T)\varphi(\mathbf{X}_i) + \lambda\mathbf{I}\right)^{-1}Y_i.\end{aligned} \tag{5}$$

By defining the coefficient $\alpha_i = \left(\varphi(\mathbf{X}_i^T)\varphi(\mathbf{X}_i) + \lambda\mathbf{I}\right)^{-1}Y_i$, the optimised weights are simply expressed as

$$\mathbf{w} = \sum_{i=1}^{N}\alpha_i\varphi(\mathbf{X}_i). \tag{6}$$

Therefore, during the training phase of the kernel-ridge regression, the aim is to calculate α_i, which are subsequently used to predict the output values. For the KRR model, we used the code from the open repository [35]. The similarity of the input vectors is determined based on the user-defined similarity function, e.g., kernel, in our case, the difference between the SOAP vectors.

2.2. Gradient Boosting Regression

The gradient boosting regression is a powerful regression firstly introduced by Freund and Schapire [36,37] through an adaptive boosting algorithm. At the beginning, this regression method was used for classification problems [38] and later on adapted for regression problems [21,22]. The aim of the gradient boosting regression method is to find a function f^* that minimises the loss function Θ [39] defined as

$$f^*(\mathbf{X}, Y) = \arg\min E_{\mathbf{X},Y}[\Theta(Y, f(\mathbf{X}))], \quad (7)$$

where $\mathbf{X}$ is the vector of input variables, Y is the output variable, and $E_{\mathbf{X},Y}$ represents the floor function applied for the vector $\mathbf{X}$ and the variable Y. In the boosted model, the function $f(\mathbf{X})$ is defined as a weighted linear combination of base learners by the following formula:

$$f(\mathbf{X}) = \sum_{i=1}^{N} \alpha_i h_i(\mathbf{X}, \beta_i), \quad (8)$$

where α_i are the real coefficients of the linear combination and β_i are the parameters of the base learners $h_i(\mathbf{X}, \beta_i)$. The minimisation of the loss function Θ is carried out via an optimisation of the function f using the recursive relation

$$f_{m+1}(\mathbf{X}) = f_m(\mathbf{X}) + \arg\min \sum_{i=1}^{N} \Theta(Y_i, f_m(\mathbf{X}_i + h_{m+1}(\mathbf{X}_i))). \quad (9)$$

The here-used gradient boosting regression method enables us to create strongly learning trees from poorly learning trees [40]. This approach utilises boosting so that the trees are created sequentially, as opposed to random forests, where the trees are generated in parallel. Each new tree is created with an effort to reduce the prediction error learning from the errors of the previous tree. The goal is to achieve the lowest possible error while keeping the predicted values as accurate as possible. We used the Anaconda distribution for Python 3.8.5, utilising the scikit-learn program package [40,41] with the GBR model, where the *random_state* hyperparameter was set to 0 and the rest of the hyperparameters were set to the default values.

2.3. SOAP Descriptors

Two data sets in comma-separated values (CSV) format were prepared using the DFT-optimised Cartesian coordinates of the zeolites and the isotropic shielding value in ppm for each atom in the zeolites. The first data set contains the Cartesian coordinates (x, y, and z) of each zeolite, the calculated σ_{iso}, the name of the chemical element, and the name of the zeolite (taken from the IZA). The second CSV file contains 3×3 tensors and the name of the corresponding zeolite. We used the DScribe package [42] to convert our data to smooth overlap of atomic positions (SOAP) descriptor vectors. Individual structures were represented as Atoms class objects from the ASE package [43,44] with the use of 3×3 tensors. We began by creating a DScribe.SOAP object, for which the parameters such as the number of basis functions, range, level l, and a list of all elements in our data were set. Subsequently, the DScribe.SOAP.create function was used to create a SOAP vector for each atom. The complete data set was split into a training and test set in a ratio of 8:2.

2.4. DFT Computational Details

A periodic DFT-based approach was used to carry out a full geometrical optimisation (atomic positions and unit-cell parameters) of all the structures in the data set. The geometrical optimisations were carried out with the Crystal17 program, based on atom-centred Gaussian orbitals [45]. All-electron basis functions of double-ζ quality were used as follows: 6-31d1 for O, N, C, and H [46]; 85-11G* for Al [47] and Pople's basis set (6–21G) with polarisation for Si. The generalised gradient-corrected PBE approximation was used as the exchange correlation (XC) functional, augmented by the empirical London dispersion (D3) term with the Becke–Johnson damping function [48]. The optimised structural parameters of zeolites with OSDA, obtained with the Crystal code and all-electron databases, were found by us to agree well with the experimental bond distances and bond angles [49–51].

For this reason, we applied the same computational protocol for the optimisation of the zeolite structures in this work.

Single-point energy calculations were carried out for these optimised geometries in order to compute the isotropic shielding values of all the atoms in the zeolites. For this, we used the open-source code QUANTUM ESPRESSO [52–54] with the GIPAW method in combination with the ultrasoft pseudopotentials with GIPAW reconstruction [55,56], from the USSP pseudopotential database [57]. The wave-function and charge density energy cut-offs were set to 60 Ry and 720 Ry, respectively. A Monkhorst–Pack grid of k-points [58] corresponding to a maximum spacing of 0.06 Å^{-1} in the reciprocal space was used. The self-consistent field (SCF) energy convergence tolerance was set to 10^{-10}.

3. Results and Discussion

3.1. Data Set and DFT Isotropic Shielding

To ensure a heterogeneity of the atomic environments giving rise to different DFT σ_{iso} values, we considered zeolites containing Al, Na, and Li cations, as well as MFI-type zeolites, containing the organic structure-directing agents (OSDAs), which are the tetrapropylammonium (TPA) and tripropylethylammonium (TPEA) cations. Among the MFI-OSDA types of structures, we considered five pure silica structures (silicalite-1), labelled as MFI-TPA and MFI-TPEA in Table S1 in the Supporting Information (SI) Section. The four MFI-ETPA structures present the location of the TPEA ethyl chain either in the direct or zig-zag channels. In silicalite-1 zeolites, the fluorine anion is the charge-compensating ion. The remaining MFI-OSDA zeolites are those with the TPA cation and one Al^{3+}, which substitutes at each of the 24 non-equivalent Si-sites of the asymmetric unit. The initial structures of the pure inorganic zeolites were the crystallographic information files (CIFs) that were collected from the IZA database. We built the MFI-OSDA structures from the available crystallographic data for TPA (ETPA) [59] and ZSM-5-TPA [60] zeolites. These structures were optimised in our previous studies [61,62] using the same level of DFT theory. We thus constructed a more heterogeneous data set that contains Si, Al, N, C, H, Li, and F atomic environments. The geometries and DFT σ_{iso} values of all atoms were used in the ML training and prediction calculations.

To access roughly the quality of the DFT σ_{iso} results, we correlated them with the experimental chemical shifts of ^{29}Si, which are available in the IZA database. The zeolites for which $\delta^{29}Si$ were collected are labelled by an asterisk in Table S1.

The linear fitting between the DFT and experimental data, illustrated in Figure 1, demonstrates that the PBE-D3 results followed reasonably well the overall experimental trend for the selected zeolites. It is worth noting that the experimental NMR data were recorded under different experimental conditions [26] and often for non-ideal zeolite structures that might contain defects, such as silanols, water, hydroxides, and in- or out-framework cations. Taking into account these factors and the linear fitting R^2 coefficient of 0.9995, as well as the root-mean-squared error (RMSE) of 2.44 ppm of the DFT values with respect to the fitted values against the experimental isotropic shieldings, we concluded a rather good correlation between the computed and experimental results.

The distributions of the calculated DFT isotropic shieldings of ^{29}Si, ^{17}O, ^{27}Al, ^{13}C, and ^{1}H are reported in Figure 2. The majority of the Si atoms have DFT σ_{iso}(^{29}Si) in the range 422–426 ppm, as follows from the maximum number of the chemical environments in this interval. Nevertheless, the predominant number of σ_{iso}(^{29}Si) is obtained in a 400–440 ppm interval, and there are few Si-sites, for which σ_{iso}(^{29}Si) < 350 ppm. The other nuclei, largely presented in the zeolites, are ^{17}O and ^{1}H. The peakin the oxygen atoms' distribution indicates that the largest number of oxygen sites has σ_{iso}(^{17}O) values at around 196 ppm. The σ_{iso}(^{17}O) values span a large interval between 150 and 250 ppm with several outliers outside this region. The isotropic shieldings of hydrogen sites are between 22 and 29 ppm, and the distribution of ^{13}C is characterised by two distinguished peaksat around 100 and 150 ppm. The hydrogen and carbon sites belong to OSDAs in MFI and ZSM-5 zeolite types.

In the studied structures, the number of Al^{3+} cations is significantly smaller, and their $\sigma_{iso}(^{27}Al)$ are spread in the 450–550 ppm interval.

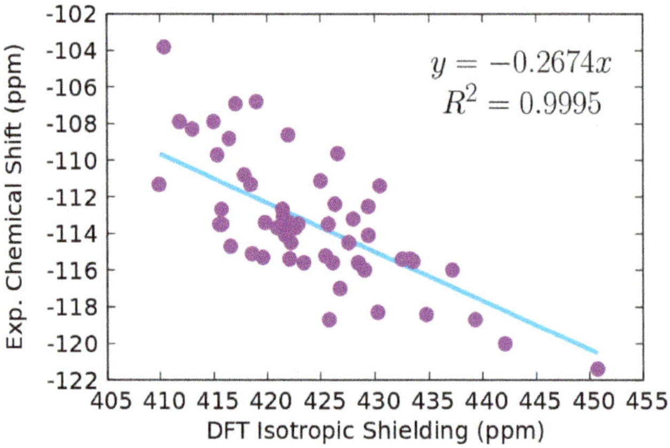

Figure 1. Comparison between DFT isotropic shielding and experimental chemical shift (both in ppm) alongside the linear fitting of the data (blue line).

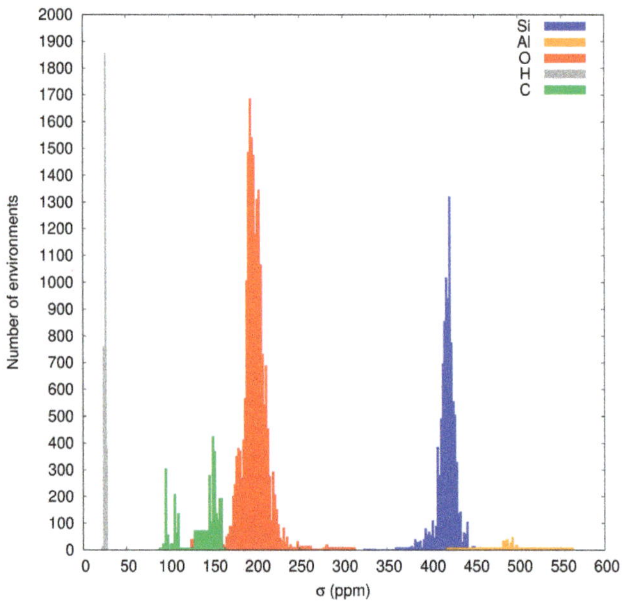

Figure 2. Distribution of the number of oxygen (O), silicon (Si), aluminium (Al), carbon (C), and hydrogen (H) atomic environments in zeolites, according to their isotropic shielding. The histograms are obtained with an interval of 2.0 ppm for C, O, Al, Si, and 1.0 ppm for H in the count of the number of atomic environment.

3.2. *KRR and GBR Models to Predict NMR Isotropic Shielding*

In this section, we discuss the performance of the KRR and GBR models. The zeolite data set, discussed above, was split into training (first 80%) and validating sets (last 20%) of zeolites. As discussed in the Methods section, we used the training set to build the SOAP descriptors. First, we considered all atoms in the zeolite structures that represent a total of 14,513 atomic environments in the KRR and GBR models. In the second part, only the silicon atoms, with their Cartesian coordinates and $\sigma_{iso}(^{29}Si)$ values, were collected in a smaller data set. The choice of Si atoms is because $\sigma_{iso}(^{29}Si)$ experimental data are most often considered as fingerprints of the local structure around Si-sites and can account for the presence of silanols, oxygen, or silicon vacancies or other defect types. The number of Si-atomic environments in this smaller data set was reduced to 3756, and among them, 3004 were used as training data and 752 as validation data.

The distribution of the differences between σ_{iso}, computed with DFT and those predicted from the KRR and GBR models, is presented in Figures 3 and 4, respectively, whereas the correlations between the predicted vs. the DFT values are reported in Figures S1 and S2. In the KRR model, the regularisation hyperparameter α was set to 0.1. Here, only one outlier value is identified in the results from the KRR application. The predicted outlier σ_{iso} = 487.7 ppm is down-shifted by about 26 ppm with respect to the "true" DFT σ_{iso} = 513.82 ppm. This outlier is in the silimanite structure with the Cartesian coordinates equal to 2.67, 1.46, and 1.10 Å. The application of both the KRR and GBR models on the smaller set containing only the Si atomic environments and their $\sigma_{iso}(^{29}Si)$ values in the interval 380–450 ppm yielded again an excellent correlation between the predicted vs. DFT computed data, as follows from the plot in Figures 5 and S3 (KRR) and Figures 6 and S4 (GBR). We obtained only one remarkable outlier $\sigma_{iso}(^{29}Si)$ value when using the KRR model. This outlier is now in the ITW zeolite. Its predicted $\sigma_{iso}(^{29}Si)$ value of 431.51 ppm is up-shifted with respect to the computed with DFT $\sigma_{iso}(^{29}Si)$ = 419.99 ppm. The coordinates of the outlier Si atom are: x = −1.31, −1.24, −2.72 Å. No outliers were identified when applying the GBR model.

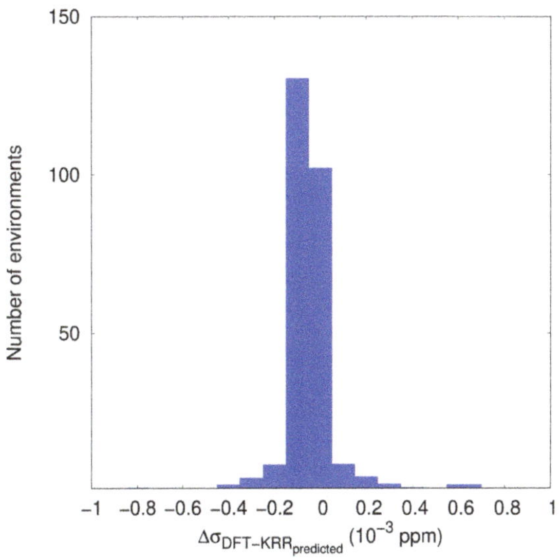

Figure 3. Distribution of the differences between the isotropic shielding values computed with DFT and those predicted with the ML-KRR method. All atomic environments are considered. The histograms are obtained with an interval of 0.1 ppm in the count of the number of atomic environments.

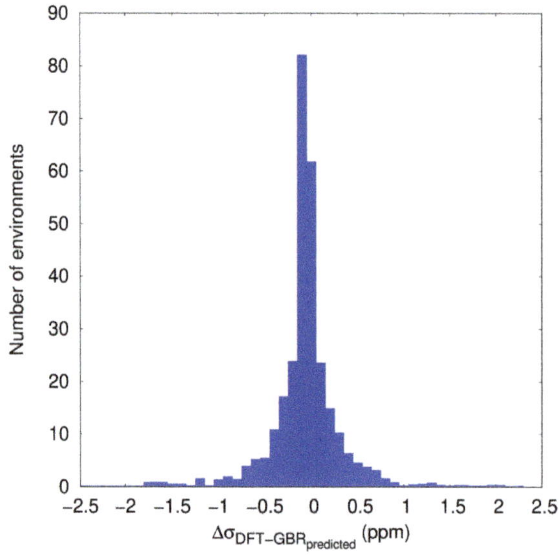

Figure 4. Distribution of the differences between the isotropic shielding values computed with DFT and those predicted with the ML-GBR method. All atomic environments are considered. The histograms are obtained with an interval of 0.1 ppm in the count of the number of atomic environments.

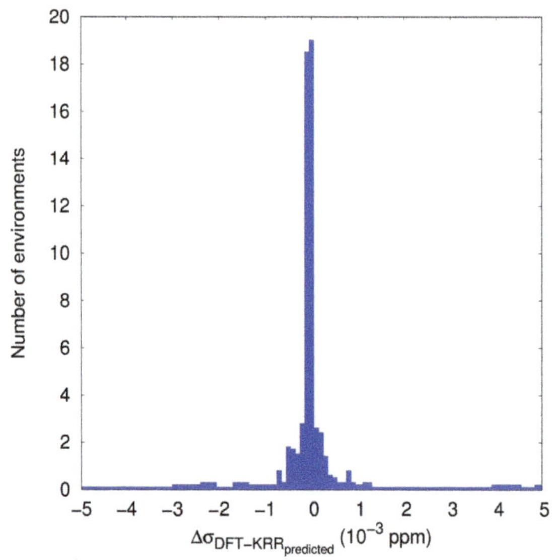

Figure 5. Distribution of the differences between the isotropic shielding values computed with DFT and those predicted with the ML-KRR method. Only silicon atomic environments are considered. The histograms are obtained with an interval of 0.1 ppm in the count of the number of atomic environments.

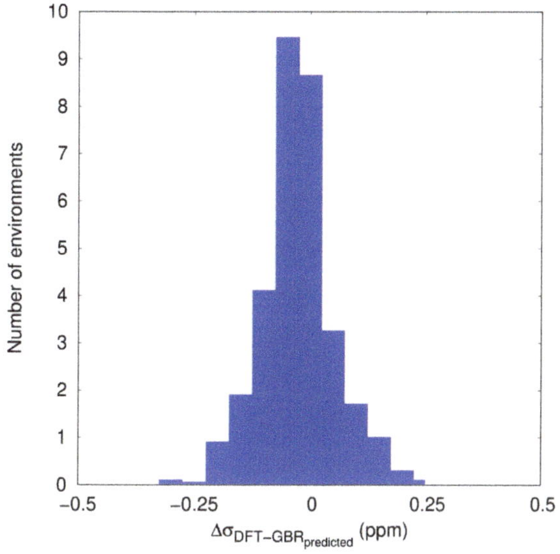

Figure 6. Distribution of the differences between the isotropic shielding values computed with DFT and those predicted with the ML-GBR method. Only silicon atomic environments are considered. The histograms are obtained with an interval of 0.05 ppm in the count of the number of atomic environments.

The mean-squared error (MSE), root-mean-squared error (RMSE), standard deviation error (STD), and the mean absolute error (MAE), as well as the R^2 coefficients are compared in Table 1. These results confirm the very good performance of both ML models leading to the R^2 coefficients of 0.997 (KRR, all atoms in the data set) and 0.999 for the other three sets of predictions. The MAE, RMSE, and MSE results are <0.6 ppm. The most notable differences between the performance of KRR and GBR models are the training and prediction time, also reported in Table 1. The GBR model appears to be faster by two orders of magnitude than the KRR model. Therefore, we concluded that the GBR model scales better than the KRR model.

Table 1. Training and prediction time, mean absolute error (MAE), root-mean-squared error (RMSE), mean-squared error (MSE), absolute and square standard deviation errors (STD AE and STD SE), and the R^2 coefficient of the KRR and GBR model predictions together with the average of both predicted values (AVG). The data shown are only for the Si atoms (Si) and all atoms (All) in the zeolite.

	Machine Learning Models					
Parameters	**KRR (All)**	**GBR (All)**	**AVG (All)**	**KRR (Si)**	**GBR (Si)**	**AVG (Si)**
Training time (s)	3796.4	49.0	-	136.7	12.2	-
Prediction time (s)	1900.8	0.6	-	74.2	0.02	-
MAE (ppm)	0.023	0.226	0.116	0.037	0.057	0.046
STD AE (ppm)	0.524	0.538	0.236	0.490	0.054	0.246
MSE (ppm)	0.275	0.341	0.069	0.241	0.006	0.062
STD SE (ppm)	12.669	8.158	1.285	5.304	0.011	1.341
RMSE (ppm)	0.524	0.584	0.262	0.491	0.008	0.250
R^2	0.999	0.999	-	0.999	0.997	-

A combination of the KRR and GBR models might remove outliers and reduce the errors. A simple estimation of the combination between both regression approaches was carried out by assuming equal weight coefficients (0.5), that is taking the mean of the predicted isotropic shielding values by the KRR and GBR ML models. The resulting distributions of the differences between the mean of the predicted and DFT values, as well as the correlation plots between the predicted vs. DFT isotropic shielding data are plotted in Figures S5–S8. The average of both ML models approached the quality of the GBR predictions. The outlier, identified in the reduced set of silicon atoms, equals 425.78 ppm; thus, it is predicted to be closer to DFT σ_{iso}(^{29}Si = 419.99 ppm). The combination of both regression methods led to a significant decrease of the STD errors, MSE, and RMSE (Table 1) in comparison to the respective errors found from the application of each ML model. It therefore follows that the combination of regression methods might be a useful approach toward the removal of errors of a single regression model.

Discussing the quality of the predicted σ_{iso} results with respect to those computed with DFT is not trivial in the case of the zeolite structures. As noted above, the rigorous comparison of the computed σ_{iso}(^{29}Si) and the experimental chemical shift data, collected from the IZA database (see Figure 1), is not straightforward. Despite this fact, considering that the RMS error of the linear fit of DFT σ_{iso}(^{29}Si) vs. the experimental δ_{iso}(^{29}Si) results (Figure 1) amounts to 2.44 ppm, we concluded that the predicted values with RMSE in the range 0.008–0.5 ppm do not worsen the quality of the DFT method used by us. This suggests a very promising application of both the KRR and GBR models, not only to predict the σ_{iso} of ^{29}Si, but also for the other nuclei in the the zeolite data set, because outliers were not identified among these nuclei. However, we note the limited number or heterogeneity of C, H, F, and Li atomic environments. It is therefore not surprising that outliers were not established among those atoms. On the other hand, the number of oxygen environments is four-times the number of Si environments in the all-atom data set. The excellent correlation between the predicted vs. DFT-computed values can be therefore also concluded for σ_{iso}(^{17}O). The combination of SOAP descriptors with simple ML regression models appears to lead to a promising predictive capability of NMR isotropic shielding of ^{29}Si and ^{17}O in the zeolites, which is in line with previous work using SOAP descriptors and regression methods for predictions of NMR parameters in the organic solids [4] and silicates [1,2].

4. Conclusions

In this paper, we studied the capability of two simple machine learning regression models, KRR and GBR, to predict the σ_{iso} values in a series of known zeolites. The DFT calculations with periodic boundary conditions were carried out to fully optimise the crystallographic zeolite structures, collected from the IZA database and the MFI-OSDA types of zeolites, and to compute the σ_{iso} values for each atom in the data set. In addition to the inorganic zeolite framework, composed by Si, O, and Al atoms, the data set contains various out-frame cations, such as Li^+, F^-, and TPA and TPEA molecular cations.

The quality of the DFT σ_{iso}(^{29}Si) was found to be reasonably good compared to the available experimental δ_{iso}(^{29}Si) in the IZA database. The SOAP descriptors, obtained from the optimised Cartesian coordinates of each atom in the DFT-based data set, were used as inputs in both machine learning regression models. Both the KRR and GBR approaches predicted isotropic shieldings with mean errors smaller than 1 ppm. The comparison between the training and predictions time gave a preference to the GBR, found to scale better than the KRR model. These results are promising for more extensive ML applications based on simple regression in combination with DFT calculations in order to accelerate the calculations of NMR parameters in various zeolitic materials.

Supplementary Materials: The following Supporting Information can be downloaded at: https://www.mdpi.com/article/10.3390/computation10050074/s1, Table S1, Figures S1–S8.

Author Contributions: Conceptualisation: T.M. and I.M.O.; construction zeolite database: R.G., H.G., B.A. and T.M.; ML software: D.D. and I.M.O.; data curation: J.N.P.-M., T.M. and I.M.O.; writing—original draft preparation: R.G., J.N.P.-M. and T.M.; supervision: T.M. All authors have read and agreed to the published version of the manuscript.

Funding: This research received no external funding.

Data Availability Statement: The optimised atomic Cartesian coordinates and isotropic shielding values are available at: https://mycore.core-cloud.net/index.php/s/FH6EGSBVBwwZZ8R (accessed on 29 March 2022).

Acknowledgments: The authors acknowledge the access to the HPC resources of CCRT/CINES/IDRIS, which was granted under the allocation A0090807369 and A0110807369 by GENCI.

Conflicts of Interest: The authors declare no conflict of interest.

References

1. Cuny, J.; Xie, Y.; Pickard, C.J.; Hassanali, A.A. Ab Initio Quality NMR Parameters in Solid-State Materials Using a High-Dimensional Neural-Network Representation. *J. Chem. Theory Comput.* **2016**, *12*, 765–773. [CrossRef]
2. Chaker, Z.; Salanne, M.; Delaye, J.M.; Charpentier, T. NMR shifts in aluminosilicate glasses via machine learning. *Phys. Chem. Chem. Phys.* **2019**, *21*, 21709–21725. [CrossRef] [PubMed]
3. Liu, S.; Li, J.; Bennett, K.C.; Ganoe, B.; Stauch, T.; Head-Gordon, M.; Hexemer, A.; Ushizima, D.; Head-Gordon, T. Multiresolution 3D-DenseNet for Chemical Shift Prediction in NMR Crystallography. *J. Phys. Chem. Lett.* **2019**, *10*, 4558–4565. [CrossRef] [PubMed]
4. Paruzzo, F.M.; Hofstetter, A.; Musil, F.; De, S.; Ceriotti, M.; Emsley, L. Chemical shifts in molecular solids by machine learning. *Nat. Commun.* **2018**, *9*, 4501. [CrossRef] [PubMed]
5. Rupp, M.; Ramakrishnan, R.; von Lilienfeld, O.A. Machine Learning for Quantum Mechanical Properties of Atoms in Molecules. *J. Phys. Chem. Lett.* **2015**, *6*, 3309–3313. [CrossRef]
6. Gerrard, W.; Bratholm, L.A.; Packer, M.J.; Mulholland, A.J.; Glowacki, D.R.; Butts, C.P. IMPRESSION—Prediction of NMR parameters for 3-dimensional chemical structures using machine learning with near quantum chemical accuracy. *Chem. Sci.* **2020**, *11*, 508–515. [CrossRef]
7. Unzueta, P.A.; Greenwell, C.S.; Beran, G.J.O. Predicting Density Functional Theory-Quality Nuclear Magnetic Resonance Chemical Shifts via Δ-Machine Learning. *J. Chem. Theory Comput.* **2021**, *17*, 826–840. [CrossRef]
8. Cordova, M.; Balodis, M.; Hofstetter, A.; Paruzzo, F.; Nilsson Lill, S.O.; Eriksson, E.S.E.; Berruyer, P.; Simões de Almeida, B.; Quayle, M.J.; Norberg, S.T.; et al. Structure determination of an amorphous drug through large-scale NMR predictions. *Nat. Commun.* **2021**, *12*, 2964. [CrossRef]
9. Aguilera-Segura, S.M.; Dragún, D.; Gaumard, R.; Di Renzo, F.; Malkin Ondík, I.; Mineva, T. Thermal fluctuations and conformational effects on NMR parameters in β-O-4 lignin dimer from QM/MM and machine learning approaches. *Phys. Chem. Chem. Phys.* **2022**, *24*, 8820–8831. [CrossRef]
10. Charpentier, T. The PAW/GIPAW approach for computing NMR parameters: A new dimension added to NMR study of solids. *Solid State Nucl. Magn. Reson.* **2011**, *40*, 1–20. [CrossRef]
11. Dib, E.; Mineva, T.; Alonso, B. Chapter Three—Recent Advances in ^{14}N Solid-State NMR *Annu. Rep. NMR Spectrosc.* **2016**, *87*, 175–235. [CrossRef]
12. Jonas, E.; Kuhn, S.; Schlörer, N. Prediction of chemical shift in NMR: A review. *Magn. Reson. Chem.* 2021, *in press*. [CrossRef] [PubMed]
13. Arun, K.; Langmead, C.J. Structure based chemical shift prediction usgin random forests non-linear regression. In Proceedings of the 4th Asia-Pacific Bioinformatics Conference, Taipei, Taiwan, 13–16 February 2006; Series on Advances in Bioinformatics and Computational Biology; World Scientific Publishing Co.: Singapore, 2005; Volume 3, pp. 317–326. [CrossRef]
14. Han, B.; Liu, Y.; Ginzinger, S.W.; Wishart, D.S. SHIFTX2: Significantly improved protein chemical shift prediction. *J. Biomol. NMR* **2011**, *50*, 43. [CrossRef] [PubMed]
15. Gao, P.; Zhang, J.; Peng, Q.; Zhang, J.; Glezakou, V.A. General Protocol for the Accurate Prediction of Molecular ^{13}C/^{1}H NMR Chemical Shifts via Machine Learning Augmented DFT. *J. Chem. Inf. Model.* **2020**, *60*, 3746–3754. [CrossRef]
16. Gao, P.; Zhang, J.; Sun, Y.; Yu, J. Toward Accurate Predictions of Atomic Properties via Quantum Mechanics Descriptors Augmented Graph Convolutional Neural Network: Application of This Novel Approach in NMR Chemical Shifts Predictions. *J. Phys. Chem. Lett.* **2020**, *11*, 9812–9818. [CrossRef]
17. Haghighatlari, M.; Li, J.; Heidar-Zadeh, F.; Liu, Y.; Guan, X.; Head-Gordon, T. Learning to Make Chemical Predictions: The Interplay of Feature Representation, Data, and Machine Learning Methods. *Chem* **2020**, *6*, 1527–1542. [CrossRef]
18. Li, J.; Bennett, K.C.; Liu, Y.; Martin, M.V.; Head-Gordon, T. Accurate prediction of chemical shifts for aqueous protein structure on "Real World" data. *Chem. Sci.* **2020**, *11*, 3180–3191. [CrossRef]
19. Gupta, A.; Chakraborty, S.; Ramakrishnan, R. Revving up ^{13}C NMR shielding predictions across chemical space: Benchmarks for atoms-in-molecules kernel machine learning with new data for 134 kilo molecules. *Mach. Learn. Sci. Technol.* **2021**, *2*, 035010. [CrossRef]
20. Gerrard, W.; Yiu, C.; Butts, C.P. Prediction of ^{15}N chemical shifts by machine learning. *Magn. Reson. Chem.* 2021, *in press*. [CrossRef]

21. Beygelzimer, A.; Hazan, E.; Kale, S.; Luo, H. Online gradient boosting. In Proceedings of the Advances in Neural Information Processing Systems 28: Annual Conference on Neural Information Processing Systems 2015, Montreal, QC, Canada, 7–12 December 2015; pp. 2458–2466.
22. Biau, G.; Cadre, B.; Rouvière, L. Accelerated gradient boosting. *Mach. Learn.* **2019**, *108*, 971–992. [CrossRef]
23. Guan, Y.; Shree Sowndarya, S.V.; Gallegos, L.C.; John, P.C.S.; Paton, R.S. Real-time prediction of ^{1}H and ^{13}C chemical shifts with DFT accuracy using a 3D graph neural network. *Chem. Sci.* **2021**, *12*, 12012–12026. [CrossRef] [PubMed]
24. Jonas, E.; Kuhn, S. Rapid prediction of NMR spectral properties with quantified uncertainty. *J. Cheminform.* **2019**, *11*, 50. [CrossRef] [PubMed]
25. Ito, K.; Xu, X.; Kikuchi, J. Improved Prediction of Carbonless NMR Spectra by the Machine Learning of Theoretical and Fragment Descriptors for Environmental Mixture Analysis. *Anal. Chem.* **2021**, *93*, 6901–6906. [CrossRef] [PubMed]
26. Baerlocher, C.; McCusker, L. Database of Zeolite Structures. Available online: http://www.iza-structure.org/databases/ (accessed on 28 March 2022).
27. Hölderich, W.; Hesse, M.; Näumann, F. Zeolites: Catalysts for Organic Syntheses. *Angew. Chem. Int. Ed. Engl.* **1988**, *27*, 226–246. [CrossRef]
28. Cejka, J.; van Bekkum, H.; Corma, A.; Schueth, F. Introduction to Zeolite Molecular Sieves. In *Introduction to Zeolite Molecular Sieves*, 3rd ed.; John Wiley and Sons: Hoboken, NJ, USA, 2007; p. 1094.
29. Evans, J.D.; Coudert, F.X. Predicting the Mechanical Properties of Zeolite Frameworks by Machine Learning. *Chem. Mater.* **2017**, *29*, 7833–7839. [CrossRef]
30. Gu, Y.; Liu, Z.; Yu, C.; Gu, X.; Xu, L.; Gao, Y.; Ma, J. Zeolite Adsorption Isotherms Predicted by Pore Channel and Local Environmental Descriptors: Feature Learning on DFT Binding Strength. *J. Phys. Chem. C* **2020**, *124*, 9314–9328. [CrossRef]
31. Helfrecht, B.A.; Semino, R.; Pireddu, G.; Auerbach, S.M.; Ceriotti, M. A new kind of atlas of zeolite building blocks. *J. Chem. Phys.* **2019**, *151*, 154112. [CrossRef]
32. Kwak, S.J.; Kim, H.S.; Park, N.; Park, M.J.; Lee, W.B. Recent progress on Al distribution over zeolite frameworks: Linking theories and experiments. *Korean J. Chem. Eng.* **2021**, *38*, 1117–1128. [CrossRef]
33. Welling, M. Kernel ridge Regression. Available online: https://web2.qatar.cmu.edu/~gdicaro/10315-Fall19/additional/welling-notes-on-kernel-ridge.pdf (accessed on 28 March 2022).
34. An, S.; Liu, W.; Venkatesh, S. Fast cross-validation algorithms for least squares support vector machine and kernel ridge regression. *Pattern Recognit.* **2007**, *40*, 2154–2162. [CrossRef]
35. ML-CSC-tutorial. Available online: https://github.com/fullmetalfelix/ML-CSC-tutorial (accessed on 24 March 2022).
36. Freund, Y.; Schapire, R.E. A Decision-Theoretic Generalization of On-Line Learning and an Application to Boosting. *J. Comput. Syst. Sci.* **1997**, *55*, 119–139. [CrossRef]
37. Freund, Y.; Schapire, R.E. A Short Introduction to Boosting. In Proceedings of the Sixteenth International Joint Conference on Artificial Intelligence (IJCAI '99), Stockholm, Sweden, 31 July–6 August 1999; Morgan Kaufmann Publishers Inc.: San Francisco, CA, USA, 1999; pp. 1401–1406.
38. Natekin, A.; Knoll, A. Gradient boosting machines, a tutorial. *Front. Neurorobot.* **2013**, *7*. [CrossRef] [PubMed]
39. He, Z.; Lin, D.; Lau, T.; Wu, M. Gradient Boosting Machine: A Survey. *arXiv* **2019**, arXiv:1908.06951.
40. Hastie, T.; Tibshirani, R.; Friedman, J. Boosting and Additive Trees. In *The Elements of Statistical Learning: Data Mining, Inference, and Prediction*; Springer: New York, NY, USA, 2009; pp. 337–387. [CrossRef]
41. Pedregosa, F.; Varoquaux, G.; Gramfort, A.; Michel, V.; Thirion, B.; Grisel, O.; Blondel, M.; Prettenhofer, P.; Weiss, R.; Dubourg, V.; et al. Scikit-learn: Machine Learning in Python. *J. Mach. Learn. Res.* **2011**, *12*, 2825–2830.
42. Himanen, L.; Jäger, M.O.; Morooka, E.V.; Federici Canova, F.; Ranawat, Y.S.; Gao, D.Z.; Rinke, P.; Foster, A.S. DScribe: Library of descriptors for machine learning in materials science. *Comput. Phys. Commun.* **2020**, *247*, 106949. [CrossRef]
43. Larsen, A.H.; Mortensen, J.J.; Blomqvist, J.; Castelli, I.E.; Christensen, R.; Dułak, M.; Friis, J.; Groves, M.N.; Hammer, B.; Hargus, C.; et al. The atomic simulation environment—A Python library for working with atoms. *J. Phys. Condens. Matter* **2017**, *29*, 273002. [CrossRef]
44. Bahn, S.R.; Jacobsen, K.W. An object-oriented scripting interface to a legacy electronic structure code. *Comput. Sci. Eng.* **2002**, *4*, 56–66. [CrossRef]
45. Dovesi, R.; Erba, A.; Orlando, R.; Zicovich-Wilson, C.M.; Civalleri, B.; Maschio, L.; Rérat, M.; Casassa, S.; Baima, J.; Salustro, S.; et al. Quantum-mechanical condensed matter simulations with CRYSTAL. *WIRES Comput. Mol. Sci.* **2018**, *8*, e1360. [CrossRef]
46. Gatti, C.; Saunders, V.R.; Roetti, C. Crystal field effects on thetopological properties of the electron density in molecular crystals: The case of urea. *J. Chem. Phys.* **1994**, *101*, 10686–10696. [CrossRef]
47. Catti, M.; Valerio, G.; Dovesi, R.; Causà, M. Quantum-mechanical calculation of the solid-state equilibrium MgO+α-$Al_2O_3 \rightleftarrows MgAl_2O_4$ (spinel) versus pressure. *Phys. Rev. B* **1994**, *49*, 14179–14187. [CrossRef]
48. Grimme, S.; Antony, J.; Ehrlich, S.; Krieg, H. A consistent and accurate ab initio parametrization of density functional dispersion correction (DFT-D) for the 94 elements H-Pu. *J. Chem. Phys.* **2010**, *132*, 154104. [CrossRef]
49. Dib, E.; Mineva, T.; Gaveau, P.; Véron, E.; Sarou-Kanian, V.; Fayon, F.; Alonso, B. Probing Disorder in Al-ZSM-5 Zeolites by 14N NMR Spectroscopy. *J. Phys. Chem. C* **2017**, *121*, 15831–15841. [CrossRef]

50. Mineva, T.; Dib, E.; Gaje, A.; Petitjean, H.; Bantignies, J.L.; Alonso, B. Zeolite Structure Direction: Identification, Strength and Involvement of Weak CHO Hydrogen Bonds. *Chem. Phys. Chem.* **2020**, *21*, 149–153. [CrossRef] [PubMed]
51. Al-Nahari, S.; Ata, K.; Mineva, T.; Alonso, B. Ubiquitous Presence of Intermolecular CHO Hydrogen Bonds in As-synthesized Host-Guest Zeolite Materials. *ChemistrySelect* **2021**, *6*, 9728–9734. [CrossRef]
52. Giannozzi, P.; Baroni, S.; Bonini, N.; Calandra, M.; Car, R.; Cavazzoni, C.; Ceresoli, D.; Chiarotti, G.L.; Cococcioni, M.; Dabo, I.; et al. QUANTUM ESPRESSO: A modular and open-source software project for quantum simulations of materials. *J. Phys. Condens. Matter* **2009**, *21*, 395502. [CrossRef]
53. Giannozzi, P.; Andreussi, O.; Brumme, T.; Bunau, O.; Nardelli, M.B.; Calandra, M.; Car, R.; Cavazzoni, C.; Ceresoli, D.; Cococcioni, M.; et al. Advanced capabilities for materials modelling with Quantum ESPRESSO. *J. Phys. Condens. Matter* **2017**, *29*, 465901. [CrossRef]
54. Quantum Espresso. Available online: https://www.quantum-espresso.org (accessed on 28 March 2022).
55. Pickard, C.J.; Mauri, F. All-electron magnetic response with pseudopotentials: NMR chemical shifts. *Phys. Rev. B* **2001**, *63*, 245101. [CrossRef]
56. Yates, J.R.; Pickard, C.J.; Mauri, F. Calculation of NMR chemical shifts for extended systems using ultrasoft pseudopotentials. *Phys. Rev. B* **2007**, *76*, 024401. [CrossRef]
57. Quantum Espresso Pseudopotentials. Available online: http://www.quantum-espresso.org/pseudopotentials (accessed on 28 March 2022).
58. Monkhorst, H.J.; Pack, J.D. Special points for Brillouin-zone integrations. *Phys. Rev. B* **1976**, *13*, 5188–5192. [CrossRef]
59. Chao, K.J.; Lin, J.C.; Wang, Y.; Lee, G. Single crystal structure refinement of TPA ZSM-5 zeolite. *Zeolites* **1986**, *6*, 35–38. [CrossRef]
60. Yokomori, Y.; Idaka, S. The structure of TPA-ZSM-5 with Si/Al = 23. *Microporous Mesoporous Mater.* **1999**, *28*, 405–413. [CrossRef]
61. Dib, E.; Mineva, T.; Veron, E.; Sarou-Kanian, V.; Fayon, F.; Alonso, B. ZSM-5 Zeolite: Complete Al Bond Connectivity and Implications on Structure Formation from Solid-State NMR and Quantum Chemistry Calculations. *J. Phys. Chem. Lett.* **2018**, *9*, 19–24. [CrossRef] [PubMed]
62. Fabbiani, M.; Al-Nahari, S.; Piveteau, L.; Dib, E.; Veremeienko, V.; Gaje, A.; Dumitrescu, D.G.; Gaveau, P.; Mineva, T.; Massiot, D.; et al. Host–Guest Silicalite-1 Zeolites: Correlated Disorder and Phase Transition Inhibition by a Small Guest Modification. *Chem. Mater.* **2022**, *34*, 366–387. [CrossRef]

MDPI
St. Alban-Anlage 66
4052 Basel
Switzerland
Tel. +41 61 683 77 34
Fax +41 61 302 89 18
www.mdpi.com

Computation Editorial Office
E-mail: computation@mdpi.com
www.mdpi.com/journal/computation

www.ingramcontent.com/pod-product-compliance
Ingram Content Group UK Ltd.
Pitfield, Milton Keynes, MK11 3LW, UK
UKHW060124010826
14089UKWH00006B/132

9 783036 549743